ANNALS OF
THE NEW YORK ACADEMY
OF SCIENCES

Volume 972

EDITORIAL STAFF

Executive Editor
BARBARA M. GOLDMAN

Managing Editor
JUSTINE CULLINAN

Associate Editors
JOHN W. KENNEDY
STEFAN MALMOLI

The New York Academy of Sciences
2 East 63rd Street
New York, New York 10021

THE NEW YORK ACADEMY OF SCIENCES
(Founded in 1817)
BOARD OF GOVERNORS, September 2002–September 2003
TORSTEN N. WIESEL, *Chairman of the Board*
JOHN T. MORGAN, *Treasurer*

Honorary Life Governors
WILLIAM T. GOLDEN   JOSHUA LEDERBERG

Governors

ELEANOR BAUM	KAREN E. BURKE	LAWRENCE B. BUTTENWIESER
PRAVEEN CHAUDHARI	BRIAN FERGUSON	GERALD FISCHBACH
JOHN H. GIBBONS	MICHAEL GOLDEN	RONALD L. GRAHAM
MARNIE IMHOFF	JACQUELINE LEO	BRUCE McEWEN
PAUL MARKS	RONAY MENSCHEL	JOHN F. NIBLACK
SANDRA PANEM	PETER RINGROSE	JOHN J. ROCHE
LEE VANCE		DEBORAH WILEY

HELENE L. KAPLAN, *Counsel* [ex officio]

VISUALIZATION AND IMAGING IN
TRANSPORT PHENOMENA

In Memory of

Horti and Larry Fairberg

(New York, USA)

ANNALS OF THE NEW YORK ACADEMY OF SCIENCES
Volume 972

VISUALIZATION AND IMAGING IN TRANSPORT PHENOMENA

Edited by
Samuel Sideman and Amir Landesberg

The New York Academy of Sciences
New York, New York
2002

Copyright © 2002 by the New York Academy of Sciences. All rights reserved. Under the provisions of the United States Copyright Act of 1976, individual readers of the Annals are permitted to make fair use of the material in them for teaching and research. Permission is granted to quote from the Annals provided that the customary acknowledgment is made of the source. Material in the Annals may be republished only by permission of the Academy. Address inquiries to the Permissions Department (permissions@nyas.org) at the New York Academy of Sciences.

Copying fees: For each copy of an article made beyond the free copying permitted under Section 107 or 108 of the 1976 Copyright Act, a fee should be paid through the Copyright Clearance Center, Inc., 222 Rosewood Drive, Danvers, MA 01923 (www.copyright.com).

∞ The paper used in this publication meets the minimum requirements of American National Standard for Information Sciences—Permanence of Paper for Printed Library Materials. ANSI Z39.48-1984.

Library of Congress Cataloging-in-Publication Data

International Symposium on Visualization and Imaging in Transport Phenomena (2002: Antalya, Turkey).
  Visualization and imaging in transport phenomena / edited by Samuel Sideman and Amir Landesberg.
    p.; cm. — (Annals of the New York Academy of Sciences; v. 972)
  "This volume is the result of the International Symposium on Visualization and Imaging in Transport Phenomena held on May 5–10, 2002, in Antalya, Turkey, and organized by the International Centre for Heat and Mass Transfer and the Visualization Society of Japan"—Contents p.
  Includes bibliographical references and index.
  ISBN 1-57331-370-X (hardcover : alk. paper) — ISBN 1-57331-371-8 (paper: alk. paper)
  1. Imaging systems in medicine—Congresses. 2. Imaging systems in biology—Congresses.
  3. Image processing—Digital techniques—Congresses. 4. Transport theory—Congresses.
  5. Energy transfer—Congresses. 6. Heat—Transmission—Congresses.
  [DNLM: 1. Energy Transfer—Congresses. 2. Biological Transport—Congresses.
  3. Image Processing, Computer Assisted—Congresses.   QU 34 1615v 2002]
  I. Sideman, S. II. Landesberg, Amir. III. Title. IV. Series.
  Q11 .N5 vol. 972
  [R857.O6]
  500 s—dc21
  [616.07

2002011259'
CIP

K-M Research/CCP
Printed in the United States of America
ISBN 1-57331-370-X (cloth)
ISBN 1-57331-371-8 (paper)
ISSN 0077-8923

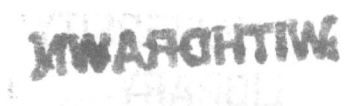

ANNALS OF THE NEW YORK ACADEMY OF SCIENCES

Volume 972
October 2002

VISUALIZATION AND IMAGING IN TRANSPORT PHENOMENA

Editors
SAMUEL SIDEMAN AND AMIR LANDESBERG

Conference Chairs
SAMUEL SIDEMAN AND TOSHIO KOBAYASHI

International Organizing Committee
SAMUEL SIDEMAN, TOSHIO KOBAYASHI, FARUK ARINÇ,
KOJI OKAMOTO, AND AMIR LANDESBERG

This volume is the result of the **International Symposium on Visualization and Imaging in Transport Phenomena**, held May 5–10, 2002, in Antalya, Turkey, and organized by the International Centre for Heat and Mass Transfer and the Visualization Society of Japan.

CONTENTS

Preface: Basic Highlights of Visualization and Imaging in Transport Phenomena.
By SAMUEL SIDEMAN xi

Part I. Enhanced Visualization

Quantitative Flow Visualization: Toward a Comprehensive Flow Diagnostic Tool.
By MORY GHARIB, FRANCISCO PEREIRA, DANA DABIRI,
AND DARIUS MODARRESS 1

Image Flows and One-Liner Graphical Image Representation.
By VADIM MAKHERVAKS, GILL BAREQUET, AND ALFRED BRUCKSTEIN. 10

Simulation and Identification of Deterministic Structures in Thermal and Magnetic Convection. *By* KEMAL HANJALIĆ AND SAŠA KENJEREŠ. 19

A New Definition of Contours in Images: Applications in Heat and Mass Transfer.
By ADIL RACHEK, JACQUES PADET, MONICA STOIAN, AND COLETTE PADET. 29

Photogrammetric and Image Processing Aspects in Quantitative Flow Visualization. *By* MATTHIAS MACHACEK AND THOMAS RÖSGEN 36

Color Interpolation Algorithms in Visualizing Results of Numerical Simulations.
    By DMITRY V. MOGILENSKIKH AND IGOR V. PAVLOV 43

Part II. Visualization of Dynamic Fields

Visualization of Molecular Dynamics by Simulation.
    By MASAHIRO OTA AND YINGXIA QI 53

Vortex Scale of Unsteady Separation on a Pitching Airfoil.
    By MASAKI FUCHIWAKI AND KAZUHIRO TANAKA 61

Imaging the Transient Boundary Layer on a Free Rotating Disc.
    By BRANIMIR MATIJAŠEVIĆ, ZVONIMIR GUZOVIĆ, AND VINKO MARTINIS ... 67

Effects of a Splitter Plate in the Near Wake of a Divergent Trailing Edge.
    By L. NEAU, J. PRUVOST, O. RODRIGUEZ, AND TA PHUOC LOC 73

Start-up Behavior of Viscoelastic Fluid Flow Near a Capillary Entry.
    By MASATAKA SHIRAKASHI, TSUTOMU TAKAHASHI,
    ATSUSHI WATANABE, AND YOHSUKE ARUGA........................ 81

Development and Validation of an X-ray Tomograph for Two-Phase Flow.
    By ERIC HERVIEU, EMMANUEL JOUET, AND LAURENT DESBAT 87

Visualization of Turbulent Wedges under Favorable Pressure Gradients
    Using Shear-Sensitive and Temperature-Sensitive Liquid Crystals.
    By TZE-PEI CHONG, SHAN ZHONG, AND HOWARD P. HODSON............ 95

Visualization and Measurement of Multiphase Flow in Porous Media Using Light
    Transmission and Synchrotron X-Rays. *By* CHRISTOPHE J.G. DARNAULT,
    DAVID A. DICARLO, TIM W.J. BAUTERS, TAMMO S. STEENHUIS,
    J.-YVES PARLANGE, CARLO D. MONTEMAGNO, AND PHILIPPE BAVEYE 103

Part III. Physiological System Dynamics

Assessment of Physiologic and Pathologic Radiative Heat Dissipation Using
    Dynamic Infrared Imaging. *By* MICHAEL ANBAR 111

Molecular Motion and Cardiac Muscle Motor Dynamics. *By* AMIR LANDESBERG,
    YOLANDA LANDESBERG, SAMUEL SIDEMAN, AND HENK E.D.J. TER KEURS . 119

Modeling the Spreading Cortical Depression Wavefront. *By* UGUR BAYSAL
    AND JENS HAUEISEN 127

Time-Varying Current Density Distributions in the Human Heart and Brain.
    By JENS HAUEISEN, MAREK ZIOLKOWSKI, AND UWE LEDER 133

Neutrophils Movement *in Vitro*. *By* ANNA KORZYŃSKA..................... 139

An Algorithm for Estimating Chaos in Mechanoemission of Blood and Magnetic
    Resonance Imaging in Patients with Gastric Cancer. *By* VALERIY E. OREL,
    ANDREY V. ROMANOV, NATALIE N. DZYATKOVSKAYA,
    AND YURI I. MEL'NIK.. 144

Part IV. Phase Change

Microscopic Study of Crystal Growth in Cryopreservation Agent Solutions and
    Water. *By* LE-REN TAO AND TSE-CHAO HUA 151

Temperature and Flow Visualization in a Simulation of the Czochralski
  Process Using Temperature-Sensitive Liquid Crystals.
    By JELENA ALEKSIC, PAUL ZIELKE, AND JANUSZ A. SZYMCZYK 158
Visualization of Droplet Boiling on Heated Transparent Solid Surfaces
  with Various Thermal Properties. By SHIGEAKI INADA 164
Ice/Water Slurry Blocking Phenomenon at a Tube Orifice. By TAKERO HIROCHI,
  SHUICHI YAMADA, TUYOSHI SHINTATE, AND MASATAKA SHIRAKASHI 171

Part V. Heat and Mass Transfer

Convective Heat Transfer and Infrared Thermography.
  By GIOVANNI M. CARLOMAGNO, TOMMASO ASTARITA,
  AND GENNARO CARDONE 177
Visualization of Flow and Heat Transfer Augmentation by Oblique Impingement
  Jets. By HIDEO KIMOTO, CHAYUT NUNTADUSIT, AND KENJI HAMABE 187
Unsteady Flow Patterns in the Vicinity of Heated Wall-Mounted Transverse Ribs.
  By GUILLAUME POLIDORI AND JACQUES PADET 193
Investigation of Transport Phenomena Inside a Microcapsule.
  By LEONID GOUBERGRITS, KLAUS AFFELD, PERRINE DEBAENE,
  AND ULRICH KERTZSCHER 200
Measurement of the Density of CO_2 Solution by Mach–Zehnder Interferometry.
  By YONGCHEN SONG, MASAHIRO NISHIO, BAIXIN CHEN, SATOSHI SOMEYA,
  TSUTOMU UCHIDA, AND MAKOTO AKAI 206

Part VI. Particle Image Velocimetry (PIV)

Particle Image Velocimetry and Thermometry for Two-Phase Flow Problems.
  By TOMASZ A. KOWALEWSKI 213
Multiphase Bubbly Flow Visualization Using Particle Image Velocimetry.
  By YASSIN A. HASSAN 223
Velocity Measurements by Particle Image Velocimetry Using a Direct
  Intercorrelation Algorithm: Application to the Interaction between a
  Water Mist and a Liquid Pool Fire. By JEROME RICHARD, ARNAUD SUSSET,
  AND JEAN-PIERRE VANTELON 229
Visualization of Bubble–Fluid Interaction by a Moving Object Flow Image
  Analyzer System. By H M CHOI, T. TERAUCHI, H. MONJI, AND G. MATSUI 235
Effect of Electrolytes on Bubble Coalescence in Columns Observed
  with Visualization Techniques. By MARÍA EUGENIA AGUILERA,
  ANTONIETA OJEDA, CAROLINA RONDÓN, AND AURA LÓPEZ DE RAMOS 242
X-ray-Based Flow Visualization and Measurement: Application in Multiphase
  Flows. By AXEL SEEGER, KLAUS AFFELD, LEONID GOUBERGRITS,
  ULRICH KERTZSCHER, ERNST WELLNHOFER, AND RENE DELFOS 247
Simultaneous Velocity and Concentration Measurements of a Turbulent Jet
  Mixing Flow. By HUI HU, TETSUO SAGA, TOSHIO KOBAYASHI, AND
  NOBUYUKI TANIGUCHI 254

PIV Measurement and Numerical Analysis of a New Refrigeration
Compartment of a Refrigerator. *By* Seong-Ho Cho, In-Seop Lee,
Jay-Ho Choi, and Young-Sok Nam 260

Part VII. Jets, Columns, and Films

Visualization of Jet Flows over a Plate by Pressure-Sensitive Paint
Experiments and Comparison with CFD. *By* Kozo Fujii,
Nobuyuki Tsuboi, and Nobuyoshi Fujimatsu 265

Transonic Injection in Interaction with Transverse Compressible Flow.
By R. Dizene, J.M. Charbonnier, E. Dorignac, and R. Lablanc 271

Visualization and Phase Doppler Particle Analysis Measurements of
Oscillating Spray Propagation of an Airblast Atomizer under Typical
Engine Conditions. *By* Peter Schober, Robert Meier,
Olaf Schäfer, and Sigmar Wittig 277

Numerical Visualization of Two-Phase Plume Formation in a Stratification
Flow Environment. *By* Baixin Chen, Masahiro Nishio,
Yongchen Song, Satoshi Someya, and Makoto Akai 285

Visualization Studies of an Acoustically Excited Liquid Sheet.
By Vayalakkara Sivadas and Manuel V. Heitor 292

Flow Characteristics of Two Immiscible Liquid Layers Subjected to a Horizontal
Temperature Gradient. *By* Satoshi Someya, Tetsuo Munakata,
Masahiro Nishio, Koji Okamoto, and Haruki Madarame 299

Part VIII. Physiological Transport and Circulation

Quantification of Myocardial Microcirculatory Function with X-ray CT.
By Stefan Möhlenkamp, Lilach O. Lerman, Željko Bajzer,
Patricia E. Lund, and Erik L. Ritman 307

Renal Handling of X-ray Contrast Media: Imaging and Exploration with
Electron Beam CT. *By* Andrew D. Rule, Željko Bajzer,
Erik L. Ritman, and Lilach O. Lerman 317

Visualization of Blood Microcirculation Parameters in Human Tissues by
Time-Integrated Dynamic Speckles Analysis. *By* Maria M. Gonik,
Alexander B. Mishin, and Dmitry A. Zimnyakov 325

Measurement of a Velocity Field in Microvessels Using a High Resolution PIV
Technique. *By* Yasuhiko Sugii, Shigeru Nishio, and Koji Okamoto 331

Biosimulation and Visualization: Effect of Cerebrovascular Geometry
on Hemodynamics. *By* Marie Oshima, Toshio Kobayashi,
and Kiyoshi Takagi 337

Index of Contributors 345

Financial assistance was received from:
- TECHNION, ISRAEL INSTITUTE OF TECHNOLOGY, HAIFA, ISRAEL
- INSTITUTE OF INDUSTRIAL SCIENCE, UNIVERSITY OF TOKYO, JAPAN
- MIDDLE EAST TECHNICAL UNIVERSITY, ANKARA, TURKEY
- SCIENTIFIC AND TECHNICAL RESEARCH COUNCIL OF TURKEY (TÜBITAK)

The New York Academy of Sciences believes it has a responsibility to provide an open forum for discussion of scientific questions. The positions taken by the participants in the reported conferences are their own and not necessarily those of the Academy. The Academy has no intent to influence legislation by providing such forums.

Preface

Basic Highlights of Visualization and Imaging in Transport Phenomena

SAMUEL SIDEMAN

*Department of Biomedical Engineering, TECHNION,
Israel Institute of Technology, Haifa, Israel*

This volume reflects the state of the art of *visualization* and *imaging* as a vehicle for quantitative analysis of transport phenomena. Catalyzed by outstanding progress in computer science, technology, and image processing during the past quarter century, we can now address, simulate, and analyze complicated three-dimensional dynamic systems involving momentum, energy, and mass transfer in the practice of engineering and medicine. These include analysis and display of scalar and vector fields, tensor data sets, and sequential dynamic events in three-dimensional space involving one or more temporal variables in single or multiphase systems. Attention is focused on visualization and imaging of dynamic transport phenomena in animate and inanimate systems. Of particular interest is the interplay between reality and imaging and the application of the quantitative information contained within the visual images of real systems. Physical description of the visualization technology and the various imaging techniques, a broad and important topic, is outside the scope and goals of this volume.

BACKGROUND AND MOTIVATION

Transport Phenomena

The transfer of momentum, energy, and mass, commonly referred to as *transport phenomena*, is essential to the continuity of the universe, and to the existence of life on Earth. The universe is maintained by the motion of the stars and the interactions between them; life on Earth depends on the radiation of solar energy and the numerous bioenergetic transformation and mass transfer processes that control the growth of plants and sustain life in all living creatures. There can be no life on Earth and no celestial omnipotence without transport phenomena! Deciphering the secrets of the various submicro- to supermacro-transport phenomena, either molecular, intracellular, or celestial, is the ultimate goal of the physical and life sciences. Acquiring the key to this knowledge is the ultimate and insatiate ambition of all the sciences. Visualization and imaging are among the latest tools available to facilitate the pursuit of understanding the nature of transport phenomena and the mechanisms that make them tick.

Visualization and Imaging

Visualization is the art of transforming normally invisible phenomena into visible measurable events. This is usually achieved by utilizing active or inert fluids or solid markers, radioactive or opaque tracers, or by modulation of wavelength refractory properties. *Imaging* is the art of creating a visual image of a normally unseen physical entity from evoked signals and data sets generated by various imaging techniques. However, since the end result of imaging is a visual picture, imaging and visualization are often synonymous terms.

The transfer of information by visual means is probably among the oldest communication techniques developed by human society. Light signals by bonfires from the top of mountains in Jerusalem some three thousand years ago served to announce the beginning of the Holy day to the people of Israel. Indian tribes used smoke signals to send messages across the land many centuries ago. Extending our vision by telescopes and electron microscopes during the past centuries led to many scientific breakthroughs that changed the world we live in. Modern devices, electronic or otherwise, that transmit and "read" invisible signals, together with new computer technology can enhance weak electromagnetic signals, allow reconstruction of two- and three-dimensional physical entities, and provide insight and understanding of many unexplored areas of our world.

ENGINEERING AND MEDICINE

Both engineering and medicine are "grass-roots" disciplines. Both grow by responding to practical needs and thrive on the progress of science and technology. Most noticeable is the huge effect that modern technology, associated with imaging and visualization, has had on extending longevity and improving the quality of life by using better diagnostics and more effective therapeutic modalities.

The development of a wide range of theoretical and practical visualization and imaging techniques has catalyzed knowledge and understanding of the invisible characteristics of the various physical and physiologic systems that play an important role in the dynamic world we live in. A large number of applications of visualization and enhanced imaging techniques are demonstrated in this volume. These range from molecular motion in biological motors, and the molecular characteristics of hydrates, to enhancing oil-well production by viscous water flooding, and to better understanding of celestial magnetohydrodynamics. Most exciting is human ingenuity and unlimited imagination, which, through the use of modeling and computers, create a virtual world by simulating real systems. Although only close approximations, at best, these theoretical models and simulations attempts allow better understanding of the investigated systems by manipulating perturbations and quantitative trial and error explorations. This exciting combination of practical tools and conceptual imaginative exploration is probably the most outstanding feature of this volume.

Flow dynamic is still attracting the scientific world, probably since it can be tackled both experimentally and theoretically. An early famous example is Reynolds' use of dyes to explore the laminar and turbulent characteristics of flowing water, which has since developed into digital image processing and time sequenced images of the motion of patterns generated by dyes, clouds, or dispersed particles. Flow dynamics is commonly studied by laser doppler velocimetry (LDV), optically passive or active dye agents (fluorescent tracers and liquid crystals). Opac (solid and liquid) or reflective particles (bubbles, aluminum dust, soot, and hot coal particles) are used in particle image velocimetry (PIV) measuring techniques, which provide insight into flow dynamics in inert and reactive micro- and macro- (clouds and plumes) single- and multiphase systems.

All optical methods, including LDV, PIV, particle tracking velocimetry (PTV), laser induced fluorescence (LIF), and other PIV related procedures depend on proper illumination (laser, visible light, and X-ray) to assure well defined reflection of phase boundaries. Image intensity, either wavelength or color, reveal density, concentration, and temperature fields and the gradients of these scalar fields define fluxes and vectorial properties of the transport phenomena involved. Obviously, although all these techniques can be used in physical and simulated physiologic systems, their application in live physiologic systems is limited to non-invasive and harmless procedures.

Medical Imaging has made huge progress in the hundred years since Roentgen discovered the "new kind of light," the X-rays, which are still the major diagnostic tool in the clinic today. The outstanding progress in computer technology now facilitates dynamic imaging, automated functional analysis, on line bed-to-office clinical information transfer and immediate access to patient files. Fluoroscopy, computed tomography (CT), infrared tomography (IRT), ultrasound (US), magnetic resonance imaging (MRI), functional MRI (fMRI), single photon emission CT (SPECT), and positron emission tomography (PET) have since been developed into clinically effective, essentially non-invasive, tools. Multiphoton microcopy and imaging of molecular dynamics permit measurement of metabolic changes in living cells in the brain. Other mind-boggling biological applications of functional imaging and molecular imaging are already on their way as proponents of biological and physiologic information and exciting genetic and molecular based therapeutic modalities for hitherto incurable human pathologies.

Particularly exciting are recent advances in optical technologies that improve our ability to detect cellular and subcellular morphologic changes in real time, thus overcoming the need for invasive biopsies. High resolution magnifying chromoendoscopy, optical reflectance spectroscopy (ORS), and optical coherence tomography (OCT) now offer sensitive and specific images for identification, screening, diagnosis, surveillance, and therapeutic monitoring of neoplastic pathology, such as the prevalent colon cancer. Furthermore, new techniques employing laser capture microdissection (LCM) allow selective procurement of targeted cells for tissue analysis; expression arrays using DNA, RNA, or proteins extracted from tumor tissues provide insights into the molecular pathogenesis of the disease. (For more details see Umar *et al.*, in **Cancer Prevention**, H.P. Osborne, Ed. *Annals of the New York Academy of Sciences* **952**: 88–108, 2001).

It is instructive at this point to note the outstanding characteristics of dynamic imaging and functional imaging and their interactions. Generally, dynamic imaging relates to the macroscale world and employs various techniques to study the temporal behavior of physical elements and physiologic organs. Functional imaging, on the other hand, is exclusively related to the physiologic world, wherein the image provides insight into the microscale cellular and molecular world. Using mathematical parlance we say that dynamic imaging is a *forward* problem, whereas functional physiologic imaging is an *inverse* problem whereby we deduce physiologic information from the physical image. Obviously, analysis of the interaction between dynamic and functional imaging in physiologic systems is rather complicated since the latter are highly nonlinear and contain adaptive feedback mechanisms between the macroscale dynamics of the system and its microscale based physiologic function.

The close and complex interplay between engineering and medical sciences promoted the symposium that lead to this volume and that aimed to explore the application of visualization and imaging techniques in the study of transport phenomena in natural and man-made dynamic systems.

CRITICAL PARAMETERS

Successful applications of visualization and imaging techniques must comply with basic operational characteristics. Following DiBianca and Ogg (*BMES Bulletin* **24**(2): 56–10, 2000) we highlight important critical parameters that affect the quality and applicability of imaging and visualization technology. (1) Spatial resolution: visualize and differentiate fine structure. (2) Contrast resolution: visualize differences in material composition (detection of contrast material, particles, radioactive agents, etc.). (3) Sensitivity: discern smallest alterations. (4) Temporal resolution: "freeze" and visualize moving elements—for example, the heart. (5) Motion and flow: follow local and spatial motion. (6) Functional changes: follow metabolic, neural, biochemical, or biophysical activity. (7) Non-invasive: visualize the desired field without disturbing it, or endangering it. (8) Safety: maintain the integrity of the system without endangering the system or the investigator. (9) Reliability and repeatability: control the imaging process and obtain consistent images. These criteria are particularly critical for meaningful understanding, quantitative evaluation, and visualization of dynamic systems involving transport phenomena, such as motion, flow, energy, and mass transfer. Obviously, other critical criteria may arise as new technologies develop in response to our needs and imagination.

ACKNOWLEDGMENTS

We gratefully thank all those who helped transform the symposium from vision to reality. First and foremost are the symposium participants whose enthusiasm and expertise made the meeting an exciting affair. Next, we thank the International Program Committee for endless suggestions and continuous vigilance, and the nameless reviewers who assured high-level presentations. Special thanks go to Professor Toshio Kobayashi, Cochairman, organizing committee, and particular thanks to Professor Faruk Arinç, who had a logical effective answer to all our problems. The International Centre for Heat and Mass Transfer (ICHMT) initiated this activity. The Visualization Society of Japan (VSJ) vigorously helped make the symposium a success. We thank our sponsors, the Technion, Israel Institute of Technology, the National Science and Research Council of Turkey (TÜBITAK), and, most significantly, we gratefully thank our departed friends, Horti and Larry Fairberg, whose Fund in the American Technion Society, New York, USA, helped facilitate this important event.

Quantitative Flow Visualization

Toward a Comprehensive Flow Diagnostic Tool

MORY GHARIB, FRANCISCO PEREIRA, DANA DABIRI, AND DARIUS MODARRESS

Graduate Aeronautical Laboratory, California Institute of Technology, Pasadena, California, USA

ABSTRACT: Quantitative flow visualization has many roots and has taken several approaches. The advent of digital image processing has made it practical to extract useful information from every kind of flow image. In a direct approach, the image intensity or color (wavelength or frequency) can be used as an indication of concentration, density and temperature field, or gradients of these scalar fields in the flow.[1] For whole-field velocity measurement, the method of choice for experimental fluid mechanicians has been digital particle image velocimetry (DPIV). This paper presents a novel approach to extend the DPIV technique from a planar method to a full three-dimensional volume mapping technique.

KEYWORDS: imaging; visualization; PIV; DPIV; diagnostics

INTRODUCTION

In general, the optical flow or the motion of intensity fields can be obtained through a time sequence of images.[2] For example, the motion of patterns generated by dye, clouds, or particles can be used to obtain such a time sequence. The main problem with using a continuous-intensity pattern, generated by scalar fields (e.g., dye patterns), is that it must be fully resolved (space/time) and contain intensity variations at all scales, before mean and turbulent velocity information can be obtained.[3] In this respect, the discrete nature of images generated by seeding particles has made particle tracking the method of choice for the entire velocimetry field. Various methods, such as individual tracking of particles or statistical techniques, can be used to obtain the displacement information and subsequently the velocity information. The spatial resolution of this method depends on the number density of the particles.

The *particle image velocimetry* (PIV) technique follows a group of particles through statistical correlation of sampled windows of the image field.[4] This scheme removes the problem of identifying individual particles, which is often associated with tedious operations and large errors in the detection of particle pairs. In terms of the spatial resolution, the velocity obtained at each window represents the average velocity of the group of particles within the window. The interrogating window in PIV is the equivalent of a grid cell in CFD. Development of the video-based digital

Address for correspondence: Mory Gharib, Ph.D., Graduate Aeronautical Laboratory, California Institute of Technology, Pasadena, CA, USA.
mory@caltech.edu

version of PIV, known as DPIV,[5,6] resulted directly from advances in charge coupled device (CCD) technology and fast, computer-based, image processing systems.

The capability of whole field measurement techniques in providing velocity vector or scalar field information in a format compatible with CFD calculations has made a major impact in defining common grounds for designing new approaches toward resolving the turbulent and two-phase flow problem. Such common grounds are difficult or impossible to define by using methods, such as LDV or hot wire anemometry, that do not address the global Lagrangian and the temporal nature of complex flows.

DPIV can be utilized to obtain three components of the velocity field. However, this extension of DPIV is limited to a few planes and cannot address the full dimensionality of turbulent flow with current video technology. *Holographic PIV techniques* are more suitable for obtaining three-dimensional (3D) distributions of the velocity vector field.[7] The photographic nature of holographic PIV techniques limits their ability to address the temporal dynamics of turbulent flows. Recent advances in 3D video-based particle tracking techniques have removed some of these shortcomings.[8] However, complexities involved in the optics, calibration, and image processing of multiple cameras and images severely limits the wide range application of multiple-camera stereo techniques.

An emerging technology that has a good potential for resolving difficulties associated with the aforementioned flow mapping techniques is the *method of defocusing imaging*.[9] *Defocusing digital particle image velocimetry* (DDPIV) is the natural extension of planar PIV techniques to a third spatial dimension. This method has shown great potential for two-phase flow studies.[10]

THE DEFOCUSING CONCEPT

The foundations of the defocusing concept were established in an early paper by Willert and Gharib.[9] We report here the most important aspects in a revised form. For clarity, we use the term *particle* when referring either to a solid particle or to a bubble.

A typical two-dimensional imaging system, consisting of a convergent lens and an aperture, is represented in FIGURE 1 to help describe the DDPIV technique. FIGURE 1A exhibits a point A, located on the object plane (or reference plane), and a point B placed in between this plane and the lens system. Point A appears focused in A', on the image plane (or sensor plane), whereas B is projected as a blurred image B'. The DDPIV technique uses a mask, with two or more apertures shifted away from the optical axis, to obtain multiple images from each scattering source, as shown in FIGURE 1B. The image shift b on the image plane, caused by these off-axis apertures, is related to the depth location of the source points, whereas the scattered light intensity combined with the blurredness is used to recover the size information.

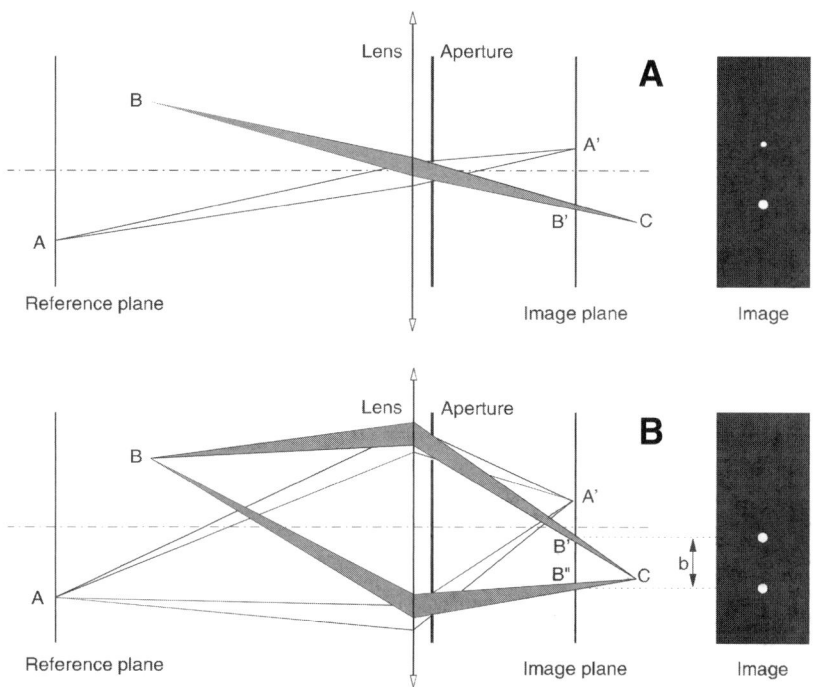

FIGURE 1. The defocusing principle: **(A)** standard g system, **(B)** defocusing arrangement.

GEOMETRIC ANALYSIS

A simplified geometric model of a two-aperture defocusing optical arrangement is represented in FIGURE 2. The interrogation domain is defined as a cube of side length a, thus, a square in the plane. The back face of this cube is on the reference plane, which is placed at a distance L from the lens plane. Let d be the distance between apertures, f the focal length of the converging lens, and l the distance from the lens to the image plane. The image plane is materialized by a photosensor (e.g., CCD) of height h. The physical space is attached to a coordinate system originating in the lens plane, with the Z-axis on the optical axis of the system. Coordinates in the physical space are designated (X, Y, Z). The image coordinate system is simply the Z-translation of the physical system onto the sensor plane, that is, at $Z = -l$. The coordinates of a pixel on the imaging sensor are given by the pair (x, y). Point $P(X, Y, Z)$ represents a light scattering source, such as particle, bubble, or a point-like dot. For $Z \neq L$, P is projected onto points $P'(x', y')$ and $P''(x'', y'')$, separated by a distance b.

The coordinates (x', y') and (x'', y'') of the images P' and P'' of $P(X, Y, Z)$ in the image plane are given by the following relations:

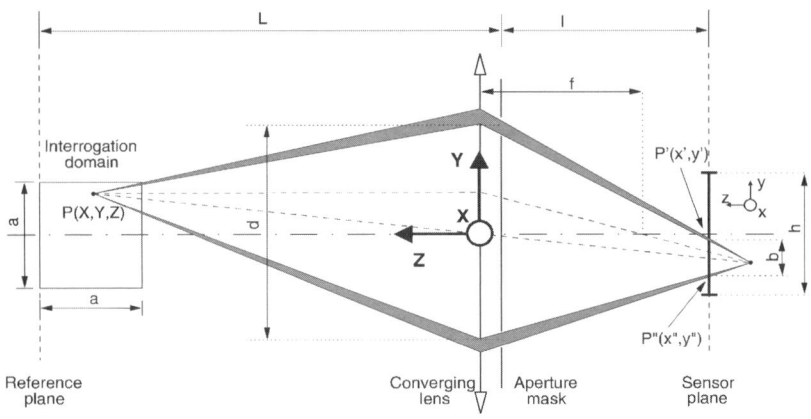

FIGURE 2. Simplified defocusing optical method.

$$\begin{cases} x' = x'' = -ML\dfrac{X}{Z} \\ y' = \dfrac{M}{2Z}[d(L-Z) - 2LY] \\ y'' = \dfrac{M}{2Z}[-d(L-Z) - 2LY] \end{cases} \quad (1)$$

where M is the optical magnification provided by the lens equation.

The image separation vector **b** represents the distance between the images P' and P''. The norm is, therefore, given by

$$b = \frac{Md}{Z}(L-Z) = \frac{1}{K}\left(\frac{1}{Z} - \frac{1}{L}\right) \text{ with } K = \frac{1}{MdL}. \quad (2)$$

Equation (2) demonstrates the extreme simplicity of the defocusing concept, which, of course, is valuable in terms of computational implementation and processing speed. In purely geometric terms, the image separation b is independent of the in-plane coordinates X and Y. Likewise, the pinhole diameter has no bearing on b and is only responsible for the amount of blurredness of any given particle image. For our prototype instrument, we use three pinholes, arranged in a triangular pattern. This configuration, shown in FIGURE 3, exhibits a flipping triangle when P moves across the reference plane and requires straightforward and fast image processing routines.

The sensitivity of the system—its ability to detect small changes of the particle location—can be evaluated through the separation gradient

$$\frac{\partial b}{\partial Z} = -\frac{1}{KZ^2}. \quad (3)$$

The coordinates of P in the global coordinate system are derived from the image coordinates of the projections P' and P'', see (1),

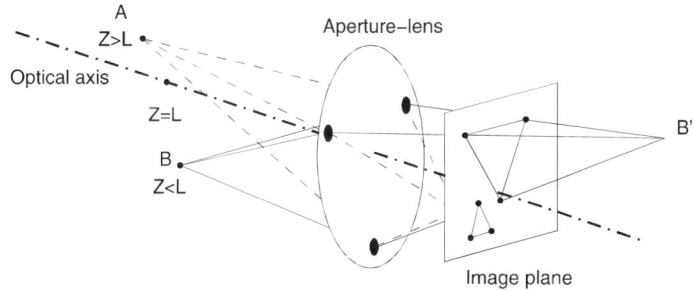

FIGURE 3. Three-aperture arrangement.

$$\begin{cases} X = -\dfrac{x_0 Z}{ML} \text{ with } x_0 = \dfrac{x' + x''}{2} \\ Y = -\dfrac{y_0 Z}{ML} \text{ with } y_0 = \dfrac{y' + y''}{2} \\ Z = \dfrac{1}{1/L + Kb}. \end{cases} \quad (4)$$

Assuming that the apertures are equidistant from the origin of the coordinate system, the image point defined by (x_0, y_0) is the image of the particle if there were a single aperture at the origin.

A camera system has been designed and fabricated based on this concept. Specific characteristics of the instrument can be found in the paper by Pereira et al.[10] The velocity vector field is obtained by local spatial cross-correlation between small volume elements (*voxels*) containing particles observed at two time steps, as shown and discussed by Pereira et al.[10]

APPLICATIONS

Two-Blade Model Propeller

A propeller was immersed in a water tank. The rotation speed was 12 rps, corresponding to a tangential velocity of 2.52 msec^{-1} at the tip of the blades. A bubble generator, placed below the propeller, produced a dense stream of rising submillimeter air bubbles. The velocity field was obtained by phase averaging.

A 3D velocity field was obtained after averaging and outlier correction. Mass-less particles were then artificially injected into the mean velocity data set, in a radial arrangement and one diameter upstream from the propeller. Paths of bubbles were determined, providing a unique insight into this complex flow, as shown in FIGURE 4. The gray level in this figure relates to the local measured velocity amplitude. The velocity reached a maximum of 2.49 msec^{-1} in the outer region of the propeller, matching closely the blade tip tangential velocity.

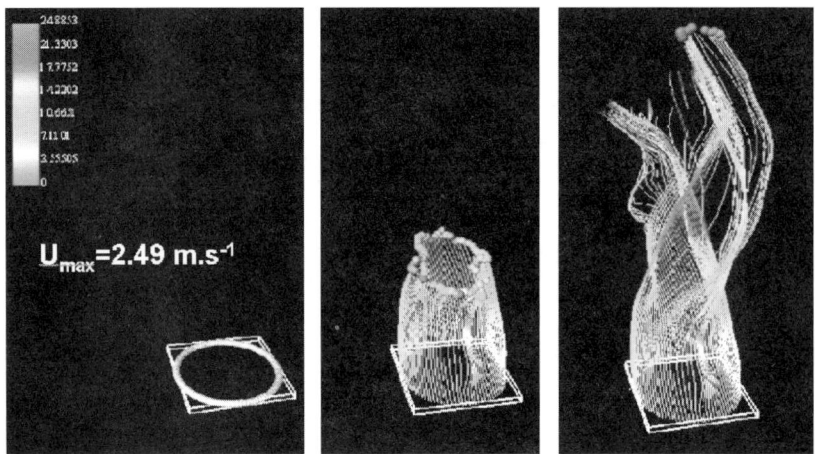

FIGURE 4. Path lines of bubbles around the propeller.

The bubble mean radius along the Y vertical axis of the flow (rotation axis of the propeller) is displayed in FIGURE 5. The mean radius increases almost linearly to nearly 325 m at $Y \approx 30$ mm, where the propeller was located. After the bubbles pass the immediate vicinity of the propeller, the radius is found to follow the opposite trend, decreasing to about 200 m. The growth of bubbles is partly due, but to a very small extent, to the decrease of the static pressure with increasing Y. In fact, bubbles experience first the low pressure in the suction side of the propeller before getting into the high-pressure region where they collapse. Included in FIGURE 5 are the histograms, calculated by taking the same volume below and above $Y = 30$ mm. The histogram peak follows the trend outlined previously and due to the pressure variations.

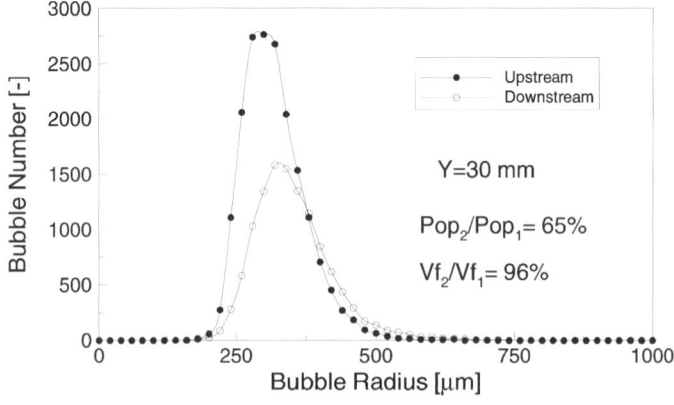

FIGURE 5. Size distribution upstream and downstream, $Y = 30$ cm.

The ratio of the upstream to the downstream populations is 65%. However, the ratio of the respective void fractions is close to 100%. These observations indicate that coalescence of bubbles is the main mechanism acting here, although breakup may occur in the propeller region.

Three-Blade Boat Propeller

The three-blade propeller had a similar configuration to the two-blade propeller discussed above. The propeller was rotated at 12 Hz. The velocity field, represented in FIGURE 6 was obtained by phase-averaging a sequence of 50 instantaneous velocity fields. Spurious vectors can be seen on the borders of the interrogation domain. A slice in the velocity field (see FIGURE 7) displays the high speed jet core along the downstream section of the propeller axis. However, the isovelocity contours, displayed in the same figure, show a viscous wake that appears as a velocity defect due to the merging of the two boundary layers from the blades. A slight contraction of the slipstream could also be detected. The wake was found to rapidly fade into the bulk flow.

SUMMARY

In this paper we seek to present some recent advances in the field of quantitative visualization. Only techniques that can provide time resolved, 3D velocity vector

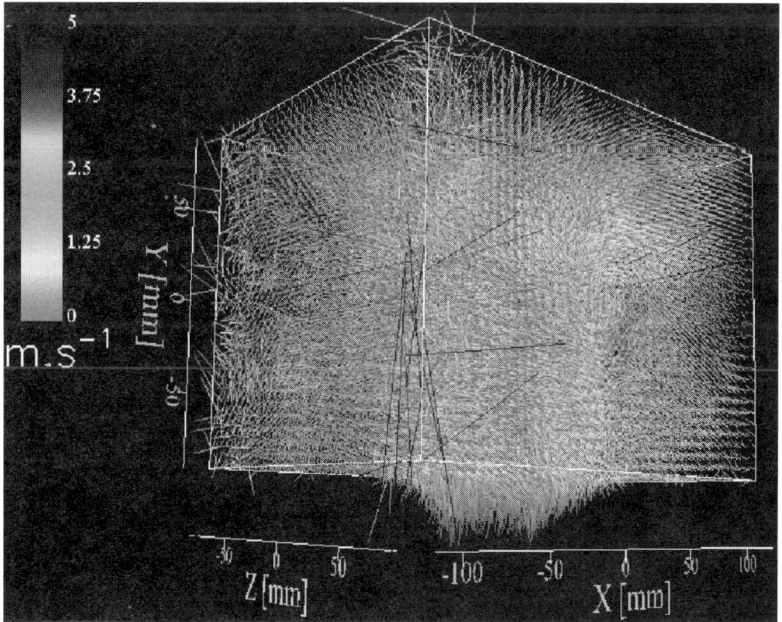

FIGURE 6. Velocity field, $200 \times 200 \times 400 \text{mm}^3$, 72,963 vectors ($33 \times 33 \times 67$ voxels).

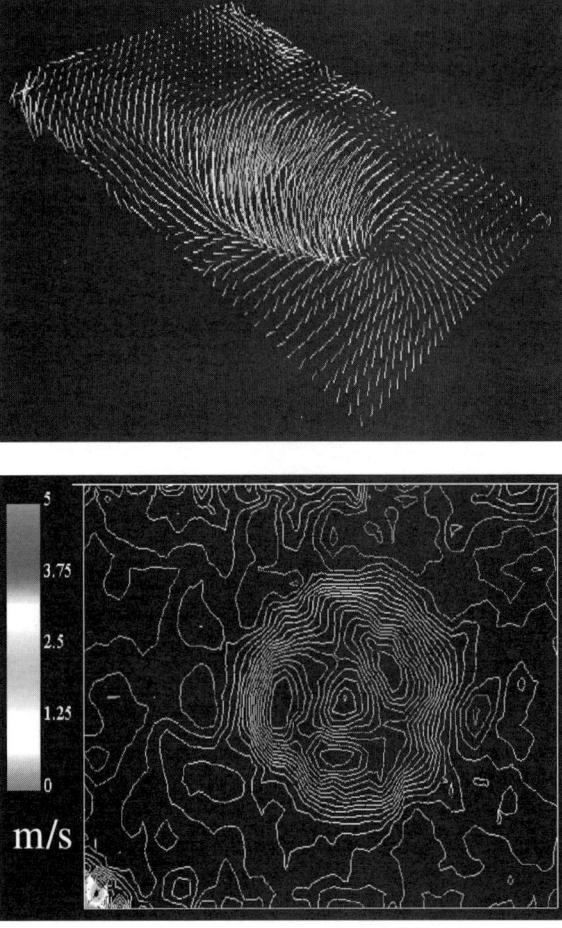

FIGURE 7. Top, velocity cross-section (downstream region, 0.5 diameter); **bottom**, corresponding isovelocity contour.

fields can offer hope in defining better common grounds with CFD. In this respect the novel method of DDPIV shows excellent potential in providing quantitative flow information comparable with that of CFD.

ACKNOWLEDGMENTS

This research was supported by the Office of Naval Research (Contract N000140010110). Defocusing digital particle image velocimetry (DDPIV) technology is protected under a U.S. patent through the California Institute of Technology.

REFERENCES

1. MERZKIRCH, W. 1987. Flow Visualization. Academic Press.
2. SINGH, A. 1991, Optic flow computation. IEEE, Computer Society Press.
3. PEARLSTEIN, A.J. & B. CARPENTER. 1995. On the determination of solenoidal or compressible velocity fields from measurements of passive and reactive scalars. Phys. Fluids **7**(4): 754–763.
4. ADRIAN, R.J. 1991. Particle-imaging techniques for experimental fluid mechanics. Annu. Rev. Fluid Mechan. **23**: 261–304.
5. WILLERT, C. & M. GHARIB. 1991. Digital particle image velocimetry. Exp. Fluids **10**: 181–183.
6. WESTERWEEL, J. 1993. Digital Image Velocimetry: Theory and Application. Delft UP, Delft, the Netherlands.
7. BARNHART, D.H., R.J. ADRIAN & G.C. PAPEN. 1994. Phase-conjugate holographic system for high resolution PIV. Appl. Optics **3**(30): 7159–7170.
8. KASAGI, N. & Y. SATA. 1992. Recent development in three-dimensional particle tracking velocimetry. Proc. of Flow Visualization Conference VI, Yokohama, Japan.
9. WILLERT C.E. & M. GHARIB. 1992. Three-dimensional particle imaging with a single camera. Exp. Fluids **12**: 353–358.
10. PEREIRA, F., M. GHARIB, M. MODARRESS & D. DABIRI. 2000. Defocusing digital particle imaging velocimetry: A 3-component 3-D DPIV measurement technique, application to bubbly flows. Exp. Fluids **29**: S78–S84.

Image Flows and One-Liner Graphical Image Representation

VADIM MAKHERVAKS, GILL BAREQUET, AND ALFRED BRUCKSTEIN

Computer Science Department, Technion—Israel Institute of Technology, Haifa, Israel

ABSTRACT: This paper introduces a novel graphical image representation consisting of a single curve—the one-liner. The first step of the algorithm involves the detection and ranking of image edges. A new *edge exploration* technique is used to perform both tasks simultaneously. This process is based on *image flows*. It uses a gradient vector field and a new operator to explore image edges. Estimation of the derivatives of the image is performed by using local Taylor expansions in conjunction with a weighted least-squares method. This process finds all the possible image edges without any pruning, and collects information that allows the edges found to be prioritized. This enables the most important edges to be selected to form a *skeleton* of the representation sought. The next step connects the selected edges into one continuous curve—the one-liner. It orders the selected edges and determines the curves connecting them. These two problems are solved separately. Since the abstract graph setting of the first problem is NP-complete, we reduce it to a variant of the traveling salesman problem and compute an approximate solution to it. We solve the second problem by using Dijkstra's shortest-path algorithm. The full software implementation for the entire one-liner determination process is available.

KEYWORDS: image representation; image flows; traveling salesman problem; edge exploration; Taylor expansion

INTRODUCTION

A one-liner representation of an image is a continuous curve passing through the important image edges. This curve is not necessarily built only from the image edges. Building a one-liner representation of a given image is a complex process, involving multiple phases and affected by multiple factors. Picasso and Calder are well-known artists that have produced beautiful one-liner images. Although a computerized system cannot compete with an artist's rendition, the work of Picasso and Calder shows that *good* image representations with one-liners are indeed possible.

The first step consists of detecting and linking the image edges. The quality of this step is crucial for success in the creation of a one-liner. Conventional edge-detection schemes include three operations: smoothing, edge detection, and edge labeling. Image derivative estimation is necessary in order to detect local grey-level changes, smoothing is performed to reduce the noise, and labeling is necessary to localize the edges and suppress false edges. Sobel[1] and Prewitt[2] have provided the most widely used operators for edge detection. They focus on detecting pixels with

Address for correspondence: Alfred Bruckstein, Computer Science Department, Technion IIT, 32000 Haifa, Israel.
freddy@cs.technion.ac.il

high gradient magnitude by convoluting the original image with a filter. The edge detectors of Sobel and Prewitt are based on thresholding the gradient modulus. This results in the need to *thin* the edges found. These operators are very sensitive to noise. To overcome those problems, more advanced methods involving smoothing operators that precede edge detection were proposed by Marr and Hildreth[3] and by Torre and Poggio.[4] Marr and Hildreth detect edges by using the zero-crossing of the Laplacian of a Gaussian convoluted with the image. Together with the positive effect of noise reduction, the smoothing operator also has a negative effect on information loss. Canny[5] proposed an algorithm that labels local maxima along the direction of the gradient vector.

Eua-Anant and Udpa[6] proposed a novel approach based on a vector field image model. They defined a new edge-tracking operator and developed a boundary-extraction algorithm based on particle motion in a force field. Our approach is similar to that of Eua-Anant and Udpa. We refer to the process of edge-detection and linking as *edge exploration*. Conventional edge detectors concentrate on looking for the points with a locally-maximal gradient magnitude. Our approach extends this idea to exploring an edge once we have found a seed portion of it. An association of the gradient vector field of the image with a hill gives us a good intuition of this approach. The height of every point of the hill is defined as the gradient magnitude in the corresponding image point. Then, an edge in the image corresponds to a ridge of this hill. The idea is to take advantage of the gradient orientation and to use both altitude and orientation to *climb* on the ridge of the hill and then to stay there. Thus, the edge-exploration process is a combination of selecting the starting point, moving toward the ridge of the hill, reaching it, and then moving in the direction that keeps the gradient magnitude locally maximal. The proposed process finds all the possible image edges without any pruning. A complex image may have many explored candidate edges, with different levels of importance. Selecting a high quality subset of representative edges is crucial, so we prioritize the edges according to image-driven information collected during the edge exploration. The selected set of image edges does not necessarily form a connected curve. We present an elegant way to resolve the problems of ordering the selected edges and connecting them into a one-liner by using image-driven curves. This problem is NP-complete in its abstract form. We reduce the problem of ordering the edges to an instance of the classic traveling salesman problem (TSP) and use a freely-available TSP solver to solve it. To complete a full one-liner representation of the image we connect the edges according to the TSP solution. We connect two endpoints of consecutive edges by using the well-known Dijkstra's shortest path algorithm to find a path passing through points with maximal gradient magnitude.

THE EDGE-EXPLORATION PROCESS

The edge-exploration step detects the image edges without any additional processing, such as thinning or skeletonization. The algorithm is based on image flows, using an edge-exploration operator in the gradient vector field. In a complex image almost every pixel has a non-zero gradient magnitude and there are many local maxima. Setting a threshold on the minimum accepted gradient magnitude may cause a significant loss of information in a very early stage of the processing. Therefore, we use another approach. We start by finding all the edges regardless of their contribution to the image representation. Then, we prioritize them using information collected during the edge exploration. Finally, we select the most important edges. A typical image consists of a large number of edges (up to hundreds of thousands). Some of the edges represent crucial parts of the image and contribute to its representation. Other edges are the result of noise or belong to the background. The classification of image edges is difficult, and the quality of the image representation greatly depends on the success of the selection process.

Image Flows

We first compute the gradient vector field of the image: magnitude and orientation for every pixel. We use the Taylor expansion to estimate the gradient vector and its low-order derivatives. This provides a system of linear equations with the image derivatives and estimation errors as unknowns. To minimize the error we use a weighted least-square method. The gradient vector field at each point consists of a vector orthogonal to the image gradient and having the same gradient as the magnitude. This vector points in a direction that preserves image intensity. The vector field defines *flows* in the image, similar to flows in fluids. The local force yielding this flow will move an imaginary particle in the direction of the field, following the same image intensity level regardless of the gradient magnitude. This will not work in complex images, where the intensity changes along image edges. Thus, we introduce an additional force that drives the flow in the direction of increasing gradient magnitude. We apply two forces: one leading an object on an equal intensity level of the image and the other pushing toward the higher gradient magnitudes. This strategy allows us, not only to explore the edge starting from an edge point, but also to discover the nearest edge from any point in the field.

Edge-Tracking Operator

The edge-exploration operator provides us with the locations of the edges, when it is applied to the gradient vector field. It works on a subpixel level to improve the quality of the resulting edges. Denote an input image by $I(x, y)$ and the corresponding gradient vector field by $\nabla I(x, y) = [p(x, y), q(x, y)]$, where p and q are the partial derivatives of the image in the pixel (x, y). The tracking-operator is given by

$$\frac{\partial P}{\partial t} = \alpha \frac{(\nabla I(x, y))_\perp}{\|\nabla I(x, y)\|} + (1 - \alpha) \frac{\nabla \|\nabla I(x, y)\|}{\|\nabla \|\nabla I(x, y)\|\|}, \quad (1)$$

where $0 \leq \alpha \leq 1$. The left-hand side of Equation (**1**) represents a shift in both coordinates, required in order to move to the next edge location. The right-hand side contains two terms. The first term represents motion in the direction that preserves image intensity. This is achieved by pointing orthogonally to the gradient direction. The second term represents motion in the direction of maximum change of the image gradient magnitude. The exploration process need not start from an image edge point. When approaching an edge, the second term points to the "fastest" way to reach the edge and has the major influence on the selection of the subsequent location. After the edge is reached, the first term takes the leading role to ensure following the edge, whereas the second term slightly fixes the direction toward a growing gradient magnitude.

The parameter α controls the proportion between the two components of the operator. A value of $\alpha = 0.1$ suits all gray-scale images with which we have experimented. For some black/white (b/w) images this parameter can be reduced even further to $\alpha = 0.05$. We explain this by the fact that b/w images have a very sharp change of the gradient magnitude near the edges, allowing us to reduce the influence of the second component. Notice that both components of the operator are normalized, enabling us to control the displacement between consecutive points and to keep it in a subpixel resolution. A large value of the gradient magnitude or a fast change in the gradient magnitude may lead to a step of multiple pixel magnitude. This may not only hamper the smoothness of the edge-tracking process, but may also cause losing the edge completely. The fine-tuning of the granularity in this step can be achieved by multiplying (1) by a scaling factor δ. We used $\delta = 0.25$ in order to enforce a quarter-pixel granularity. The real coordinate space is used to minimize the estimated errors. Due to the nature of Equation (1), the movement is not assumed to be discrete and it allows a smooth tracking of the edges. However, the first and second derivatives of the image altitude are calculated for discrete pixels. To approximate the derivatives at a subpixel level we used a linear interpolation method based on estimated values at the four corners of a pixel.

The Full Exploration Algorithm

We use a scan-line algorithm to explore and prioritize all image edges. For each image pixel, the algorithm finds a candidate edge corresponding to it. The algorithm moves along the edge by using our edge-exploration operator, the resolution of the operator being at least a single pixel. The flow along the tracked edge has three termination conditions: (1) hitting a "white" pixel (a pixel with zero gradient magnitude); (2) entering a stream of locations belonging to an already explored edge; and (3) exiting the image, that is, being led to a place that is located outside of the image plane.

After the candidate edge is found, a host of heuristic algorithms can improve the quality of its tracing process. The edge is considered to be a *good* or a *stable* edge if it runs over places having a relatively smooth distribution of the overall gradient magnitude. If this is not so, we implement a subdivision of a candidate edge into several parts that are considered separately in the sequel.

During the exploration process the algorithm also collects the information necessary for edge prioritization. This involves two parameters calculated for each edge: the number of exploration streams flowing into the edge (indicating its importance), and the average gradient magnitude along it (indicating its intensity). The edges are prioritized according to both parameters, and the user can select the most important edges.

THE EDGE-LINKING PROCESS

At this stage we want to connect the selected edges into a continuous curve—a one-liner. This goal can be separated into two separate tasks: (1) order the set of edges so that their connected version is *nice* and intuitive; and (2) compute a set of curves that connect the edges. We accomplish each task separately and combine the solutions together.

Ordering the Edges

We seek an *optimal* ordering of the edges in which they will be connected. Since, formally, the one-liner is not required to not intersect itself, the one-liner problem can be reduced to the problem of connecting a given set of edges into one curve with an upper bound on the length of the connectors. We were able to prove that this abstract setting of the problem is NP-complete by a reduction from the Hamiltonian path/circle problem. Thus, it is unlikely to find efficiently the optimal order to connect the edges. Instead we convert the problem to an instance of the TSP and use a publicly-available TSP solver to find an approximate solution.

The one-liner problem differs slightly from the classic TSP: Our problem not only contains $2n$ vertices and all possible weighted edges connecting them, but it is also constrained by a set of n edges that must be part of the solution. To satisfy this we add a new point in the middle of every such edge, and assign distances (edge weights) between the points so as to force the TSP solution to contain all the preselected edges. Although the TSP problem is NP-complete,[7,8] many special cases of it, in particular, our one-liner problem, can be approximated efficiently. We used the publicly-available LK program (by Neto,[9] which implements the Lin–Kernighan heuristic. It mostly follows the design outlined by Johnson and McGeoch.[10]

Connecting the Edges

To produce a one-liner fully adapted to the given image, we should use image-driven curves to smoothly connect the edges. To connect endpoints of different edges, we seek a curve with maximal average gradient magnitude, which does not have loops and is located in the neighborhood of the connected endpoints. To this end, we called upon the shortest-path graph algorithm of Dijkstra.[11] Given the two connector endpoints, we build a graph that has a vertex for each pixel in the neighborhood and an edge connecting the pixel with its neighboring pixels. The weight of the edge is the reciprocal of the gradient magnitude of the neighboring pixel. We use an implementation of the algorithm found in SPLIB (Cherkassky, Goldberg, and Radzik[12]).

EXPERIMENTAL RESULTS

We have implemented the processes described above using two external packages: the LK program, for solving the Euclidean TSP problem, and an implementation of Dijkstra's shortest-path algorithm. We then experimented extensively with the software. We present below a series of images that demonstrate the various stages of the algorithm.

FIGURE 1 demonstrates the operation of the edge-exploration step on some simple images. Very few edges need to be selected to obtain quite nice representations. FIGURE 2 shows the application of the same algorithm on a few more complex images.

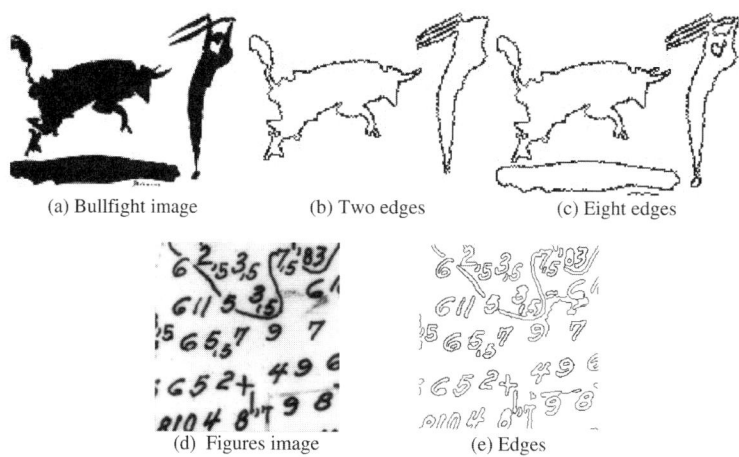

FIGURE 1. Main edges determined by the edge-exploration process in simple images.

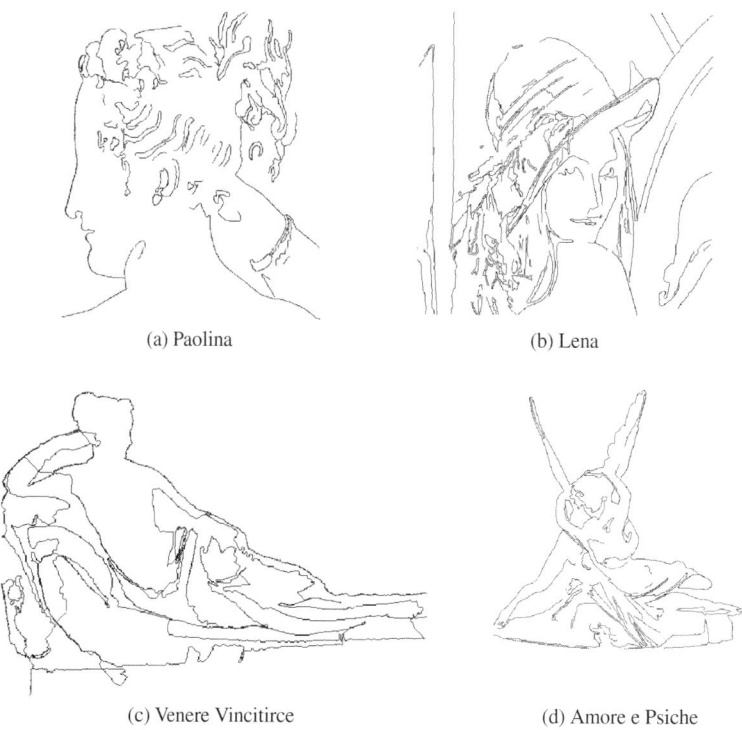

FIGURE 2. Main edges in more complex images.

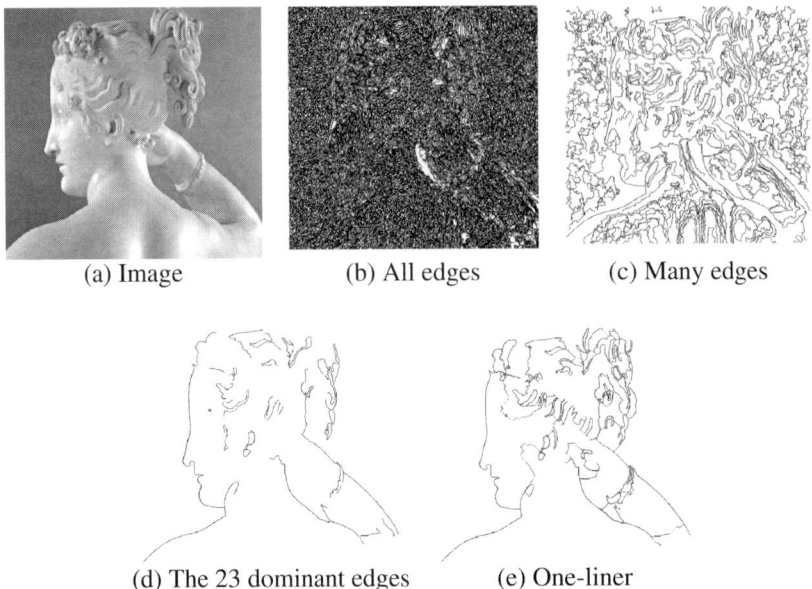

FIGURE 3. Paolina, one-liner image representation.

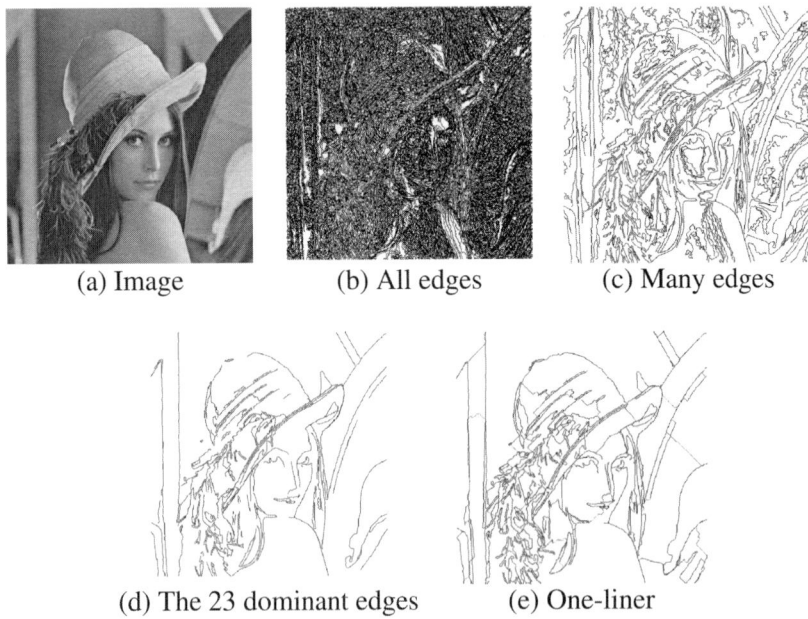

FIGURE 4. Lena, one-liner image representation.

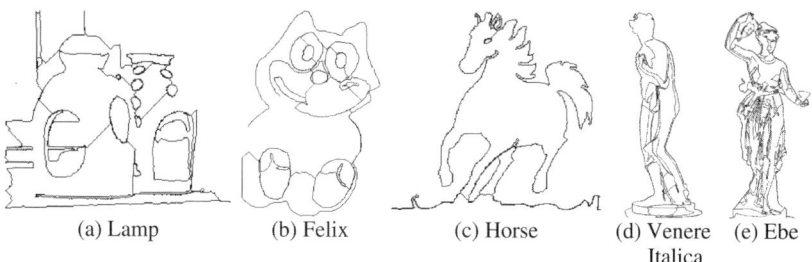

(a) Lamp (b) Felix (c) Horse (d) Venere Italica (e) Ebe

FIGURE 5. Various one-liners.

FIGURES 3 and 4 show various steps of the algorithm, demonstrating the effect of varying the edge-prioritization parameters and show one-liner representations of each image.

FIGURES 3b and 4b show the edges found when neither prioritization parameter (average gradient and number of streams flowing into the edge) were constrained. FIGURES 3c and 4c show the effect of requiring at least 40 incoming streams.

Finally, FIGURE 5 shows one-liner representations of various gray-scale and b/w images.

We ran our experiments on a P-III 600Mhz personal computer with 256MB of SRAM. TABLE 1 shows some running times, measured for three gray-scale images differing in their dimensions and complexities. For all the images we used the parameter value $\alpha = 0.1$ for the edge-exploration step. The statistics for this step are shown in the first section of the table. The second section shows statistics for the TSP and Dijkstra's implementations.

TABLE 1. Performance of the algorithm

Image	Lamp	Paolina	Felix
Edge exploration process			
Dimensions	256 × 256	512 × 480	240 × 346
Derivative estimation (MS)	841	3,135	1,072
Path exploration (MS)	7,190	21,531	9,954
Explored paths	7,703	31,865	10,515
Connecting selected edges			
Selected edges	122	574	182
TSP vertices	183	861	273
TSP execution (MS)	140	1,422	250
Graph vertices (average)	2,301	2,248	2,428
Dijkstra's algorithm (MS)	531	2,213	801

ACKNOWLEDGMENT

This research was supported in part by the Fund for the Promotion of Research at the Technion, IIT.

REFERENCES

1. LYVERS, E.P. & O.R. MITCHELL. 1988. Precision edge contrast and orientation estimation. IEEE Trans. Pattern Analysis Machine Intelligence. **10:** 927–937.
2. PREWITT, J.M.S. 1970. Object enhancement and extraction. *In* Picture Processing and Psychopictorics. B.S. Lipkin & A. Rosenfeld, Eds. Academic Press, New York.
3. MARR, D. & E.C. HILDRETH. 1980. Theory of edge detection. Proc. Roy. Soc. Lond. **207:** 187–217.
4. TORRE, V. & T.A. POGGIO. 1980. On edge detection. IEEE Trans. Pattern Analysis Machine Intel. **8:** 147–163.
5. CANNY, J.F. 1986. A computational approach to edge detection. IEEE Trans. Pattern Analysis Machine Intel. **8:** 679–698.
6. EUA-ANANT, N. & L. UDPA. 1999. Boundary detection using simulation of particle motion in a vector image field. IEEE Trans. Image Process. **8:** 1560–1572.
7. GAREY, M.R., R.L. GRAHAM & D.S. JOHNSON. 1976. Some NP-complete geometric problems. Proc. ACM Symp. Theory Comput, 10–22.
8. PAPADIMITRIOU, C.H. 1977. Euclidean TSP is NP-complete. Theor. Comput. Sci. **4:** 237–244.
9. NETO, D. 1999. Efficient cluster compensation for Lin-Kernighan heuristics. Ph.D. Thesis, Department of Computer Science, University of Toronto.
10. JOHNSON, D.S. & L.A. MCGEOCH. 1997. The Traveling Salesman Problem: A Case Study. Local Search in Combinatorial Optimization. John Wiley & Sons.
11. DIJKSTRA, E.W. 1959. A note on two problems in connection with graphs. Numer. Math. **1:** 269–271.
12. CHERKASSKY, B.V., A.V. GOLDBERG & T. RADZIK. 1996. Shortest paths algorithms: theory and experimental evaluation. Math. Program. **73:** 129–174.

Simulation and Identification of Deterministic Structures in Thermal and Magnetic Convection

KEMAL HANJALIĆ AND SAŠA KENJEREŠ

Department of Applied Physics, Delft University of Technology, Delft, the Netherlands

ABSTRACT: Large-scale vortical structures, found in many flows in nature, in laboratory or industrial equipment, have often a deterministic character. Simulation, identification, visualization, and interpretation of such structures is essential for understanding the physics of turbulence and transport processes, and for their control, but still poses a challenge especially in complex flows at high Reynolds and Rayleigh numbers. We show that such structures can be simulated by a transient Reynolds-averaged Navier–Stokes (T-RANS) approach. Various structure–identification methods show that the captured deterministic structures resemble closely those obtained from experiments or direct numerical simulations (DNS). The potential of the T-RANS method is illustrated by examples of classical and magnetic Rayleigh–Bénard (R-B) convection in a range of Rayleigh numbers that are at present inaccessible to experiments or other simulation techniques.

KEYWORDS: deterministic structures; identification; thermal convection; magnetic convection

INTRODUCTION

Many flows, laminar or turbulent, contain well-organized, coherent flow structures that play major role in transporting momentum, heat, and species. The dynamics of entrainment, mixing, chemical reaction, combustion, and noise generation is often directly governed by such structures. In many internal flows they provide the major communication link between the bounding walls, thus controlling friction and drag, heat and mass transfer. In turbulent flows they may represent the well-organized eddies (*true* turbulence), or a form of secondary motion with inherent, but deterministic (organized) unsteadiness, coexisting and interacting (production, transport and dissipation of turbulence) with the random fluctuations. Such structure may appear in the early stage of turbulence generation (e.g., vortex shedding and transition to turbulence along a solid wall), or are continuously present in some flow areas (e.g., near-wall streaks, ejections and sweeps, or the rolling vortices at the edges of free shear layers). Whatever their origin, the identification of such structures and their morphology is the key prerequisite to understanding their role in controlling the flow and transport processes. Properly identified, coherent structures can be

Address for correspondence: Kemal Hanjalić, Ph.D., Department of Applied Physics, Delft University of Technology, Lorentzweg 1, 2628 CJ Delft, the Netherlands.
hanjalic@ws.tn.tudelft.nl

subjected to various methods of flow control and tested for optimum performance for the preset optimization criteria. Comparison of the structure morphology can also serve as a basis for comparing various simulation methods, as is done in this work.

Several techniques for structure identification have been proposed and used successfully, but they all require detailed information about the instantaneous velocity and scalar fields. Such information can be provided by direct or large-eddy numerical simulations (DNS, LES), or by extensive and tedious experiments. However, all these techniques have serious limitations: because of excessive demand on computational resources, DNS and LES are still restricted to simple geometries and low-to-moderate Reynolds and Rayleigh numbers. Similarly experiments are still limited in the extent of information they can provide, and are very time-consuming. Handling complex geometries and, especially, resolving the flow very close to solid walls bounding the flow, are the challenges that remain. On the other hand, the RANS (Reynolds-average, Navier–Stokes) approach, used widely for computational prediction of complex turbulent flows, cannot recognize any specific eddy structure because of inherent one-point and single-scale assumptions.

In order to overcame the cost and capacity problems and still resolve large-scale structures—at least in flows where they play a dominant role—several combined or hybrid methods have been proposed recently, aimed at taking advantages of the simplicity and computational efficiency of RANS, and the potential of LES to fully resolve the large-scale part of the turbulence spectrum. One such approach, called transient RANS (T-RANS) and its application to the simulation of classic and magnetic Rayleigh–Bénard convection is discussed here. The simulated coherent roll-cell structures obtained by T-RANS are first compared with DNS results to illustrate the potential of T-RANS and the method is then applied to predict cases at higher Rayleigh numbers, with bottom wavy walls and with simultaneous imposition of a magnetic field.

COHERENT STRUCTURE IDENTIFICATION AND VISUALIZATION

A variety of coherent structural shapes have been reported in the literature in various flows (hence, different origin) and using different tools for their identification. These structures show a range of features that are difficult to describe in a unique manner, so that a complete and unique definition of coherent motion is still missing despite a number of definitions proposed in the literature.[1] Robinson[2] defined coherent motion as "three-dimensional region over which at least one fundamental flow variable exhibits significant correlation with itself or with another variables over a range of space and/or time that is significantly larger than the smallest local scales of flows". Robinson identified no less than eight classes of coherent motion, of which only *one* class are the "*vortical* structures in various forms". In contrast to this general formulation, most other definitions proposed associate coherent structures with vortices; for example, as "closed circular or spiral pathlines or instantaneous streamlines mapped onto a plane normal to the vortex core…in reference frame moving with the vortex center,"[2] or "connected turbulent fluid mass with instantaneous phase correlated vorticity over its spatial extent" (Hussain[3]). Jeong and Hussain[4] stated simply that coherent structures in turbulent flows are now commonly regarded as vortices, and modified slightly the previous definition by replacing "vorticity" by "domains of phase-correlated negative λ_2" (see below). Lesieur *et al.*[5] defined *coherent vortices* in turbulent flows as regions satisfying three (qualitative) conditions: (1) that the vortex concentration should be high enough so that a local roll up of the surrounding fluid is possible, (2) that they should keep their shape longer than the local turnover time, and (3) that they should be unpredictable! The latter criteria signifies the application of the definition to turbulent flow, thus distinguishing common patterns, such as vortex shedding in turbulent (unpredictable) flow, from the supposedly predictable patterns in laminar flow.

Once the coherent structure are identified with vortices, the identification criteria can be defined as isosurfaces of characteristic variables defining the strength of vorticity $\omega = \text{rot}\,u$ either in term of its absolute or relative modulus, or relative to strain-rate intensity. Most definitions are directly or indirectly related to low pressure areas (vortex tube in a moving frame), as follows from Euler equation

$$\frac{\partial u}{\partial t} + \omega \times u = -\frac{1}{\rho}\nabla P, \qquad (1)$$

where $P = p + \rho u^2/2$ is the dynamic pressure and ρ is fluid density. If the first term in Equation (1) can be neglected (Lesieur's second criterion), Euler equation reduces to the cyclostrophic balance $\rho\omega \times u = -\nabla P$. Although both vorticity magnitude or pressure lows can be used, each has some advantages and disadvantages, depending primarily on the type of flow and structures considered.

Jeong and Hussain[4] argue that three common intuitive indicators of vortices—pressure minimum, closed or spiraling streamlines and pathlines, and isovorticity surfaces—are inadequate in detecting vortices in an unsteady flow in general. Indeed, regions with low dynamic pressure can also be found in irrotational flows, or can misrepresent a vortex due to the non-local character of the pressure, which may have a larger scale than the vortex core. Circular or spiraling pathlines may not be closed during the vortex lifetime due to non-linear vortices interactions and breakdown. The use of streamlines is also questionable in moving frames since they are not Galilean invariant. Vortex intensity $|\omega|$, considered as a useful criterion for free shear flows, may be inadequate for high-shear near-wall regions because of contamination with the background shear strain.

Recently, several more general definitions of a coherent vortices, all Galilean invariant, have been proposed, aimed at separating or filtering the vorticity from the strain rate. The basis is the velocity gradient tensor, which is decomposed into strain rate and rotation rate, $A_{ij} = \partial u_i/\partial x_j = S_{ij} + \Omega_{ij}$. Hunt *et al.*[6] argued that the region with positive second invariant Q of the characteristic equation for A_{ij} defines an eddy. The Q criterion can be expressed in various forms: for example,

$$Q = \frac{1}{2}(\Omega_{ij}^2 - S_{ij}^2) = \frac{1}{4}(\omega_i^2 - 2S_{ij}^2) = \frac{1}{2\rho}\nabla^2 p. \qquad (2)$$

As seen from Equation (2), Q represents, in fact, the excess of rotation rate over the strain rate, but corresponds also to the pressure minimum. It is noted that another criterion, the kinematic vorticity number N_k, defined as vorticity modulus normalized with the strain-rate, introduced by Truesdall[4] as early as 1953, is related to Q:

$$N_k = \left(\frac{\omega_i^2}{S_{ij}^2}\right)^{1/2} = \left(1 + \frac{2Q}{S_{ij}^2}\right)^{1/2}. \tag{3}$$

As shown below, N_k and Q often give very similar results. Chong et al.[7] defined the vortex core as a region with complex eigenvalues of the characteristic equation for A_{ij}, corresponding to the closed or spiraling streamline pattern in moving frame, which occur when the discriminant Δ is positive: that is,

$$\Delta = \left(\frac{Q}{3}\right)^3 + \left(\frac{R}{2}\right)^2 > 0. \tag{4}$$

Yet another criterion has been proposed by Jeong and Hussain,[4] based on the evolution equation for the strain rate (the gradient of the Navier–Stokes equations):

$$\frac{D}{Dt}S_{ij} + \Omega_{ij}^2 + S_{ik}S_{kj} = -\frac{1}{\rho}\frac{\partial^2 p}{\partial x_i \partial x_j}. \tag{5}$$

If the unsteady term is neglected, the local pressure minimum occurs if the Hessian of pressure has two positive eigenvalues, or $S^2 + \Omega^2$ has two negative eigenvalues. This condition can be reduced to the criterion that the second eigenvaue λ_2 be negative.

Applications of all of the criteria listed above in various flows have been reported in the literature. Although in some flows different criteria may give very similar outcome, in other flows (e.g., close to a solid wall) this may not be the case and a careful selection is needed to reach clear identification of coherent structures. We show below some results of their application in classic and magnetic R-B convection.

VERY LARGE EDDY SIMULATIONS BY T-RANS

The methods that combine the RANS and LES strategies can generally be classified as very large eddy simulations (VLES). The name implies a form of LES (not necessarily based on grid-size filtering) with a cut-off filter at much lower wave number, or simply solving ensemble averaged equations. The basic rationale behind VLES is: resolve only very large, coherent, or deterministic structures and model the rest! Modeling a larger part of the spectrum requires a more sophisticated model than the standard subgrid-scale model for LES; that is, a form of RANS model that is not related to the size of the numerical mesh. On the other hand, as compared with the conventional RANS, the model is required only for the incoherent random fluctuations, during which the large scales are resolved. The solution of the resolved part of the spectrum can follow the traditional LES practice, using grid size as a basis for defining the filter (hence the name hybrid RANS/LES), or solve ensemble- or conditionally averaged Navier–Stokes equations. We follow the latter approach, named T-RANS[8,9] and demonstrate its application to confined turbulent flows subjected separately or jointly to thermal buoyancy and magnetic field. The T-RANS resembles the semideterministic method (SDM)[10] in the sense that the instantaneous flow can be decomposed into unsteady ensemble-averaged (organized) motion and random (incoherent) fluctuations, so that the instantaneous flow property $\hat{\Psi}(x_i, t)$ can be written as a sum of time-mean, $\bar{\Psi}(x_i, t)$, deterministic $\tilde{\Psi}(x_i, t)$ and random $\psi(x_i, t)$ parts. That is,

$$\hat{\Psi}(x_i, t) = \overline{\Psi}(x_i) + \tilde{\Psi}(x_i, t) + \psi(x_i, t) = \langle \Psi(x_i, t) \rangle + \psi(x_i, t), \qquad (6)$$

where $\langle \Psi(x_i, t) \rangle$ is an ensemble-averaged quantity that is fully resolved by numerical computation. The ensemble-average (denoted by $\langle \cdot \rangle$) equations for momentum, energy and electric potential can, therefore, be written as follows:

$$\frac{\partial \langle U_i \rangle}{\partial t} + \langle U_j \rangle \frac{\partial \langle U_i \rangle}{\partial x_j} = \frac{\partial}{\partial x_j}\left(\nu \frac{\partial \langle U_i \rangle}{\partial x_j} - \tau_{ij} \right) - \frac{1}{\rho} \frac{\partial (\langle P \rangle - P_{\text{ref}})}{\partial x_i}$$

$$+ \beta g_i (\langle T \rangle - T_{\text{ref}}) + \underbrace{\frac{\sigma}{\rho}\left(-\varepsilon_{ijk} \langle B_k \rangle \frac{\partial \langle \Phi \rangle}{\partial x_j} + \langle U_k \rangle \langle B_i \rangle \langle B_k \rangle - \langle U_i \rangle \langle B_k^2 \rangle \right)}_{\langle F_i^L \rangle} \qquad (7)$$

$$\frac{\partial \langle T \rangle}{\partial t} + \langle U_j \rangle \frac{\partial \langle T \rangle}{\partial x_j} = \frac{\partial}{\partial x_j}\left(\frac{\nu}{Pr} \frac{\partial \langle T \rangle}{\partial x_j} - \tau_{\theta j} \right) \qquad (8)$$

$$\nabla^2 \langle \Phi \rangle = \nabla \cdot [\langle U \rangle \times \langle B \rangle] \qquad (9)$$

where $\langle F_i^L \rangle$ is the Lorenz force, $\langle \Phi \rangle$ is the electric potential, $\langle B_i \rangle$ is the imposed magnetic field (all ensemble averaged), τ_{ij} and $\tau_{\theta j}$ are the stress and flux contribution due to the incoherent motion, provided from a subscale (RANS) model.

It is noted that the total long-term averaged second moments consist of the resolved (deterministic) and incoherent (random) part, which are assumed non-interacting

$$\overline{\hat{\Psi}\hat{Y}} = \overline{\overline{\Psi}\overline{Y}} + \overline{\tilde{\Psi}\tilde{Y}} + \overline{\psi\gamma} = \overline{\langle \Psi \rangle \langle Y \rangle} + \overline{\psi\gamma}. \qquad (10)$$

Both parts are expected to be of the same order of magnitude in general, with the modelled contribution prevailing in the near-wall regions, where the deterministic motion is weak.

Verification of the T-RANS Method by Structure Identification

The primary criteria for verifying a solution method in fluid mechanics and transport phenomena involve integral parameters (friction, heat, and mass transfer coefficient) and averaged velocity, temperature, and concentration fields. However, such properties give no indication about vortical structures and reasonable agreement with reference data (e.g., provided from experiments) can sometimes be obtained even if the structures are fully ignored (e.g., RANS solutions) or not properly captured. Either the coherent structures play no role in spatial and temporal nonuniformity (e.g., in wall attached—non-separating flows, dominated by pressure gradient), or the errors in models of various terms in equations compensate each other. Needless to say, in the latter case strong spatial and temporal variations in flow properties with possible excessive values (e.g., extreme in friction or heat transfer, hot spots) may either remain hidden or be predicted incorrectly.

Since VLES is an approximate method that is aimed at capturing large coherent or deterministic vortical structures—where they exist, the best verification of the approach is to compare such identifiable structures obtained by VLES with those computed by DNS or LES. Here, one has a choice between the several identification techniques and criteria defined above: the relative or maximum vorticity modulus, kinematic vorticity number N_k, the Q criterion, the λ_2 criterion, and others. All these criteria are essentially related to the local pressure minimum, which is assumed to be the major indication of the existence of a coherent vortical structure. However, because the T-RANS captures only the large deterministic structures, care is needed to define proper thresholds for various criteria1.[1,12] This is illustrated in the next section.

EXAMPLES OF CLASSIC AND MAGNETIC RAYLEIGH–BÉNARD CONVECTION

We present results from the application of various methods for the education and identification of distinct structure morphology in examples of classic Rayleigh–Bénard (R-B) convection over walls with flat and corrugated topography, and in the case when the system is also subjected to a magnetic field (magnetic R-B convection). First, we use these education methods to validate the VLES approach and then provide some insight into the mechanism and consequences of the structure reorganization on the overall flow properties and wall heat transfer. These structures are known to exist in such flows as a consequence of self-organization of convective roll cells. Large-scale deterministic structures captured with VLES can be compared with those obtained by DNS and LES for the same Rayleigh number, using several identification criteria. Next, results of structure reorganization in R-B convection at very high Rayleigh numbers are presented, illustrating the potential of VLES to handle such flows and to get insight into eddy structure morphology in conditions that are inaccessible to either DNS, LES, or even experiments. The last example involves combined effects of thermal buoyancy and magnetic field in magnetic R-B convection, which is featured by a strong reorganization of large-scale vortical structures, with a profound effect on wall heat transfer. The figures shown below illustrate some of the findings.

FIGURE 1 shows a quantitative comparison of the captured instantaneous vortical structures in classic R-B convection using, in parallel, randomly selected single realizations from DNS and T-RANS for the same Rayleigh and Prandtl numbers. In the background the numerical grid used in each method is presented. The structures are identified by the kinematic vorticity number N_k. The picture depends on the selected threshold value, and for $N_k = 2$ the figure illustrates the basic difference. As expected, the T-RANS captures only the large structures, but their shape, morphology, and distribution in both DNS and T-RANS are very similar.

FIGURE 1. Qualitative comparison of captured vortical structures ($N_k = 2$) for DNS and T-RANS realizations: $Ra = 6.5 \times 10^5$, $Pr = 0.71$; perspective view with underlying mesh.

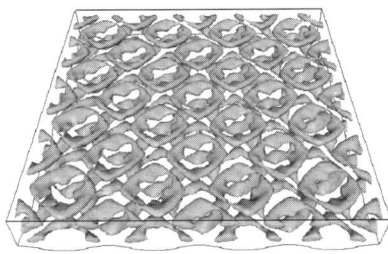 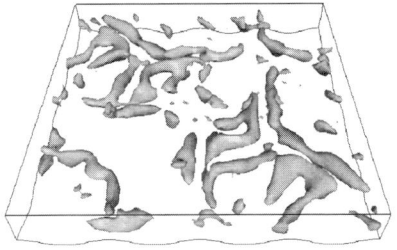

FIGURE 2. Time evolution and spatial organization of large coherent structures defined and identified by $N_k = 2$ at $\tau^* = 50$ (**left**) and $\tau^* = 200$ (**right**), three-dimensional waviness. (COLOR PLATE 1.)

The subsequent figures illustrate the potential of T-RANS for studying morphological changes and reorganization of coherent structure when subjected to external perturbations. We consider first the effects of changes in boundary configuration and then the effect of a magnetic field. These two cases can be considered as examples of *boundary control* and *body-force control* of flow and transport processes. FIGURE 2 shows two sequences in the time evolution of the organized deterministic structures in R-B convection over a three-dimensional wavy bottom wall. The structures, again identified by $N_k = 2$ as in FIGURE 1, strikingly reflect the bottom wall topography in the initial phase, but the pattern gradually fades away due to intensive structure interaction and mixing. FIGURE 3 shows effect of a strong, uniform, vertically oriented magnetic field (Hartmann number $Ha = 100$) applied to R-B convection at $Ra = 10^7$.

The Lorentz force generated by the magnetic field acts in horizontal plane and suppresses horizontal movement and turbulence fluctuations. The momentum due to buoyancy force is, thus, consumed mainly in vertical motion of plumes. The effect can be visualized simply by isosurfaces of the vertical components of the instantaneous velocity W, as shown in FIGURE 3 (normalized by buoyancy velocity), showing the reorganization of typical polygonal structures into a multitude of small concentrated cylindrically shaped plumes. Another view of the same process is shown in FIGURE 4, where isosurfaces of the vorticity kinematic number $N_k = 1.5$ for non-magnetic and magnetic convection, the latter for two Hartmann numbers ($Ha = 20$ and

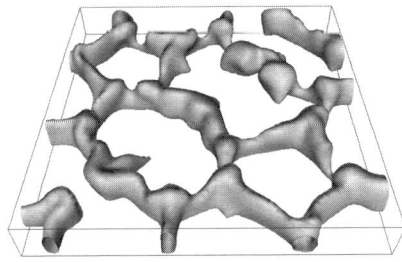 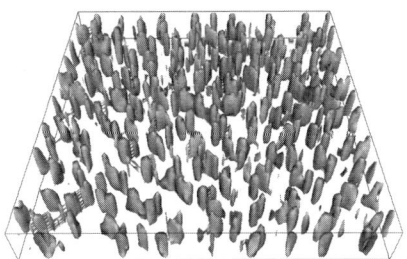

FIGURE 3. Vertical velocity component, $W^* = W/\sqrt{\beta g_i \Delta T H} = 0.1$, $Ha = 0$ (**left**) and $Ha = 100$ (**right**).

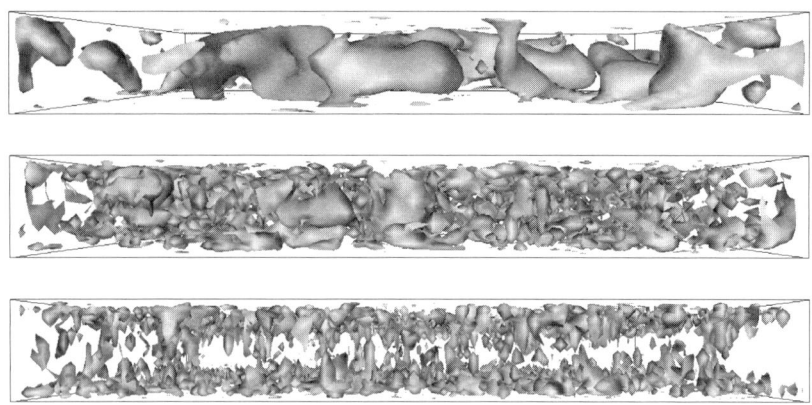

FIGURE 4. Side view of kinematic vorticity number distributions for various intensities of magnetic field, $Ha = 0$ (**top**), $Ha = 20$ (**middle**), $Ha = 100$ (**bottom**), $Nk = (\omega_i^2/2S_{ij}^2)^{1/2} = 1.5$.

100) corresponding to moderate and strong magnetic fields. A very similar picture is obtained using other identification criteria; for example, the Q parameter (as shown in FIGURE 5), or discriminant Δ, or λ_2 (Hanjalić and Kenjereš[9]). Although these criteria are regarded as superior in identifying vortical structure because they separate vorticity from the strain rate, apart from obvious reduction in scales FIGURE 4 does not show clearly the real process of structure reorganization, as does FIGURE 3. Even better insight is obtained with isosurfaces of absolute vorticity, with a carefully selected threshold, here $|\omega| = 5\% |\omega_{max}|$, FIGURE 6, which captures the essence of the structure reorganization. A typical plume structure in a classic R-B convection, asso-

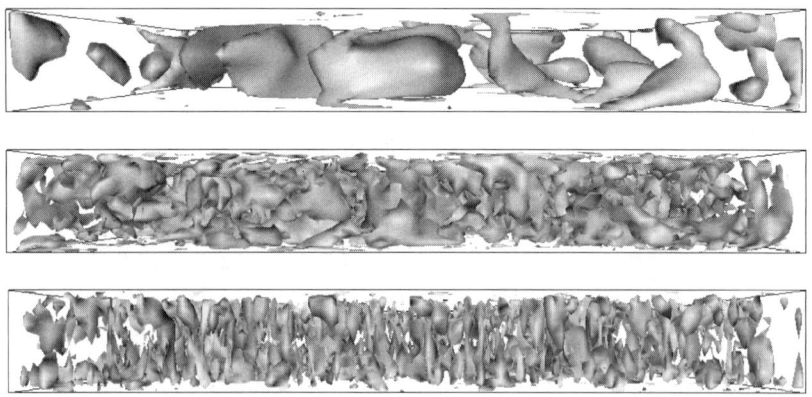

FIGURE 5. Second invariant of the velocity gradient tensor distributions for various intensities of magnetic field, $Ha = 0$ (**top**), $Ha = 20$ (**middle**), and $Ha = 100$ (**bottom**), $Q = 0.15$.

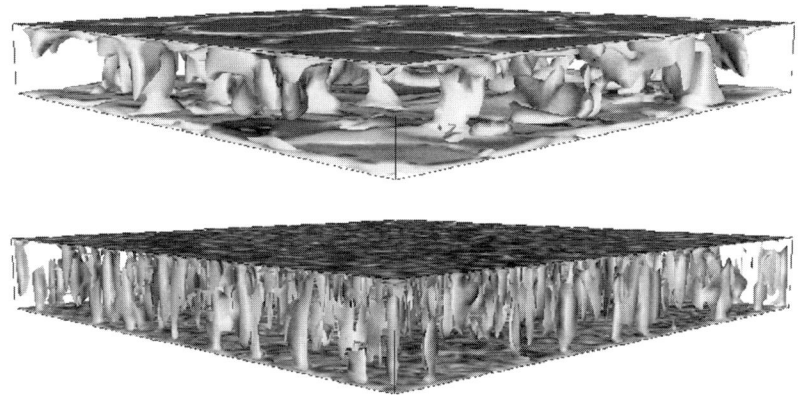

FIGURE 6. Effect of magnetic field on spatial reorganization of vortical structures, $Ra = 10^7$, $Ha = 0.100$, modulus of vorticity $|\omega_i| = \sqrt{\omega_x^2 + \omega_y^2 + \omega_z^2} = 5\% |\omega_{max}|$. (COLOR PLATE 2.)

ciated with constant threshold vorticity, is depicted in FIGURE 6(top), showing randomly distributed mushroom- and finger-like shapes. Imposing a strong magnetic filed leads to structure transformation into a more regular pattern with almost cylindrical vortical plumes stretching from the bottom to the top wall and providing direct communication (heat transfer) between them, FIGURE 6(bottom).[9] Because of a strong contribution of shear strain in the near-wall region, we selected a very small value of the threshold vorticity ($5\%|\omega_{max}|$), corresponding to the very core of the vortical plumes.

SUMMARY

Education, visualization, and imaging of coherent vortical structures in complex turbulent flows have become essential tools for understanding the physics of turbulence and transport processes, and for their control. We demonstrated this in a study of classic and magnetic Rayleigh–Bénard convection at a range of Rayleigh numbers that are featured by well-organized large-scale roll-cell patterns with a deterministic character. Various methods of structure education proposed in the literature have been applied to identify structure morphology, all showing similar outcome in the flows considered. The kinematic vorticity number N_k and the second invariant of the velocity gradient tensor Q provide very good insight and are very convenient for application. These criteria show dramatic morphology changes in plume structures in R-B convection when subjected to additional body force—here Lorentz force, or non-plane configuration of the bounding walls.

The prerequisite for this is the availability of data on instantaneous velocity, vorticity, and scalar (temperature and concentration) fields that need to be provided by experiments or numerical simulations. Major difficulties appear in handling flows in complex geometries and at high Reynolds and Rayleigh numbers, where the available techniques face serious constraints. We demonstrated that the T-RANS approach is a feasible and convenient method for simulating flows with deterministic vortical structures especially at very high Rayleigh numbers that are inaccessible to the conventional DNS and LES techniques.

ACKNOWLEDGMENT

The research of Dr. Saša Kenjereš was made possible by a fellowship of the Royal Netherlands Academy of Arts and Sciences (KNAW).

REFERENCES

1. LUGT, H.J. 1979. The dilemma of defining vortex. *In* Recent Developments in Theoretical and Experimental Fluid Mechanics. U. Mueller, K.G. Roesner & B. Schmidt, Eds.: 309–321. Springer.
2. ROBINSON, S.K. 1991. Coherent motions in the turbulent boundary layer. Ann. Rev. Fluid Mech. **23:** 601–638.
3. HUSSAIN, F. 1986. Coherent structures and turbulence. J. Fluid Mech. **173:** 303–328.
4. JEONG J. & F. HUSSAIN. 1995. On the identification of a vortex. J. Fluid Mech. **285:** 69–94.
5. LESIEUR, M., P. BEGOU, P. COMTE & O. METAIS. 2000. Vortex recognition in numerical simulations. ERCOFTAC Bull. **46:** 25–28.
6. HUNT, J.C.R., A.A. WRAY & P. MOIN. 1988. Eddies, stream and convergence zones in turbulent flows. Center for Turbulence Research Report, CTR-S88.
7. CHONG, M.S., A.E. PERRY & B.J. CANTWELL. 1990. A general classification of three-dimensional flow field. Phys. Fluids A **2:** 765–777.
8. HANJALIĆ, K. & S. KENJEREŠ. 2002. Simulation of coherent eddy structure in buoyancy-driven flows with single point turbulence closure models. *In* Closure Strategies for Transitional and Turbulent Flows. B.E. Launder & M. Sandham, Eds. Cambridge Univ. Press, Cambridge, U.K. 659–684.
9. HANJALIĆ, K. & S. KENJEREŠ. 2000, Reorganization of turbulence structure in magnetic Rayleigh–Bénard convection: a T-RANS study. J. Turbulence (electronic journal), IOP Publishing **1**(008): 1–22.
10. HA MINH, H. & A. KOURTA. 1993. Semi-deterministic turbulence modeling for flows dominated by strong organized structures. Proc. 9th Symp. Turbulent Shear Flows, Kyoto, Japan. 10.5.1–10.5-6.
11. KENJEREŠ, S. & K. HANJALIĆ. 1999. Identification and visualization of coherent structures in Rayleigh–Bénard convection with a time-dependent RANS. J. Visualization **2**(2):169–176.
12. DUBIEF, Y. & F. DELCAYRE. 2000. On coherent-vortex identification in turbulence. J. Turbulence (electronic journal) IOP Publishing. **1**(011): 1–22.

A New Definition of Contours in Images
Applications in Heat and Mass Transfer

ADIL RACHEK,[a] JACQUES PADET,[a] MONICA STOIAN,[b] AND COLETTE PADET[a]

[a]*Laboratoire de Thermomécanique-UTAP, Université de Reims-Champagne Ardenne, Reims, France*

[b]*University Politehnica Bucuresti, Splaiul Independentei, Bucuresti, Romania*

> ABSTRACT: A new definition for the contours of objects used in experimental visualization is proposed and used for edge detection and image smoothing in infrared thermography. This work is a contribution to image analysis, in the search for outlines of the features and their structures, aimed at identifying the trends they represent. A good example is the discovery of structures in the flow of liquids.
>
> KEYWORDS: edge detection; trends; image smoothing; contours; infrared; thermography

INTRODUCTION

Visualization through hierarchical development allows access to a vast amount of information of otherwise obscured phenomena in thermodynamics, fluid mechanics, heat and mass transfer, and other fields of science. The discovery of the outlines of objects in a noisy environment is a major element in the determination of the boundaries of the object for which we are searching, and in various techniques of segmentation, transmission, and storage of the complete image. Existing methods lead to rather arbitrary choices, by permitting their user to choose one or more parameter values to fit his idea about the results he may wish to obtain. For this reason, we endeavor to construct an objective self-adapting method that does not lead to exceptional cases or arbitrary decisions. We then use the original features of our method to propose a new definition of the outlines in an image. Finally, we demonstrate its application to the images obtained by infrared thermography in the thermal study of cooling following soldering.

BACKGROUND

We commence by reviewing the main published methods for detecting outlines.[1–5] The definition of outlines usually corresponds either to areas of rapid change in the level of grayness of an image (lines of maximum contrast) or to the lines of constant value. In the majority of existing methods, the extracted parameters are invariably

Address for correspondence: Jacques Padet, Ph.D., Laboratoire de Thermomécanique-UTAP, Université de Reims-Champagne Ardenne, B.P.1039, 51687 Reims, France.
adil.rachek@univ-reims.fr

optimized only after the completion of a process subject to visualization. There are essentially two types of contour detection methods: those using operators of differentiation and those relying on models of the contours. Through the use of operators that approximate first derivatives (e.g., gradient operators of Sobel and Kirsh[3,6]) or second derivatives (e.g., Laplace operators or the level of grayness[7,11]) one can bypass local discontinuities. In practice, to make contours explicit through differentiation, one looks for the extrema of the gradient or zeros of the Laplacian. These operators bring out the contour by detecting sudden variations in grayness. Those methods that rely on models of contours permit, in the process of their detection, the use of *a priori* information about the structure. Several such models have been proposed. The use of Markov fields leads to a family of stochastic models that employ the notion of neighborhood and neighborliness.[1] Other methods use models of active contours.[5] Finally, one can detect contours by applying basic operations (opening and closure) to the elementary operators (erosion and dilatation) of mathematical morphology.[2]

As noted above, available methods frequently lead to rather arbitrary choices, by permitting their user to choose one or more parameter values to fit his idea about the results he may wish to obtain. These preconceived ideas may bear little relationship to the specific features of the analyzed images.

THE NEW AUTOADAPTIVE METHOD

In the following, we concentrate on detecting contours of numerical images. Here, an image consists of a matrix of pixels with K rows and L columns. A given function $I(k,L)$, which represents the intensity of luminosity at point (k,l), $k = 1,...,K$ and $l = 1,...,L$, is a discrete function or a discrete signal. It gives a spatial representation of an object with a two-dimensional background. We write it as a $K \times L$ matrix.

The proposed method considers each row, column and diagonal of the image data as a one-dimensional (1D) data set and finds its trend, following the method proposed by Roche[9,10] and described below. The trends of each row or column form a surface, termed *the trend of the image*. The intersections of various trends of this type and of the original image determine the contours.

Low-Pass Filter and Moving Average

Let $I(i)$ be a 1D data set together with a low-pass filter and a cut-off period K_N. The moving average of this collection, following the method of Hamming window,[3] $a(n)$, is defined by

$$\overline{I(i)} = \sum_{n=-N/2}^{N/2} I(i+n) h_{K_N}(n) a(n), \qquad (1)$$

where

$$h_{K_N} = \begin{cases} 2/K_N & n = 0 \\ \dfrac{1}{n\pi}\sin\left(\dfrac{2n\pi}{K_N}\right) & n \neq 0 \end{cases}$$

$$a(n) = \begin{cases} 0.54 + 0.46\cos\left(\dfrac{2n\pi}{N}\right) & |n| \leq \dfrac{N}{2} \\ 0 & |n| > \dfrac{N}{2}. \end{cases}$$

The value N is both the length of truncation and the length of the window of the moving average. The problem that arises is one of determining the length N of the window at the beginning of the period of truncation. Roche obtained the relationship $N = K_N$, which we use in the following. Note that in all the methods of smoothing, there is a problem of boundary location. The points corresponding to $i + n$ in Equation (1) could be located outside the domain. We decided to extend the range of points of the domain by assigning constant values on the boundary.

Index of Non-Linearity

The index non-linearity (INL) is a parameter that permits objective determination of the size of the filter to be used for finding the trends and other features of the phenomena.[10] It is defined by

$$INL = \left(\dfrac{\sum_{i=1}^{N-1} |I_{i+1} - I_i - I_N + I_1|}{\sum_{i=1}^{N-1} |I_{i+1} - I_i|} \sum_{i=1}^{N-1}\left(1 + \left(\dfrac{I_{i+1} - I_i}{I_{max} - I_{min}}\right)^2\right) \right)^{1/2} \quad (2)$$

Trends in a Data Set

To establish a criterion to determine trends in a data set, Roche *et al.*[8,9] further suggested how to combine the two techniques outlined above. Following (1), the low-pass filter $\bar{I}$ is a mapping

$$\bar{I}: I(i), K_N \to I_{K_N}(i). \quad (3)$$

If we vary the length of the window, a collection of sequences $\overline{I_{K_N}(i)}$ arises by applying the filter (1) to the initial sequence. The filter $\bar{I}$ decreases and smooths the perturbations in the sequence $I(i)$, according to the increase in the length K_N. For each window cut off period K_N, we evaluate the sequence of values $\overline{I_{K_N}(i)}$ by computing its INL. This is a mapping, applied to $\overline{I_{K_N}(i)}$ and K_N:

$$INL: \overline{I_{K_N}(i)}, K_N \to INL(K_N, \overline{I_{K_N}(i)}). \quad (4)$$

The mapping can be simplified if we assume the same initial sequence $I(i)$

$$INL: K_N \to INL(K_N). \quad (5)$$

A comparison of the data fluctuations after each smoothing is then performed. A graph INL-GRAMME can be associated with the index of nonlinearty. This graph demonstrates the global decrease in INL when K_N increases. The INL-GRAMME will also demonstrate, depending on the specific case, the decrease in slope of flat regions or of minima. The values of K_N that correspond to minima in the index of nonlinearity can provide criteria for choosing a length of the window and, thus, of the low-pass filter that produces the trends of $I(i)$, Any such criteria is a form of the principle of *minimum action*. These minima recognize the weakest possible fluctuations in the trend. The collection of values K_N, determined in this way represent levels of characteristic observations of the studied phenomenon. Each of these values can be associated with a low-pass filter whose window has length $2K_N$. We define partial trends of order j of the sequence $I(i)$ as the trend resulting from the filter associated with the jth value K_N. We denote them by $T_j(i)$. The trend corresponding to the largest K_N value (the last one) is called an ultimate trend and denoted by $T_u(i)$.

Trend of an Image

Let $I(k,l)$ be an arbitrary image. The values k and l represent the number of rows and columns in the image. Each row can be considered as a 1D curve. We apply the low-pass filter to such a curve while varying the length of the window from 1 to the value given as one half of the length of the row. To each smoothed curve that results, we associate its INL value.

The collection of these values is represented by the INL-GRAMME. Using previous criteria, we obtain partial and ultimate trends of the curve. The process can be applied to each row of the image that needs to be processed. The trends of order j of all the rows form an image $\overline{I_j(k, l)}$. We call this the partial trend of order j of the original image. In the same way, the ultimate trend, $\overline{I_u(k, l)}$, is constructed from those of the separate rows. For each row, the trend of order j is computed on the basis of its own moving window. The dimensions of these windows can vary from row to row. Consequently, we have defined the partial and ultimate trends of the image, using the concept of the index of nonlinearity.

Contours

We now proceed to use the trends of the image to determine the contour of an image. We call this a local 1D method of detection. We know that an image, $\overline{I(k, l)}$, can be viewed as a surface in the ambient space. The pair (k,l) represents the coordinates in the base plane and $\overline{I(k, l)}$ is the value of the function defined over this plane.

However, the partial and ultimate trends that result from the 1D processing along rows, columns, or diagonals themselves form the separate images. Each of them is represented by a surface. Recall that the intensities of luminosity within an image vary for different objects. When we traverse, simultaneously, an image and one of its trends within the same space, the trend of the image crosses the fluctuations of the original image. The resulting intersections and their projections form the contour of the image that we are processing. We refer to these intersections as passing through zero of the difference between the original image and one of its trends. To a projection, one can assign a labelling of contours by giving value 1 to the points of the contour and 0 otherwise. In this way, starting from trends of separate rows, each pair of rows of $\overline{I(k, l)}$ and $\overline{I_j(k, l)}$ forms, along the horizontal axis l, two crossing curves. The points of intersections in the space project onto those on the axis. The mathematical condition for the relevant coordinates is that the points to be found are those pairs (k,l) for which

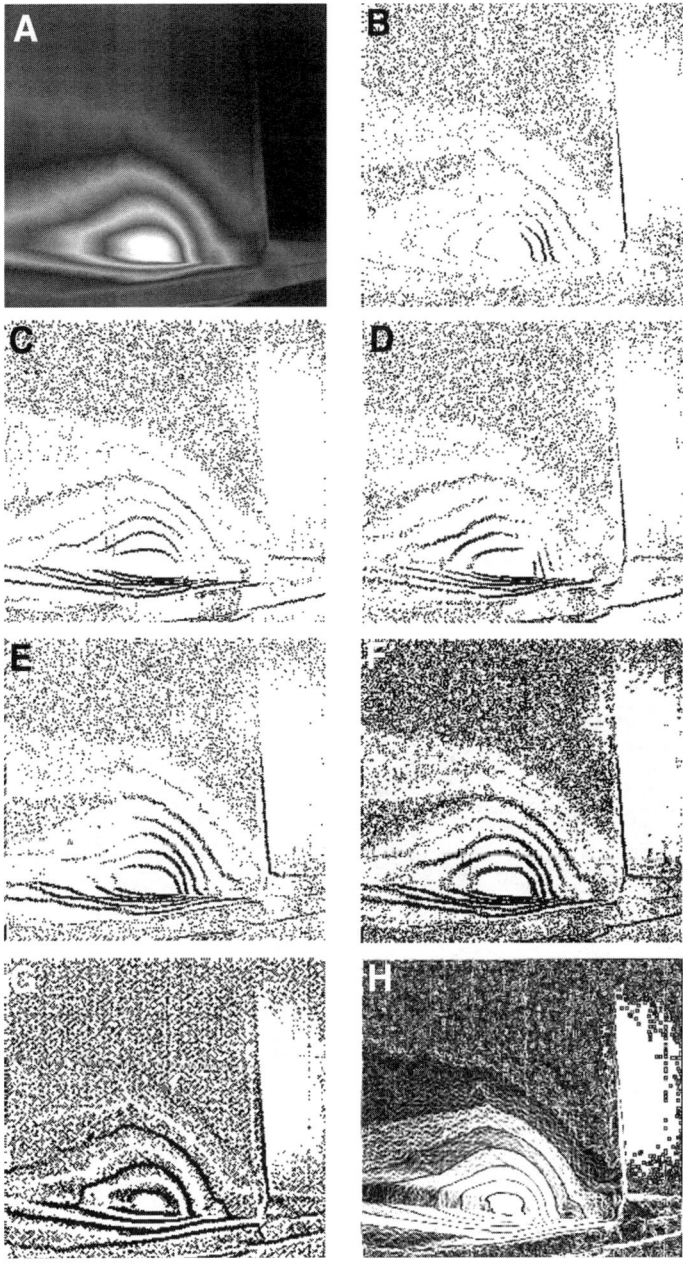

FIGURE 1. **A** Original image, **B** horizontal intersection, **C** vertical intersection, **D** left diagonal intersection, **E** right diagonal intersection, **F** superimposition, **G** Laplacian operator method, **H** contour obtained from commercial software.

$$\left| I(k, l-1) - \overline{I_j(k, l-1)} \right| \times \left| I(k, l+1) - \overline{I_j(k, l+1)} \right| < 0. \tag{6}$$

The complete contours are obtained by superposition of several different intersections, thus permitting extraction of maximum information.

APPLICATION

We applied our method to two images that resulted from the thermal study of soldering during its cooling stage (see FIGURES 1 A and 2 A). The images were obtained by infrared thermography at Universitatea Politehnica Bucuresti, Romania. They represent the temperature field of one soldering during its cooling. The first image was taken when the soldering started to cool and the second at the end of the cooling process. The size of the images is 250 × 230 pixels. Since our method treats the image in gray levels, the color images were transformed into gray levels. We processed, separately, each row, column, and diagonal to first find the trends and then the intersections with the original image. The resulting contours are shown in FIGURE 1 B–E. Comparing the results of such separate 1D analysis, we observe that some points appear only in one of the horizontal, vertical, or diagonal direction, as in FIGURE 1 B–E. To maximize the information, we superimposed the intersections

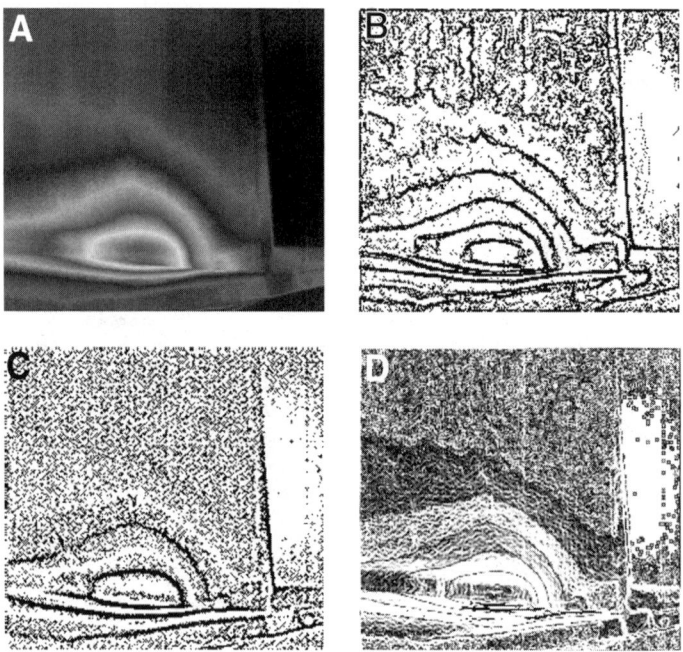

FIGURE 2. **A** Original image, **B** superimposition, **C** Laplacian operator method, **D** contour obtained by commercial software.

obtained from these three directional treatments (FIGS. 1F and 2B). We conclude that the contours can be superimposed and completed almost perfectly. The method proposed here gives satisfactory results for the search of contours in the images. Moreover, an improvement of the detection and the quality of contours was observed with our method when comparing our results with those obtained using other commercial software (FIGS. 1G and H, and 2C and D).

SUMMARY

A new procedure was applied successfully to the images of the soldering during its cooling. The superposition of the contours obtained by analyzes along rows, column, and diagonals allows one to obtain complete contours and other relevant information about the structure of the image.

REFERENCES

1. COCQUEREZ, J.P. & S. PHILIPP. 1995. Analyse d'images : filtrage et segmentation. Masson.
2. COSTER, M. & J.L. CHERMANT. 1989. Précis d'analyse d'image. Presses du CNRS.
3. COULON, F.T. 1984. Théorie et traitement des signaux. DUNOD.
4. DUDA, R.O. & P.E. HART. 1973. Pattern Classification and Scene Analysis. Wiley, NewYork.
5. KASS, M., D. WITHIN & D. TERZOPOULOS. 1988. Snakes: active contour models. Int. J. Computer Vision **1:** 312–331.
6. KIRSH, R. 1971. Computer determination of the constituent structure of biological images. Comput. Biomed. Res. **4:** 315–328.
7. KUNT, M., G. GRANLUND & M. KOCHER. 1973. Traitement numérique des images. 2 Diffusion. Cent-Ernst, Paris.
8. ROCHE, J.M. 1995. Tendances d'un phénomène irrégulier: définition et caractérisation. Thèse, Université de Reims.
9. ROCHE, J.M., C. PADET & J. PADET. 1992. Non-linearity index and trends of irregular phenomena. Proc. 4th Int. Conf. IPMU, (Palma de Majorque, Espana). 197–203.
10. ROCHE, J.M. & C. PADET. 1996, New methods in data analysis: potentials offered for the study of variable phenomena. Proc. Int. Symp. on Transient Convective Heat Transfer, (Cesme, Turkey). 369–379.
11. SONG, H. 1998. Analyse de données bidimensionnelles: Tendances, Structures et Contours. Thèse, Université de Reims.

Photogrammetric and Image Processing Aspects in Quantitative Flow Visualization

MATTHIAS MACHACEK AND THOMAS RÖSGEN

Institute of Fluid Dynamics, ETH Zürich, Switzerland

ABSTRACT: The development of a measurement system for the visualization, topological classification, and quantitative analysis of complex flows in large-scale wind tunnel experiments is described. A new approach was sought in which the topological features of the flow (e.g., stream lines, separation and reattachment regions, stagnation points, and vortex lines) were extracted directly and preferably visualized in real-time in a virtual wind tunnel environment. The system was based on a stereo arrangement of two CCD cameras. A frame rate of 120 fps allowed measurements at high flow velocities. The paper focuses on the problem of fast and accurate reconstruction of path lines of helium filled soap bubbles in three dimensions (3D). A series of simple algorithmic steps was employed to ensure fast data processing. These included fast image segmentation, a spline approximation of the path lines, a camera model, point correspondence building, calculation of path line points in 3D and creation of a three-dimensional spline representation. The path lines, which contained both velocity and topological information, were analyzed to extract the relevant information.

KEYWORDS: complex flows; streamlines; separation and attachment regions; topological classification; wind tunnel; vortex lines

INTRODUCTION

Classical visualization methods (e.g., smoke wire, laser vapor screen, and tufts) provide an overview of the flow topology but quantitative data is often not available and depends strongly on interpretation by the researcher. An alternative is to take classical qualitative visualization methods and combine them with modern computer vision technology to generate quantitative data. The intuitiveness of the visualization approach is maintained by representing the results in a *virtual wind tunnel* environment with an interactive user interface. For the equivalent approach using numerical data see Reference 1.

The camera and illumination were configured so that the images of the moving particles represent path lines (see FIGURE 1). The spatial processing of the acquired image data is done based on the principles of the photogrammetry theory.[2]

Address for correspondence: Matthias Machacek, Dipl. Ing., Institute of Fluid Dynamics, Sonneggstrasse 3, ETH Zürich, Switzerland.
machacek@ifd.mavt.ethz.ch

FIGURE 1. A typical image with helium-filled soap bubble tracers, prior to any image processing.

EXPERIMENTAL SETUP

The visualization system consisted of two progressive-scan interline CCD cameras recording images with a frame rate of 120 fps and a resolution of 640 × 480 pixels. Lighting was provided by four halogen spot lamps with a total power of 6,000 W. The helium-filled soap bubbles for seeding were introduced into the test section through a strut mounted nozzle.[3] The bubbles had an approximate diameter of 2 mm. The components were set up in a wind tunnel with a cross section of 2 × 3 m and a flow velocity of up to 60 m/sec.

Camera Model and Calibration

The camera model used was the pinhole model adopted from photogrammetry theory.[2] The model parameters were the focal length f, the principal point (x_0, y_0), the stretching of the x-axis s_x, and the skew θ. In addition, lens distortion was accounted for by two second order polynomials for the radial distortion (k_1, k_2) and for the tangential distortion (p_1, p_2). The camera position and orientation with reference to a global coordinate system is described by the position parameters (X_0, Y_0, Z_0) and orientation angles $(\varphi, \omega, \kappa)$. Thus, the camera model includes a total of 15 parameters that are unknown for any given experimental set up and must be estimated by a calibration routine.[2,3]

The major problem for the camera calibration in the wind tunnel test section is the large size of the measurement volume. An accurate calibration requires calibration markers distributed throughout the whole field of view, hence a calibration target of a similar size is needed. Such a target is difficult to manipulate and rather expensive. Therefore, a two-step calibration procedure was developed in which the inner parameters $(f, x_0, y_0, s_x, \theta, k_1, k_2, p_1, p_2)$ were calibrated separately from the external parameters $(X_0, Y_0, Z_0, \varphi, \omega, \kappa)$. Since the inner parameters are independent of the experimental setup and measurement volume size, they can be estimated with a calibration target of any convenient size. The external parameters were then estimated with an eight-point algorithm with normalized correspondences.[2] This method uses the information on point correspondences between two views and images of point pairs with a precisely known distance. This was accomplished by using a stick with two precisely mounted point source LEDs at the tips as a calibration target. For the calibration procedure the stick was moved by hand randomly in the test section, generating a cloud of calibration point pairs with a known relative distance.

TWO-DIMENSIONAL PATH RECONSTRUCTION

A three-dimensional reconstruction requires the representation of a path line in the various camera views. As a first step, the finite thickness bubble tracks were reduced to a representative line close to the center line (skeleton) of the track (see FIGURE 2). Because of several desirable mathematical properties (e.g., smoothness and differentiability) least-squares parametric splines based on cubic B-splines[4] were used to represent the path lines.

A crucial aspect in the processing chain was the configuration of the digital cameras that recorded the bubble images. The bubbles were illuminated with a continuous light source and recorded with the maximum exposure time (reciprocal frame rate) by an interline CCD chip with a negligible read-out time. This led to a set of consecutive images in which a moving particle left a continuous track of connected path segments, forming a complete path line (FIG. 2). Previous attempts[3] have shown that a path line reconstruction based on the analysis of independent path line segments (i.e., independent image frames) can lead to topologically inaccurate solutions. This is due to the difficulty of processing the end points of the segments in a consistent manner when the information of the previous and following segments is not taken into account. Hence, the high level processing was based here on multi-frame path lines as reconstructed from a connected set of path line segments.

The initial low level processing was achieved by the following series of steps: background subtraction, binarization with an appropriate thresholding operator, and

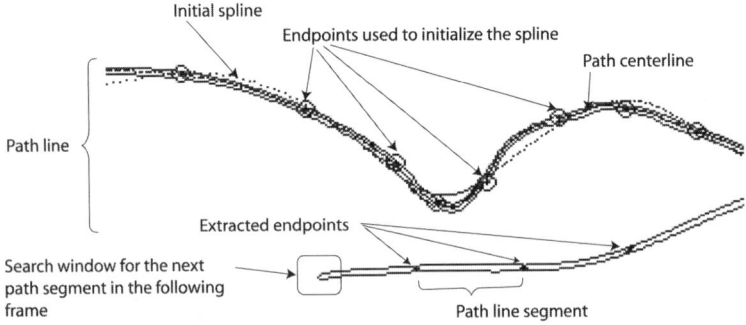

FIGURE 2. Processing a path line. The upper path line is finished and, thus, processed to derive the path representation. The circles indicate the path segment end points used for the initial spline representation of the particle path.

median filtering to reduce the residual noise. The current image was then scanned in the immediate vicinity of the path line segment endpoint detected in the previous frame (FIG. 2). In cases of a successful match the new segment boundary pixels and its two endpoints were extracted and corrected for the distortion (k_1,k_2,p_1,p_2) and the image mapping (f,x_0,y_0,s_x,θ) with the use of a look-up table containing the distortion corrected value (the so-called normalized coordinate[2]) for each image pixel. This approach provided a significant speedup for the nonlinear part of the overall coordinate transform and reduced the subsequent processing to linear operations despite the nonlinear formulation.

The procedure was continued until a path line is completed; that is, the particle left the camera field of view. Thereafter, its representation in spline form could be calculated. For that purpose, the spline representation was initialized with the end points of the path segments. A least-squares spline with a fixed and equidistant knot placement was used. The spline was then improved in an iterative process until a sufficiently accurate representation of the center line was found.

THREE-DIMENSIONAL PATH RECONSTRUCTION

To calculate the three-dimensional path line, corresponding point pairs on corresponding path lines had to be found in both camera views. Corresponding path lines were found by projecting the end points of a path line segment into the second camera image, creating so called epipolar lines (see FIGURE 3). If the distance between the epipolar line projection and an endpoint was below a critical value, the endpoints, and thus, the path lines were considered to be corresponding. Since the epipolar condition for two cameras is ambiguous, all candidate end points were tested for correspondence until a unique solution was found.

Path line points were defined by equidistant spacing on the spline model in the first view. These points were then projected into the second view, again creating a set of epipolar lines. These lines were intersected with the corresponding second image spline approximation, to produce corresponding points. All corresponding two-dimensional point pairs were finally used to calculate the coordinate points in the

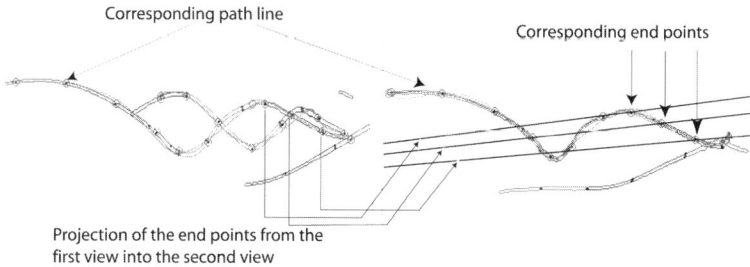

FIGURE 3. The image shows the relation between the two camera views. Correspondence is found through the point projections (epipolar lines).

three-dimensional space. The path line was thus reconstructed as a sequence of points. Finally these three-dimensional points were used to rebuild a three-dimensional spline representation of the path line. This spline curve was also based on cubic B-splines.

RESULTS

The two step calibration, based on a stick with two LEDs as localized point markers to estimate the external parameters, gave accurate results. At the same time the calibration procedure was greatly simplified and the time needed significantly reduced. In addition, accurate manufacture of a two point calibration target proved straightforward. The visualization method was demonstrated on the wake of a delta wing flow in a large scale wind tunnel as can be seen in FIGURES 4, 5, and 6.

SUMMARY

A method for the accurate and fast calculation of three-dimensional path lines was described, wherein the measurement data was derived from two independent camera views. It was found that by using approximating splines based on cubic B-splines the center line of a compound multiframe path line could be accurately modeled. The use of a non-linear camera model, in combination with a look-up table for the normalized pixel coordinates, allowed the implementation of accurate and fast operations for the photogrammetric reconstruction. The final three-dimensional path lines were also represented with a model based on cubic B-splines. These analytical

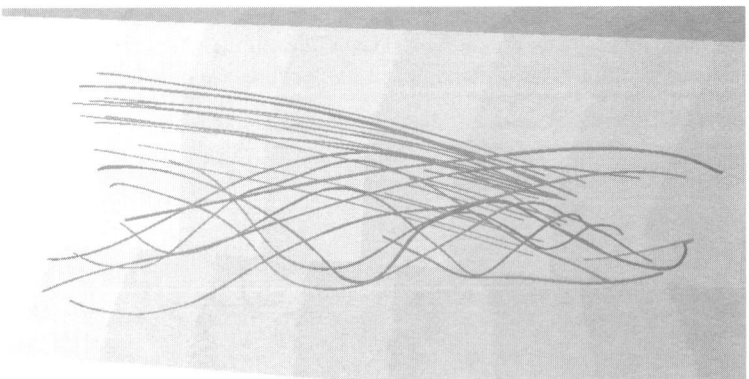

FIGURE 4. Visualization of the wake flow of a delta wing. The characteristic delta wing vortex in the lower part of the image and a less affected part of the flow in the upper part can be identified. The measurement time was 2.5 sec., which corresponds to 300 frames. The delta wing had a span width of 0.5 m.

FIGURE 5. A different view of the wake flow shown in FIGURE 4. A VRML environment is used for the data representation. The observation window of the wind tunnel where the two CCD cameras were placed is shown in the back.

curves provide accurate and continuous representations of the path lines and form a solid base from which topological information on the flow can be extracted. Compared with three-dimensional particle tracking velocimetry, the present method does not require elaborate tracking algorithms and provides continuous information on the particle tracks.

FIGURE 6. The vortex as seen from the back. The measurement time was 8.3 sec., which corresponds to 1,000 frames. The path lines were approximated with linear segments.

REFERENCES

1. BRYSON, S. & C. LEVIT. 1991 The virtual wind tunnel: an environment for the exploration of three-dimensional unsteady flows. NAS Technical Reports, RNR-92-013.
2. HARTLEY, R. & A. ZISSERMAN. 2000, Multiple View Geometry in Computer Vision. Cambridge University Press.
3. MACHACEK, M. & T. RÖSGEN. 2001, Development of a Quantitative Flow Visualization Tool for Applications in Industrial Wind Tunnels, Proceedings of the 19th Int. Congress on Instrumentation in Aerospace Simulation Facilities, Cleveland, Ohio.
4. DIERCKX, P. 1993, Monographs on Numerical Analysis: Curve and Surface Fitting with Splines. Oxford Science Publ.

Color Interpolation Algorithms in Visualizing Results of Numerical Simulations

DMITRY V. MOGILENSKIKH AND IGOR V. PAVLOV

Russian Federal Nuclear Center—the All-Russian Scientific and Research Institute of Technical Physics Named after Academician E.I. Zababakhin (RFNC-VNIITF), Snezhinsk, Chelyabinsk Region, Russia

ABSTRACT: The paper addresses the application of graphical replenishment (GR) color interpolation algorithms for visualizing numerical simulation of physical processes in order to enhance the information density of the graphic interpretation of these processes. The GR algorithms presented here may be successfully applied to two- and three-dimensional grids, and grids of various structures.

KEYWORDS: color interpolation; graphic replenishment; finite differences

INTRODUCTION

Techniques and algorithms for generating smooth changes of color inside a polygon are typically applied in visualizing three-dimensional (3D) geometries for formation of a photorealistic effect of a smooth object surface, with illumination taken into account. Such techniques include the methods of Gourot,[1–5] Phong,[3–6] and the algorithms proposed by Pletniov.[7] Similar color interpolation algorithms (illumination) inside polygons are seldom applied in scientific visualization applications for presentation of physical content of mathematical models, although this method of visualization is very informative for two-dimensional (2D) model analysis. This study presents algorithms and approaches that may be considered, in part, as variants of known algorithms. The process of interpolation or smooth change of colors inside polygons during visualization of scientific and experimental data is called *graphical replenishment* (GR)[8] of the initial data. The strict mathematical term *interpolation* fits more accurately here, but it is fundamental in the context of error estimation.[5, 8–10] The present study uses the special notion *graphical replenishment* in order to reflect the application area and to stress the fact that accurate estimates are important during visualization, as well as fast approximation, in order to obtain satisfactory results, with accuracy comparable with that obtained by the classical techniques of interpolation and visual perception.

The major goal of the difference and finite-difference techniques is a grid with a vector of physical characteristics specified on it. Cells and points are the grid

Address for correspondence: D.V. Mogilenskikh, Scientific Visualization Laboratory, E.I. Zababakhin Russian Federal Nuclear Center (RFNC-VNIITF), 456770 P.O. Box 245, Snezhinsk, Chelyabinsk Region, Russia.
 d.v.mogilenskikh@vniitf.ru

components. In other words, the scalar and vector fields specified at the sets of cells or points are the numerical simulation results.

STATEMENT OF PROBLEM

The problem of replenishment of discretely specified functions of two variables is well studied.[5,8–10]. Here, the emphasis is shifted toward the creation of adequate algorithms of color interpretation of the function change when passing points of a cell. The application of splines requires great computing power. The dimensionalities of grids in modern problems, as, for example, in continuum mechanics, are huge and require very large systems of equations for applying classical interpolation methods. Here, we describe the algorithms on models that are specified with 2D regular difference grids. This supposition does not restrict the generality. The geometry of the initial data—that is, the regular 2D grid—may be presented as 2D arrays $X[M,N]$, $Y[M,N]$, where M,N is the grid dimensionality. We assume that physical characteristics are specified at each point. We denote one of the physical characteristics by $U[M,N]$. Let U_{max} and U_{min} be the maximum and minimum values of U, respectively, over the entire grid.

It is necessary to analyze and compare results that yield a discrete color cell picture or a picture (see FIGURE 1 A) with a smooth physical characteristics changed inside the cells (FIG. 1 B). In this method of visualization of the model physical content, the following general changes in characteristics can be displayed: vortices, turbulence, fine transitions of state, special objects such as shock waves, lines of discontinuity, interfaces, isolines, and lines of flow. GR is a good instrument for interpretation when it is combined with other means of visualization, such as construction of isolines (FIG. 1 C), lines of flow, and grid superposition.

The problem is to obtain appropriate GR algorithms for color presentation of the changes in the physical values inside the cells. Several methods exist for estimating the correctness of this replenishment, or for comparing two replenishment algorithms. A description of the methods for estimating the GR quality is beyond the frameworks of this paper.[11]

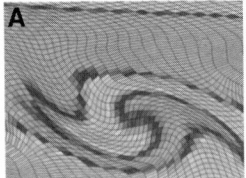 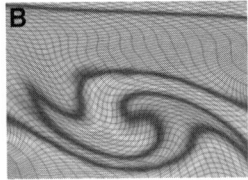 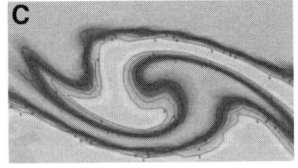

FIGURE 1. Shows three images of the same fragment of the numerical model with the computational grid. **A.** The cell-like manner of color filling, that is, the cells are colored with one color without replenishment inside the cells. **B.** Application of Algorithm 2 GR of the physical value in the form of color replenishment inside the cells. **C.** Application of Algorithm 2 GR, but in this case the image is presented without a grid with visualization of isolines. One can see the match of isolines with the lines of color transition in algorithm 2 GR.

COLOR SOLUTIONS FOR THE PROBLEM OF REPLENISHMENT

The Palette

The most widely used technique for color interpretation in a model with physical content is the linear law of correspondence (LLC) between the physical characteristic and the color space. One of the basic concepts in this subject is the color palette. Two definitions are useful: (1) The derived red, green, blue (RGB) colors are called an additive system for RGB color mixing, or RGB space. (2) A color palette is a sequenced set of colors from the RGB space represented by a one-dimensional (1D) array, where each color corresponds to an integer-valued index—its sequence number in the palette.

Consider a palette consisting of $(NC+1)$ colors. In this indexed representation of colors, each index corresponds to a color from the RGB color space. As a rule, a palette is created by smooth transitions from one color to another. (see FIGURE 2). According to the LLC, the color for the grid cells is calculated as follows:[12] the range of the value change U is $IN_u = [U_{max}, U_{min}]$. The range of discrete change of colors in the palette is $IN_c = [0, MC]$.

We calculate the coefficient of proportionality k of the segment length from $k = IN_u/IN_c$, where k is the length of the division segment IN_u with the peculiarity that the cells of values U from one division segment would are filled with one color. Consider the current cell of value $U_{ij} = U_{ij}[i,j]$, where $i \in [0,M]$ and $j \in [0,N]$. We calculate for this cell the color index from the palette

$$I_{i,j} = \left\lfloor \frac{U_{i,j}}{k} + 0.5 \right\rfloor, \tag{1}$$

where $\lfloor x \rfloor$ denotes the floor (integer part) of x. The result is that each cell, or point of the grid, has a specific color from the palette (FIG. 1A).

Modes of Replenishment

Replenishment in the Palette

The essence of the replenishment mode of color visualization lies in using only colors from the given palette, with each color corresponding to the range of change in the U value. In this case the interpolation takes place at the level of numbers for the colors in the palette. This technique is convenient since, for any rule of correspondence between a color and a physical value, GR guarantees that the color is obtained rigorously from this palette (see FIGURE 3A).

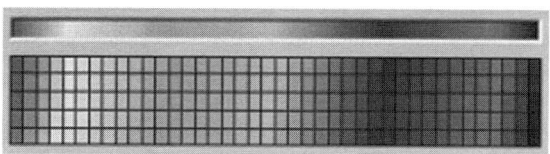

FIGURE 2. Two variants representing the palette.

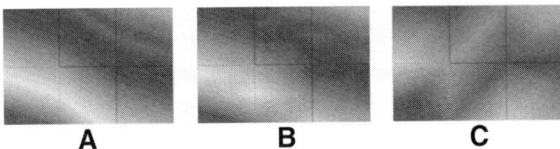

FIGURE 3. The same test grid with specified values at the grid points. **A.** Application of Algorithm 2 GR of the value inside a cell in the area of the palette colors. **B.** Application of Algorithm 2 GR of the value inside a cell within the RGB color space area; that is, by all the three components. One can observe shortfalls of GR in RGB; some "dirty" colors are outside the palette, pseudo-transparency, and ghosts appear. **C.** Application of GR with the help of OpenGL functions. The quality of the GR value in the cells is evidently bad. (COLOR PLATE 3.)

Replenishment in the RGB Color Space

There is another interpolation technique between the palette colors whereby the replenishment is performed individually with RGB components. This results in bad agreement and inadequate correspondence between the colors and the physical content. In FIGURE 3B we applied the replenishment by the RGB space. The image is less accurate and less informative than FIGURE 3A due to the appearance of colors that are not from the palette. Note the appearance of "dirty" effects, such as quasi-transparency and ghosts, in FIGURE 3B. Finally, more calculations are needed in this interpolation technique, since three components must be interpolated rather than one for replenishment in the palette.

Replenishment by Application of OpenGL

It is possible to implement GR using the functions of the standard graphical library OpenGL. This is a graphic standard developed and approved in 1992 by the leading software developers and manufacturers of graphic accelerators. FIGURE 3C presents a variant of replenishment of the same data as in FIGURE 3A and B, but with OpenGL. The result is obviously unsatisfactory.

ALGORITHMS FOR GENERATING A RASTER SCAN OF A CELL

The area of GR is a raster screen consisting of a 2D set of discrete points—pixels. All the graphic objects displayed on a screen require the process of raster scan generation. Typically, the methods of raster scan are hidden from a user behind graphic functions, but there are some problems for which this process should be performed explicitly. Raster scan algorithms are closely connected with GR algorithms. Here we consider one of the optimal raster decomposition algorithms (Algorithm 1) of the grid cell for GR Algorithms 2–4.

Algorithm 1. Application of the Brezenheim Raster Scan Algorithm

To obtain a cell raster scan it is possible to apply the classical integer raster scan algorithms for a line segment of Brezenheim,[4] using the following steps:

1. Calculate the raster scan of a cell ribs in integer screen coordinates by the Brezenheim algorithm.
2. Sort the points (pixels) of the raster scan of the ribs, for example, vertically, by coordinate Y.
3. Eliminate from the list of points of the ribs raster scan the neighboring points having the same Y coordinates and different X coordinates, leaving the leftmost and rightmost raster points.
4. Search for pairs that determine the beginning and the end of the inner scanning segment.
5. Obtain the raster ordered scan of the cell boundary.

For each inner pixel in a line, it is then possible to calculate the value U by linear interpolation, or the color number I from the palette. Next, it is possible to use different algorithms to replenish the characteristic U inside the cell. Application of the Brezenheim algorithm is fast when generating a raster scan. Furthermore, this approach to raster scan generation shows very good results in visualizing curvilinear grids.

Generalization for Description of GR Algorithms

As a result of applying Algorithm 1, we locate the inner pixel, A (see FIGURE 4) that belongs to the current cell with the screen coordinates (X_a, Y_a). The GR algorithms operate similarly for other pixels inside the current cell. Practical implementation of various algorithms for the cell raster scan has shown that, on average, for any grid type (especially when their dimensionality is enlarged), the application of the Brezenheim algorithms to the ribs provides a considerable advantage in visualization speed.

Algorithm 2. Bilinear Interpolation

We describe the replenishment algorithms, without restricting the generality, by assuming that inner pixels have been already localized with the help of Algorithm 1.

Algorithms 2 and 3 may be considered as 2D modifications of the methods of Gouraud[1] and Phong,[2] as they apply to a discrete scalar function of two variables specified on a grid for 2D scientific visualization.

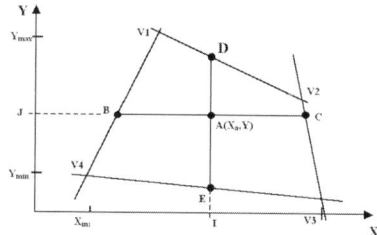

FIGURE 4. Representation of the current cell with the localized inner point **A** and with lines |BC| and |DE| as scanning segments.

The First Interpolation

Scanning line J yields the cross-points B and C on the cell ribs (FIG. 4). In these points it is necessary to interpolate the values of the characteristic U from the top points using linear interpolation.

$$U_B = (1-t)U_{v_1} + tU_{v_4}$$
$$U_C = (1-t^*)U_{v_2} + t^*U_{v_3} \qquad (2)$$

where U_B, U_C, U_{v_1}, U_{v_2}, U_{v_3}, and U_{v_4} are the values of the characteristic U in the corresponding points and the parameters t and t^*,

$$t = \frac{|V_1B|}{|V_1V_4|}, \quad 0 \leq t \leq 1;$$
$$t^* = \frac{|V_2C|}{|V_2V_3|}, \quad 0 \leq t^* \leq 1. \qquad (3)$$

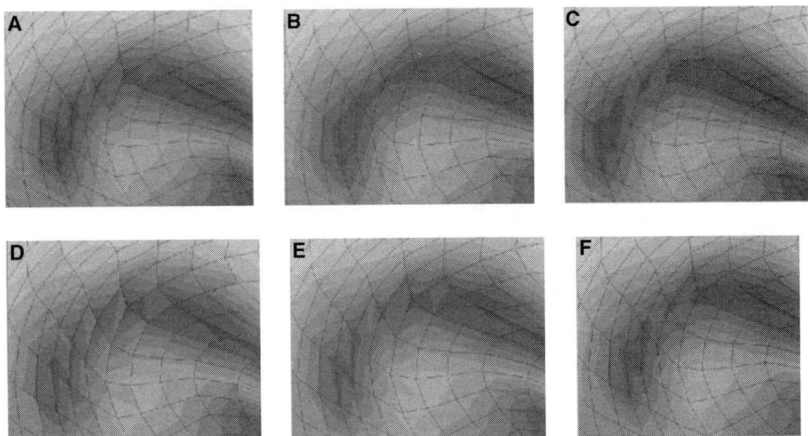

FIGURE 5. Comparative illustrations of the results of the operation of GR algorithms modifications of the same fragment of the numerical model. **A.** Application of the Algorithm 2 GR, scanning the cell insides by a column that is a vertical to the second interpolation value (FIG. 4, segment DE). **B.** Application of Algorithm 2 GR scanning the cell insides by a line that is horizontal—the second interpolation value (FIG. 4, segment BC). **C.** Application of Algorithm 3 GR with double scanning of the cell insides by a column and by a line that are a double bilinear interpolation (FIG. 4, segments BC and DE). **D–E.** Results of Algorithm 4 modification replenishment by height. **D.** Shows technique 1 of Algorithm 4 GR with the assumption that the cell nodes belong to one plane. **E.** Application of technique 3 of Algorithm 4 GR, partitioning the cell into two triangles and determining the cell inner points belonging to a certain triangle. **F.** Application of Method 4 of Algorithm 4 GR, partitioning the cell into four triangles with a common apex in the geometric center of gravity and determining the cell inner points belonging to a certain triangle.

The Second Interpolation

It is necessary to interpolate the value U into a pixel, or the point A, from points B and C. Similar to the previous step

$$U_A = (1-p)U_b + pU_c, \quad p = \frac{|AB|}{|BC|}, \quad 0 \le p \le 1. \tag{4}$$

Color Calculation

Color can be calculated by using the LLC I_{ij} formula, Equation (**1**). The example for this algorithm operation is shown in FIGURE 3A. FIGURE 5A and B mark the model fragment and show the difference in replenishment (scanning by columns and by lines). Typically, the representation of a model as a whole depends only slightly on the scanning type.

Note that the step of color calculation may be made much earlier. One can calculate the color at the cell points at once and then perform calculations relative to the palette indices. For example, when generating a raster scan of the cell ribs with Algorithm 1, rather than calculate the value U for each scan point, calculate for the color number. This approach is applied in the Gouraud[1] method, when the illumination intensity is calculated for the polygon point; at this time the intensity is interpolated. However, in order to preserve the maximum possible accuracy it is recommended that the move to color be made as late as possible. Note that early transition to integer calculations leads to increased visualization speed.

Algorithm 3. Double Bilinear Interpolation

This algorithm is a modification of Algorithm 2 and is more time-consuming. The idea lies in the simultaneous application of interpolation both by column and by row. The algorithm is completely similar to Algorithm 2. Let the value U_A at point A be calculated by Algorithm 1 with interpolation by row. The value U_A^* is calculated similarly, with interpolation by column (FIG. 4). To calculate the value of U at point A, we propose, for example, to use the arithmetic average, $U_A = (U_A + U_A^*)/2$. Other methods of value combination are also possible. The color calculation step is similar to Algorithm 2.

Application of double interpolation is seldom justified because of the large time cost for insignificant improvement in the quality of the result. The result of the given algorithm operation is presented in FIGURE 5C, which shows the same model fragment as FIGURE 5A and B.

Algorithm 4. Replenishment by Height

We assume that the inner pixels are localized with the help of Algorithm 1, and represent the value U as a function of the two variables $U = U(X, Y)$. The grid cell represents part of the 3D surface. In the general case, the points of a rectangular grid do not belong to one planar surface. One may propose several ways to implement Algorithm 4.

First Method

We assume that the points belong to one planar surface. When a pixel is localized in a cell (point A in FIG. 4), then coordinates X and Y are known, and it is necessary to find the coordinate Z, or the value U_A in our case. By LLC, we determine the corresponding number of color in the palette. This method provides good results when all the points belong to the plane. The result is comparable to Algorithm 2 in quality and speed. The main shortfall of this method is that the image is not consistent and is uneven (FIG. 5D). The model is the same as used in FIGURE 5A–C.

Second Method

Assume that the cells do not necessarily belong to one plane. It is necessary to partition the cell into two triangles and find general equations of the planes that correspond to the triangles. We obtain two values of U for the current inner pixel, U_A^1 and U_A^2. The variant of determination of the resulting value is $U_A = (U_A^1 + U_A^2)/2$. The result is slightly better than that for the first method.

Third Method

Similar to Method 2, it is necessary to partition a cell into two triangles. The difference lies in the fact that, for each inner pixel, it is necessary to determine the triangle that contains it. Then, when searching for the value U_A, apply the equation of one plane or another. The result of the replenishment algorithm operation is as good as the replenishment by Algorithm 2. However, it is slower than Algorithm 2 by 10–15%. This method provides accurate correspondence of color transitions at the cell boundaries (FIG. 5E). In some cases, however, this algorithm presents a worse general picture. Partitioning a rectangular cell into two triangles, for example, is not ambiguous.

Fourth Method

Partition a cell into four triangles, calculate the geometric center of the cell and obtain four equations of the planes. Next, joining the top points at the geometrical center forms the triangles. Practical implementation shows that this method has worse quality and has time rate indices larger than Algorithm 2. FIGURE 5F presents a qualitative example of this algorithm operation on a problem similar to that in FIGURE 5A–E. The time rate indices worsen when enlarging the grid dimensionality.

POTENTIAL APPLICATIONS OF THE GR ALGORITHMS

1. The results of raster scan of cell ribs may be used for adjacent cells.
2. An image may be formed in the main memory in the form of a buffer, for example, as a bitmap image. One can then display the image entirely on the screen with, for example, OpenGL functions.
3. It is possible to apply Algorithms 2–4 once and to record the result as an image in bitmap format. Furthermore, if it is necessary to enlarge the model parts, then these algorithms need not necessarily be applied. Faster algorithms or standard capabilities of graphic libraries may be applied to enlarge the size of a bitmap image.

4. It is possible to apply Algorithms 2–4 to 3D grid surfaces with a plane raster scan, using, for example, OpenGL functions for work with textures.
5. The algorithms are very convenient for hardware implementation.

CONCLUSIONS AND RECOMMENDATIONS

1. Generally, the best results are obtained with Algorithm 2. It is the fastest of the algorithms working without OpenGL. The quality and adequacy of image is the best.
2. It is preferable and more effective to apply Algorithm 1 for generating a raster scan.
3. Application of OpenGL is justified for the quick preliminary review of an image.
4. It is possible to apply OpenGL for work with bitmap images, when enlarging, or when using the textures, prepared by Algorithms 2–4.
5. It is recommended to perform graphic replenishment within the frameworks of the color range of the palette, and not in RGB space.
6. It is recommended to perform a late transition to palette color numbers, and thus eliminate the staircase effect when moving from color to color.

The main point is that GR of one cell Algorithms 2–4 do not require information about other cells in the model, in contrast to the application of splines or interpolating polynomials for GR. It should be particularly noted that in the application of GR the splines or interpolating polynomials for finite-element grids is generally a complex and time-consuming problem. Consequently, Algorithms 2–4 are very practical for GR grid fields.

ACKNOWLEDGMENT

We are grateful to the members of the Department of Mathematics of RFNC-VNIITF for their careful reading of this report, for their important advice and remarks on the study content, and for their help in the implementation of the illustrations. The approaches and algorithms stated in this study were investigated and implemented within the frameworks of the work on creating a system of scientific visualization of the results of numerical simulation and experiments VIZI2D.[13–15]

REFERENCES

1. GOURAUD, H. 1971. Computer Display of Curved Surfaces. Doctoral Thesis, University of Utah, 1971. Also as Comp. Sci. Dept. Rep. UTEC-CSc-71-113 and NTIS AD 762 018. A condensed version in IEEE Trans. **C-206:** 623–628.
2. Proc. of the Int. Conferences on Computer Graphics. Graphicon 1995–2000.
3. ROGERS, D. 1989. Algorithmic bases of computer graphics. M. Mir. 391–399.
4. SHIKIN, E.V. & A.V. BORESKOV. 1997. Computer Graphics Dynamics, Realistic Images. Dialogue-MIFI. Moscow. 318–320.
5. IVANOV, V.P. & A.S. BATRAKOV. 1995. 3D Computer Graphics. Radio and Communication. Moscow.

6. PHONG, B.-T. 1975. Illumination for Computer Generated Images. Doctoral Thesis, University of Utah, 1973. Also as Comp. Sci. Dept. Rep. UTEC-CSc-73-1296 NTIS ADA 008 786. A condensed version is given in CACM **18:** 311–317.
7. PLETNIOV, F.A. 1997. Fast methods of filling in realistic graphics. Issues of Atomic Science and Engineering (VANT). Mathematical Simulation of Physical Processes **4:** 39–50.
8. BAIAKOVSKIY, YU.M., V.A. GALAKTIONOV & T.M. MIKHAILOVA. 1985. Graphical Expansion of FORTRAN. Nauka, Moscow.
9. BAKHVALOV, N.S. 1973. Numerical Methods. Nauka, Moscow.
10. KOBKOV, V.V. & YU.I. SHOKIN. 1983. Spline-Functions in Numerical Analysis. Study Guide, Novosibirsk.
11. CHAN, SH.K. 1994. Principles of Designing Visual Information Systems. M. Mir.
12. MOGILENSKIKH, D.V. 2000. Nonlinear color interpretation of physical processes. Proc. Int. Conf. on Computer Graphics. Graphicon 2000. 201–211.
13. MOGILENSKIKH, D.V., I.V. PAVLOV, V.V. FYODOROV, *et al.* 2000. Principles of constructing and functional content of visualization system for analysis of scalar and vector fields specified at 2D regular grids. Preprint RFNC-VNIITF No. 172. <http://www.ch70.chel.su>.
14. MOGILENSKIKH, D.V., V.V. FYODOROV & I.V. PAVLOV. 1998. Systems of scientific visualization "VIZI" for graphical representation of mathematical simulation results. Third Siberian Congress on Applied Industrial Mathematics (INPRIM-98). Abstracts Part III. Novosibirsk,. Institute of Mathematics SB RAS. 17–18.
15. MOGILENSKIKH, D.V., I.V. PAVLOV & E.E. SAPOZHNIKOVA. 1998. Methods of 3D graphical presentation of 2D data of results of mathematical physics problems. V Zababakhin Scientific Talks. Proc. Int. Conf, Snezhinsk.

Visualization of Molecular Dynamics by Simulation

MASAHIRO OTA AND YINGXIA QI

Department of Mechanical Engineering, Tokyo Metropolitan University, Minami-osawa, Hachi-oji, Tokyo, Japan

ABSTRACT: The mechanism of formation of methane hydrate is investigated here at a molecular level. The key to whether or not methane hydrates can be formed is the stability of the hydrate structure. This paper deals with a computer simulation of methane hydrate type 1 formation by a molecular dynamics method and an accurate description of the crystal structure. Computer simulation results show that very stable type 1 methane hydrates can be formed below 275 K, in which 46 water and 8 methane molecules are contained in a single cubic unit cell of length 12.0 Å. Above 275 K, the stability of the hydrates increasingly degrades, logarithmically as a power function, up to 300 K, with no stability at 350 K. Pressure, as a parameter of formation conditions, increases with temperature and becomes tremendously high at 350 K. Hydrates with various hydration numbers (at least one guest molecule in a small cage) have roughly the same degree of stability, but higher pressures are required for hydrates with fewer guest molecules. Empty hydrates dissolve above 275 K.

KEYWORDS: methane hydrate; structure; molecular dynamics; simulation

INTRODUCTION

Extensive accumulations of natural gas hydrates in permafrost regions and in sediments on deep ocean floors have recently been explored. Sloan[1] suggests that the total resource may surpass the energy content of the total fossil fuel reserves by as much as a factor of two. Since methane gas (the main constituent of natural gas) could be recovered by the dissolution of methane hydrates, natural gas hydrates are a potential future energy supply. On the other hand, hydrates have been suggested as a natural gas storage medium[2] because of their high storage efficiency. Furthermore, compared to liquid natural gas (LNG), hydrates have been shown to have the advantages of lower storage pressure and temperature, at a lower cost.[3] The application of natural gas hydrates as an energy resource and as means of energy storage requires detailed knowledge of the thermodynamic, kinetic, and mechanical parameters that affect the formation and dissolution of hydrates. Extensive data on hydrate thermodynamic properties is available and the phase diagram is well known.[4] The crystal structure of clathrate hydrate type has been accurately determined by measuring a single crystal of ethylene oxide hydrate by X-ray and neutron diffraction, at 248 K

Address for correspondence: Masahiro Ota, Ph.D., Department of Mechanical Engineering, Tokyo Metropolitan University, Minami-osawa, Hachi-oji, Tokyo 192-0397, Japan.
ota-masahiro@c.metro-u.ac.jp

and 80 K, respectively.[5,6] The hydration number of solid methane hydrates has been tested at 193 K by ^{13}C NMR.[7] Fleyfel et al.[8] experimentally investigated methane hydrates in their metastable/nonequilibrium stage, also by ^{13}C NMR. Tse[9] studied atomic pair distribution functions and frequency spectra of methane hydrate at 145 K by a molecular dynamics method using the measured structure data as the initial configuration. Hirai et al.[10] simulated the formation process of CO_2 hydrate at 270 K by a molecular dynamics method, stipulating that the CO_2 guest molecules were fixed at the center of cages. Tests on the production rate and "memory effect" of methane hydrate formation have recently been carried out in our laboratory.[11] Here we discuss our attempt to form methane hydrate type 1 by allowing all the molecules to move freely at a series of temperatures, so as to gain detailed information on the crystal structure and compare the results to experimental data. Based on these results, we can determine whether methane hydrates form or dissolve by determining if the crystal structure can be maintained. The hydration number is also investigated.

SIMULATION METHOD

A computer simulation using a molecular dynamics method requires a model system, initial and boundary conditions, an equation of motion and its method of solution, and a specified interaction potential between particles. For a crystal solid, a single unit cell is selected as the model system. For type 1 methane hydrate, the unit cell is a cube of length 12.0 Å containing 46 water molecules and 8 methane molecules. In general, hydrate formation simulation should be commenced from the initial condition of a random configuration. However, to save computational time, the initial configuration of water molecules is first artificially designed as a regular pentagonal dodecahedron composed of the center-of masses of water molecules at the center of the cubic cell, but the orientations of water molecules are random. The methane molecules are placed in positions that are roughly similar to that in an ideal hydrate structure.

A periodic boundary condition and a minimum image convention are imposed for surface effects. The *leap-frog* algorithm solves the equations of motion. The constant-NVT method is used, in which the number of molecules, the volume size, and the temperature remains constant throughout the simulation. The total system potential energy Φ is a sum of pairwise potential energies over all pairs of particles, given by

$$\Phi = \sum \varphi_{ij}, \qquad (1)$$

where i and j denote the interaction sites.

For interactions between water molecules, the analytical potential model of Matsuoka, Clementi, and Yoshimine[12] (MCY) is used. This model consistently gives more accurate predictions of the ice crystal structures than other potentials, such as those of ST2, BNS, Rowlinson, RSL, RSL2, Watts, LS, TI4P, and SPC.[13] The MCY potential model involves four interaction centers in a rigid molecule with the geometry of a O–H bond length of 0.9572 and an angle H–O–H of 104.52°. The function has two components: an electrostatic interaction between arrays of fractional point charges and a set of short-range (exponential) atom–atom potentials as follows:

$$\varphi(m_1, m_2) = q^2\left(\frac{1}{r_{13}} + \frac{1}{r_{14}} + \frac{1}{r_{23}} + \frac{1}{r_{24}}\right) + \frac{4q^2}{r_{78}} - 2q^2\left(\frac{1}{r_{18}} + \frac{1}{r_{28}} + \frac{1}{r_{37}} + \frac{1}{r_{47}}\right)$$
$$+ a_1 e^{-b_1 r_{56}} + a_2[e^{-b_2 r_{13}} + e^{-b_2 r_{14}} + e^{-b_2 r_{23}} + e^{-b_2 r_{24}}]$$
$$+ a_3[e^{-b_3 r_{16}} + e^{-b_3 r_{35}} + e^{-b_3 r_{45}}]$$
$$+ a_4[e^{-b_4 r_{16}} + e^{-b_4 r_{26}} + e^{-b_4 r_{35}} + e^{-b_4 r_{45}}]. \tag{2}$$

The parameters used in the MCY model are given by Matsuoka et al.[12]

For interactions between water and methane molecules, via methane–methane molecules, the Lennard–Jones (L-J) potential is

$$\Phi^{LJ}(r) = 4\varepsilon\left[\left(\frac{\sigma}{r}\right)^{12} - \left(\frac{\sigma}{r}\right)^{6}\right]. \tag{3}$$

The parameters for the L-J potentials are given elsewhere.[14] Methane molecules are treated as spherical particles that interact with water molecules at the center of mass of water molecules. Using the venerable Lorentz–Berthelot mixing rule allows us to approximate the L-J parameters between unlike molecules,

$$\sigma = \frac{1}{2}[\sigma_1 + \sigma_2] \tag{4}$$

$$\varepsilon = [\varepsilon_1 + \varepsilon_2]^{1/2}. \tag{5}$$

For a long-range electrostatic interaction, Ewald's method[15] is used. For a short-range interaction, such as the exponential terms in MCY and the terms in L-J, the potential calculation is truncated with a spherical cutoff distance of half the length of the cube edge.

RESULTS AND DISCUSSION

According to a previous study,[13] the O–O atomic pair distance predicted by the MCY potential is about 7% too large. However, the geometry of the crystal lattice is not influenced by this error. To resolve the problem, the O–O atomic pair distance is scaled here by the factor of 1.07, but only for the MCY potential energy calculation between water molecules. As a result, the practical O–O pair distance in the simulation is equal to the normal value.

Crystal Structure

The system potential at 0 K is nearly minimum and stable during the latter part of the simulation; that is, the system is in an equilibrium state, every molecule has moved to an appropriate position, and the interaction forces among all of the molecules have achieved equilibrium. Consequently, every molecule is nearly motionless. The structure obtained at this time is defined to be the *static structure* and is shown in FIGURE 1. This is the structure of hydrate type 1. The drawings were made using RasMac v2.6. FIGURE 1 shows the hydrogen-bonded network of water molecules. The general picture is that of tetrahedrally hydrogen-bonded water molecule network. Two kinds of cages, a large tetrakaidecahedron cage and a small pentagonal

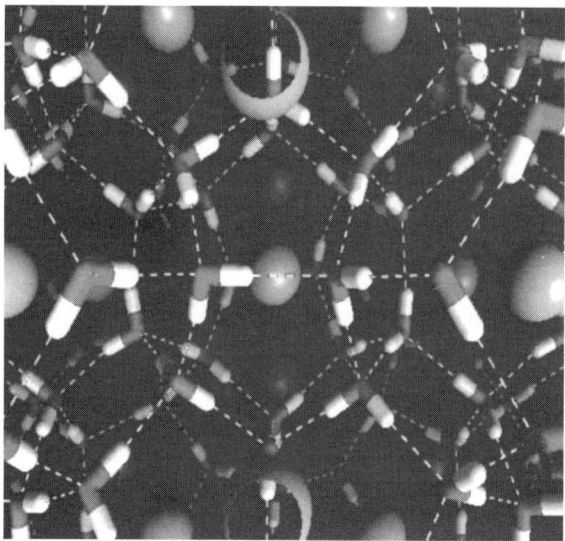

FIGURE 1. Hydrate lattice structure shown in three-dimensions.

dodecahedron cage can be clearly seen in FIGURE 1. Statistically, there are two small cages and six large cages in a unit cell. Methane molecules locate approximately at the center of the cages. Practically, except for one methane molecule in the central cage, the methane molecule in the small cage, being at the vertex of the cube, is shared by eight unit cells and the remaining six in the large cages are shared by two unit cells. Therefore, these methane molecules should be at the vertices or on the surface of the cube. Since the model system used here is only a single cubic cell, their positions in the static structure are close to the surface or the vertices of the cube. When the temperature is above 0 K, it can be observed that these molecules

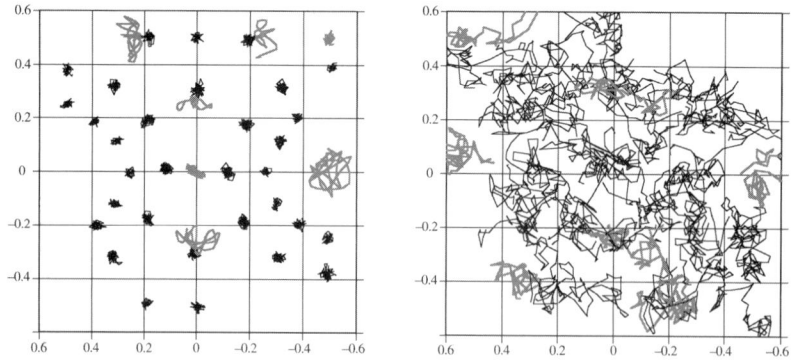

FIGURE 2. Trajectories of motion of molecular center-of-mass. *Gray* represents methane molecules, *black* represents water molecules: **left** 100 K, **right** 350 K.

move about, through the surface, sometimes in the cube, sometimes outside the cube. FIGURE 2 shows the trajectories of molecule motion at 100 K and 350 K in two dimensions (2D), monitored every 100 steps over the last 5,000 steps. The figure shows the trajectories of 46 water molecules and 8 methane molecules in the cell. The drawings show projections on the YZ plane. It is clear that the motion ranges of molecules increase with increasing temperature. However, up to 300 K, each molecule only moves about its equilibrium position, which is the position in the static structure at 0 K. Any deviation from the equilibrium position could cause the molecule to be pulled away by the intermolecular forces. That is, up to 300 K, the hydrate structure can be maintained even though there is stronger thermal molecular motion. It is, therefore, concluded that methane hydrate type 1 can be formed stably up to 300 K. At 320 K, however, the wider range of molecular motion causes a structural instability. According to tests of Marshall et al.,[16] the formation pressure of methane hydrates is 397 MPa at 320 K. Consequently, it is practically very difficult to form methane hydrates at 320 K. It is shown, that the hydrate structure is completely destroyed at 350 K, and it seems impossible to form methane hydrate type at 350 K. Another point of interest is that the range of motion of a methane molecule in a large cage is greater than that in a small cage. This implies that a methane molecule in a large cage escapes more easily from the hydrates than a molecule in a small cage. Quantitative descriptions of the hydrate structure are given below.

Pair Distribution Function

The definition of a pair distribution function is

$$g(R) = \frac{V}{N^2} \langle \sum_i \sum_{j \neq i} \delta(R - r_{ij}) \rangle. \quad (6)$$

FIGURE 3 exhibits simulated $g_{HH}(R)$ and $g_{MW}(R)$ pair distribution functions at various temperatures, where the bin width is 0.1. Each curve is the average of data collected over the last 1,000 steps. At 0 K, the first peak position in $g_{HH}(R)$ is at about 2.4 Å. This means the distance between the two closest hydrogen atoms belonging to the two closest water molecules, is 2.4 Å. Up to 300 K, changes in $g_{HH}(R)$ are confined to a reduction in peak heights and an increase in peak width as the temperature increases. According to the present simulation studies, there is a distinct feature in the hydrate structure stability. When the hydrates become unstable, even to the point of being thoroughly dissociated, the valley between the first peak and the second

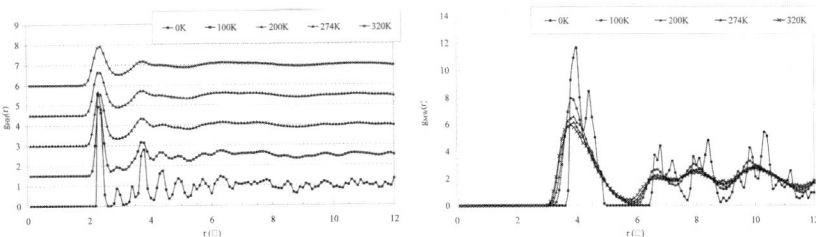

FIGURE 3. Pair distribution functions: **left** hydrogen–hydrogen, **right** methane–water molecules.

peak gradually rises until the second peak disappears. The position of the unique peak in $g_{MW}(R)$ at 0 K is about 3.8 Å, which corresponds to the distance between the centers of most of methane and water molecules. When the temperature increases, the top of the peak becomes flat. However, similar curves are obtained for temperatures in the range 200 K–275 K. This means that a methane molecule in a large cage has some free-motion space, that is to say, its equilibrium position in a large cage is not seriously restricted. Our calculated results for atomic pair distribution functions are similar to those obtained by Tse et al.[9] based on the SPC potential model.

Hydrogen Bond Angle

FIGURE 4 shows the pair distribution function, where the bin width is 1°, for a hydrogen bond angle of O–H⋯O that is formed between nearest-neighboring water molecules with a O–O distance less than the position of the first minimum (3.5 Å) in the oxygen–oxygen pair distribution function. Each water molecule is hydrogen-bonded to four others, and each donates and accepts two hydrogen bonds. Hence, another condition that must be satisfied is that the distance between the hydrogen atom and another oxygen atom is less than the position of the first minimum (2.5 Å) in the oxygen–hydrogen pair distribution function. For the static structure at 0 K, the peak position is about 176°. Most of the hydrogen atoms are nearly on the line connecting two of the nearest-neighboring oxygen atoms, as shown in FIGURE 1. With increasing temperature, the peak position shifts to a smaller value, and the curve becomes flatter.

Tetrahedral Angle

As mentioned above, every water molecule is hydrogen-bonded to four others in a tetrahedral structure. According to Uttormark et al.,[17] an angular order parameter, F, can be used to evaluate the deviation from perfect tetrahedral bonding,

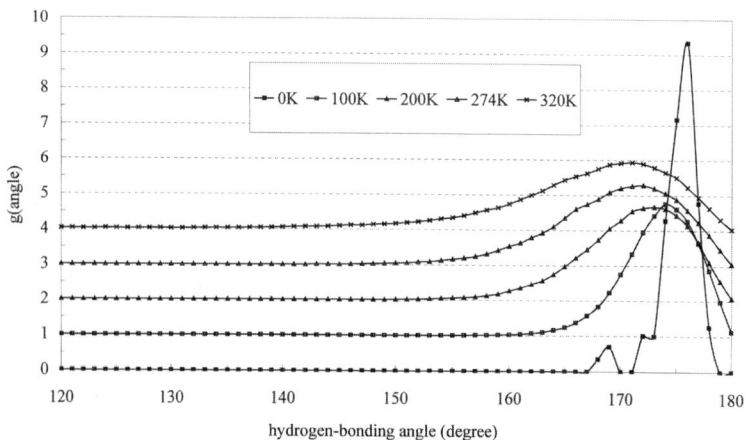

FIGURE 4. Distribution for hydrogen bond angle O–H–H.

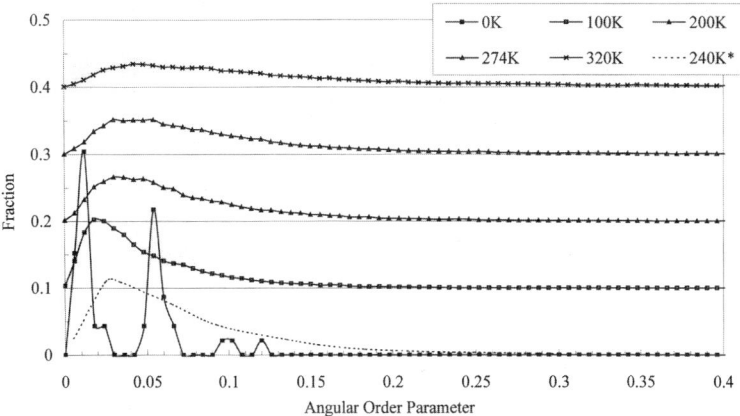

FIGURE 5. Distribution of angular order parameter in methane hydrate (reproduced from Baez[18] with permission).

$$F = \sum_{i=1}^{6} (|\cos x_i| \cos x_i + 0.11)^2, \qquad (7)$$

where x_i is the bond angle formed by the *test* oxygen atom with pairs of oxygen atoms from among its nearest-neighboring water molecules and $6 = 4(4-1)/2$ is the number of independent angles. The nearest-neighboring molecules are considered to be those within a distance less than the position of the first minimum in the oxygen–oxygen pair distribution function. F has a value of zero for the perfect tetrahedral structure, and positive nonzero values for all other structures. The calculated results for the distribution of the angular order parameter, using a bin width of 0.006, are shown in FIGURE 5. For the static structure at 0 K, the first sharp peak position is 0. This means that the hydrate structure is hydrogen-bonded by means of a nearly perfect tetrahedron. With increasing temperature, the peak positions shift to larger values, and the curves become flatter. For comparison, the distribution of F of the hydrates at 240 K, taken from Baez and Clancy[18] is also presented in FIGURE 5. The slight differences in peak height may be due to the different methods of collecting data. It seems that at 350 K, the curve has almost changed to a horizontal line and the tetrahedral structure has disappeared completely.

CONCLUSIONS

The molecular mechanism of hydrate formation was investigated by using molecular dynamics simulations conducted at various temperatures under constant NVT conditions.[18–20] This revealed that type 1 methane hydrate is more stably formed below 275 K. Although it can also be formed up to 300 K, it cannot be formed at 350 K. The crystal structure of hydrate type 1 generated by our simulations is in very good agreement with that measured by Hollander and Jeffery.[6] In fact, the measured crystal structure should be referred to the static structure obtained form the simulations at 0 K, because the molecules above 0 K vibrate only around their equilibrium positions at very high frequency. It was also inferred that for methane hydrate type 1, the guest molecule in a large cage escapes from the hydrate more easily than one in a small cage. Moreover, hydrates with all of the cages filled with guest molecules are stable at a lower equilibrium pressure than partially filled hydrates, which are stable only at much higher pressures. Empty hydrates cannot exist at all above 275 K.

REFERENCES

1. SLOAN, E.D., JR. 1994. Conference overview. Ann. N.Y. Acad. Sci. **715:** 18–19.
2. OTA, M., H. KOMATSU, K. MURAKAMI & M. SAKAMOTO. 1997. Storage of natural gas hydrates as energy and water resources. Int. Conf. on Fluid and Thermal Energy Conversion, Yogyakarta, Indonesia. 413–417.
3. GUDMUNDSSON, J.S. & A. BORREHAUG. 1996. Frozen hydrate for transport of natural gas. Second Int. Conf. on Natural Gas Hydrates, Toulouse, France. 415–422.
4. SLOAN, E.D., JR. 1989. Clathrate Hydrates of Natural Gases. Marcel Dekker, Inc., New York.
5. MCMULLAN, R.K. & G.A. JEFFREY. 1965. J. Chem. Phys. **42:** 2725–2731.
6. HOLLANDER, F. & G.A. JEFFREY. 1977. Neutron diffraction study of the crystal structure of ethylene oxide deuterohydrate at 80 K. J. Chem. Phys. **66**(10): 4699–4705.
7. RIPMEESTER, J.A. & C.I. RATCLIFFE. 1988. Low-temperature cross-polarization/magic angle spinning ^{13}C NMR of solid methane hydrates: structure, cage occupancy, and hydration number. J. Phys. Chem. **92:** 337–339.
8. FLEYFEL, F., K.Y. SONG, A. KOOK, et al. 1994. ^{13}C NMR of hydrate precursors in metastable regions. Ann. N.Y. Acad. Sci. **715:** 210–224.
9. TSE, J.S. & M.L. KLEIN. 1983. Molecular dynamics studies of ice Ic and the structure I clathrate hydrate of methane. J. Phys. Chem. **87:** 4198–4203.
10. HIRAI, S., et al. 1996. Study for the stability of CO_2 clathrate-hydrate using molecular dynamics simulation. Energy Convers. Mgmt. **37**(6–8): 1087–1092.
11. WAGATSUMA, Y., M. OTA, Y.X. QI, et al. 1998. On production of clathrate-hydrate. National Thermal Engineering Conference, No. 98-7. JSME 151–152. (Japanese)
12. MATSUOKA, O., E. CLEMENTI & M. YOSHIMINE. 1976. CI study of the water dimer potential surface. J. Chem. Phys. **64**(4): 1351–1361.
13. MORSE, M.D. & S.A. RICE. 1981. Tests of effective pair potentials for water: predicted ice structures. J. Chem. Phys. **76:** 650–660.
14. REID, R.C., J.M. PRAUSNITZ & B.E. POLING. 1987. The properties of Gases and Liquids. McGraw-Hill.
15. ALLEN, M.P. & D.J. TILDESLEY. 1987. Computer Simulation of Liquids. Oxford Science Publications, Oxford.
16. MARSHALL, D.R., S. SAITO & R. KOBAYASHI. 1964. AIChE J. **10**(2): 202–205.
17. UTTORMARK, M.J. 1992. Melting Kinetics of Small Crystalline Cluster in the Liquid by Molecular Dynamics. Ph.D. Dissertation, Cornell University, Ithaca, N.Y.
18. BAEZ, L.A. & P. CLANCY. 1994. Ann. N.Y. Acad. Sci. **715:** 177–186.
19. QI, Y. & M. OTA. 2000. Molecular transfer processes in the formation of methane hydrate. Proc. NHTC'00, 34th National Heat Transfer Conference, Pittsburgh. 1–8.
20. OTA, M. & Y. QI. 2000. Numerical simulation of nucleation process of clathrate hydrates. JSME Int. J. B **43**(4): 719–726.

Vortex Scale of Unsteady Separation on a Pitching Airfoil

MASAKI FUCHIWAKI[a] AND KAZUHIRO TANAKA[b]

[a]*Osaka Science & Technology Center, Osaka, Japan*
[b]*Kyushu Institute of Technology, Iizuka, Fukuoka, Japan*

> ABSTRACT: The streaklines of unsteady separation on two kinds of pitching airfoils, the NACA65-0910 and a blunt trailing edge airfoil, were studied by dye flow visualization and by the Schlieren method. The latter visualized the discrete vortices shed from the leading edge. The results of these visualization studies allow a comparison between the dynamic behavior of the streakline of unsteady separation and that of the discrete vortices shed from the leading edge. The influence of the airfoil configuration on the flow characteristics was also examined. Furthermore, the scale of a discrete vortex forming the recirculation region was investigated. The non-dimensional pitching rate was $k = 0.377$, the angle of attack $\alpha_m = 16°$ and the pitching amplitude was fixed to $A = \pm 6°$ for $Re = 4.0 \times 10^3$ in this experiment.
>
> KEYWORDS: unsteady separation; vortex; flow visualization; pitching airfoil; reattachment

INTRODUCTION

A number of studies on the unsteady flow in the low Reynolds number (Re) region have been catalyzed in recent years by an interest in microelectromechanical systems (MEMS) based on the concept of flow control[1] and the development of the human power vehicles.[2] Sunada *et al.*,[3] Fuchiwaki *et al.*,[4] and Sato *et al.*[5] reported the characteristics of steady forces acting on several kinds of airfoils in the low Reynolds number region. Fuchiwaki *et al.*[6] visualized the discrete vortices shed from the leading and trailing edges of a pitching airfoil for $Re = 4.0 \times 10^3$, and reported the frequency of vortex shedding during one pitching cycle and the arrangement of unsteady separation of a pitching airfoil. Lai *et al.*,[7] Jones *et al.*,[8] and Peter *et al.*[9] reported the structure of wake behind a pitching and plunging airfoils for the low Re region. Moreover, Joseph *et al.*[10] showed the possibility of controlling separation by a setting a small plunging airfoil in the separation region behind the back step or a bluff body. However, the detailed structure and the scale of a vortex forming the recirculation region in unsteady separation on a pitching airfoil for the low Re region is not been sufficiently understood.

The objective of this paper is to clarify the scale of a discrete vortex forming the recirculation region and the relationship between the discrete vortex and the

---

Address for correspondence: Masaki Fuchiwaki, Department of Mechanical Systems Engineering, Kyushu Institute of Technology, 680-4 Kawazu, Iizuka-city, Fukuoka 820-8502, Japan. Voice: +81-948-297763; fax: +81-948-297751.
 futiwaki@mse.kyutech.ac.jp

streakline of unsteady separation on a pitching airfoil. Comparing the results produced by two visualization methods allow to achieve this. The dye flow method was utilized to visualize the streakline of unsteady separation and the Schlieren method was another way to visualize the discrete vortex shed from the leading edge. Furthermore, two kinds of airfoils, NACA65-0910 and a blunt trailing edge airfoil, were tested in order to study the influence of airfoil configuration.

EXPERIMENTAL

Dye Flow Visualization

Dye flow visualization showed the streakline of unsteady separation on the pitching airfoil surface. The experimental apparatus for dye flow visualization consisted of a water tunnel, a tested airfoil, equipment for driving the pitching motion, a halogen light sheet source, a plane mirror, a 30-frames/second digital video camera and two kinds of dyes.[4] The pitching motion was driven by a stepping motor, which was directly connected with the pitching shaft of a test airfoil.

The tested airfoils were NACA65-0910, which had a round leading edge and a sharp trailing edge, and a blunt trailing edge (BTE) airfoil, which had a sharp leading edge and a thick trailing edge. Their chord length c and span length l were 0.04 m and 0.12 m, respectively. The tested airfoils had some pinholes on the surface for the dye flow visualization. The pinhole diameter was 0.5 mm. They had slightly different locations in the spanwise direction so as to maintain dye flow independence. Two kinds of dyes, uranine and rhodamin B, with respective colors green and yellow in the halogen light sheet, flowed out from the pinholes located on the leading and trailing edge, respectively.

Schlieren Visualization

Schlieren visualization showed a discrete vortex shed from the leading edge and moving along the suction surface. An experimental apparatus for this visualization consisted of a low-speed wind tunnel, oscillating equipment, a tested airfoil, and a Schlieren visualizing system with a high-speed camera.[6] The system consisted of a He–Ne laser (5 mW) as a light source, a condensing lens, a pinhole, two reflectors, a concave mirror of 0.3 m diameter and 3.0 m focal length, a plane mirror for reflection, a knife edge, and a camera lens. The high-speed camera photographed the optical images every 4,500-frames/second. In order to create a difference in density in the air flow around the tested airfoil, the central part of the surface was painted black and was heated to 373 K by a halogen light and two illumination lights. The chord length c and span length l of the tested airfoils were 0.06 m and 0.14 m, respectively.

Experimental Conditions

The pitching motion with a sinusoidal wave was performed around its mid-chord axis. The non-dimensional pitching rate was defined by $k = 2\pi f (c/2)/V_0$, where f is the pitching frequency (Hz), and V_0 is the main flow velocity (m/sec). The rate, $k = 0.377$ was realized with $V_0 = 0.1$ m/sec, $f = 0.3$ Hz in the dye flow visualization experiment, and with $V_0 = 1.0$ m/sec, $f = 3.0$ Hz, in the Schlieren visualization experiment, respectively. The mean angle of attack during one pitching cycle was $\alpha_m = 16°$ where complete separation occurred from the leading edge under stationary conditions. The pitching amplitude was fixed to $A = \pm 6°$. Flow visualizations were performed for chord $Re = 4.0 \times 10^3$.

RESULTS AND DISCUSSIONS

Dynamic Behavior of the Streakline and Recirculation Region on a Pitching Airfoil Surface

FIGURE 1 shows the streaklines during one pitching cycle at $\alpha_m = 16°$ and $k = 0.377$ around the pitching NACA65-0910 (FIG. 1A–1C) and BTE (FIG. 1D–1F). Pictures (A)–(C) show the results at the bottom dead position ($\alpha = 10°$), the position after moving clockwise ($\alpha = 18°$), and the top dead position ($\alpha = 22°$) during one pitching cycle, respectively. FIGURE 1(D)–(F) show the results at the same positions as FIGURE 1(A)–(C).

In the case of NACA65-0910, a green-colored streakline was attached to the front half of the suction surface and an orange-colored dye reversed from the trailing edge toward the leading edge at the bottom dead position, (FIG. 1A). As a result,

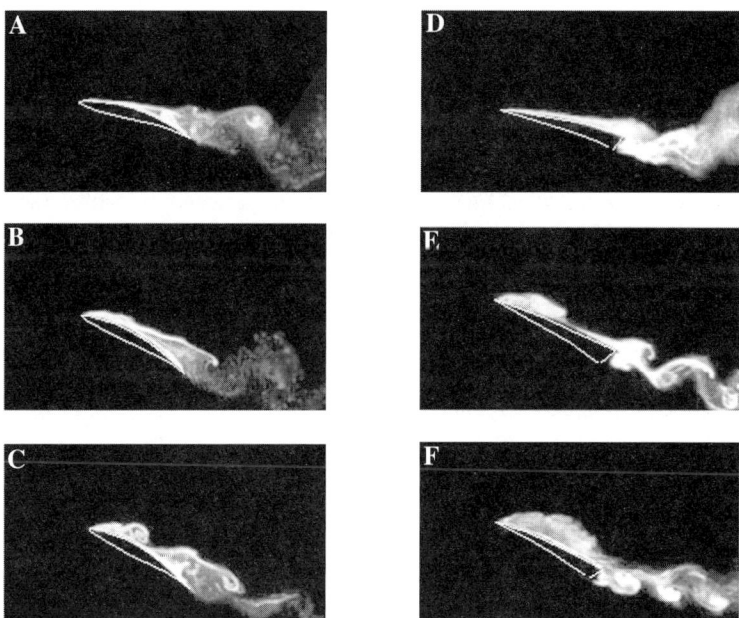

FIGURE 1. Streaklines during one pitching cycle at $\alpha_m = 16°$ and $k = 0.377$ around the pitching NACA65-0910 and BTE; **(A)** NACA65-0910, bottom dead position ($\alpha = 10°$); **(B)** NACA65-0910, moving upward ($\alpha = 18°$); **(C)** NACA65-0910, top dead position ($\alpha = 22°$); **(D)** BTE, bottom dead position ($\alpha = 10°$); **(E)** BTE, moving upward ($\alpha = 18°$); **(F)** BTE, top dead position ($\alpha = 22°$). (COLOR PLATE 4.)

separation occurred on the rear half of the suction surface. Increasing the angle of attack from the bottom dead position, the green-colored streakline became unstable (FIG. 1B). The vortex with a strong clockwise rotation was then shed from the leading edge into the separation region and caused the separation streakline to instantly reattach to the suction surface. As a result, the recirculation region was formed on the suction surface. It grew until the angle of attack reached the top dead position, to cover as far as the center of suction surface (FIG. 1C). Decreasing the angle of attack from the top dead position, the streakline separated from the surface and flew out toward the wake.

In the case of BTE, the streakline attached completely to the suction surface (FIG. 1D). When the angle of attack increased, the discrete vortex with a strong clockwise rotation suddenly appeared from the leading edge and formed the recirculation region (FIG. 1E). The separation streakline was attached as far as the trailing edge behind the reattachment point. Increasing the angle of attack caused the recirculation region to grow gradually and cover the suction surface fully at the top dead position (FIG. 1F). The state of the region was then maintained against a decreasing angle of attack. However, the region separated from the surface at $\alpha = 20°$ and large scale separation occured from the leading edge.

It is well known that the phenomenon of reattachment occurs generally with a transition from the laminar boundary layer to the turbulent boundary layer and that the laminar separation occur in the low Re region. However, the phenomenon of reattachment appeared around a pitching airfoil with higher pitching rate, even in the low Re region and at a higher angle of attack.

Comparing the state of the streaklines visualized in FIGURES 1D and 2C, the phenomenon of reattachment similarly occurred at the different angle of attack. The feature of this NACA case was that the streakline on the surface changed gradually from a stable state to the reattached state through the unstable state. The feature of the BTE case was that the streakline on the surface changed suddenly from the stable state to the reattached state, which lasted longer than in the NACA case. These results suggest that the dynamic behavior of the streakline depends on the airfoil configuration.

For both airfoils, reattachment appeared only at the higher angle of attack with higher non-dimensional pitching rate. This fact shows that a certain threshold in rotational speed is needed for a discrete vortex shed from the leading edge in order to realize the phenomenon of reattachment.

Dynamic Behavior of the Discrete Vortices on the Suction Surface

FIGURE 2 shows the dynamic behavior of the discrete vortex shed from the leading edge of a pitching NACA65-0910 (FIG. 2A–2C) and BTE (FIG. 2D–2F). The respective pictures in FIGURE 2 have the same angle of attack during one pitching cycle as the illustrations at the same level in FIGURE 1.

In the case of NACA65-0910, one discrete vortex was generated near the center of suction surface at the bottom dead position (FIG. 2A). The location of the vortex coincided with that of the separation point shown in FIGURE 1A. Increasing the angle of attack, the location approached the leading edge, and the vortex moved along the surface. During these conditions the streakline was unstable (FIG. 2B). These facts showed that the instability of the streakline meant the generation of discrete vortex.

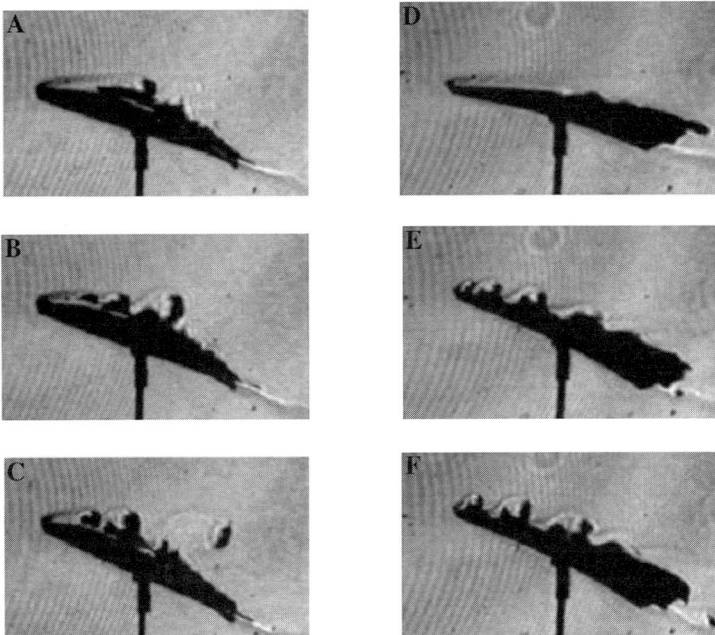

FIGURE 2. Dynamic behavior of the discrete vortex shed from the leading edge of a pitching NACA65-0910 and BTE; **(A)** NACA65-0910, bottom dead position ($\alpha = 10°$); **(B)** NACA65-0910, moving upward ($\alpha = 18°$); **(C)** NACA65-0910, top dead position ($\alpha = 22°$); **(D)** BTE, bottom dead position ($\alpha = 10°$); **(E)** BTE, moving upward ($\alpha = 18°$); **(F)** BTE, top dead position ($\alpha = 22°$).

FIGURE 2C shows the discrete vortices clearly at the top dead position. Comparing this result with FIGURE 1C shows that the recirculation region consisted of two discrete vortices. Decreasing the angle of attack, these vortices separated from the surface and flew out to the wake.

In the case of BTE, FIGURE 2D shows the generation of discrete vortices visualized on the rear half of suction surface at the bottom dead position. As the angle of attack increased, a discrete vortices were generated, one after another, from the leading edge. These vortices were attached to the surface and moved toward the trailing edge (FIG. 2E). Comparing this figure with FIGURE 1E, the recirculation region consisted of a few discrete vortices. Comparing FIGURE 2F with FIGURE 1F shows four or five discrete vortices in the recirculation region on the surface at the top dead position.

Two or three discrete vortices formed the recirculation region, the length of which was half of the chord length in FIGURE 2C of the NACA case and in FIGURE 2E of the BTE case. Four or five discrete vortices formed the recirculation region, the length of which was the full chord length in FIGURE 2F of the BTE case. From these results, it was clear that the scale of discrete vortex shed from the leading edge was about one fourth of the chord length and it did not depend on the airfoil configuration. The length of the recirculation region to the chord length determined the number of discrete vortex existing there.

SUMMARY

The unsteady separation around pitching airfoils, such as NACA65-0910 and BTE was visualized with higher angle of attack and higher non-dimensional pitching rate for $Re = 4.0 \times 10^3$ by dye flow (for streaklines) and Schlieren (for discrete vortex shedding) visualization techniques.

The dynamic behavior of streaklines depended on the airfoil configuration and the instability meant the generation of a discrete vortex. However, looking at the discrete vortex, the scale shed from the leading edge was about one fourth of the chord length and it did not depend on the airfoil configuration.

The length of the recirculation region to the chord length determined the number of discrete vortices generated there.

ACKNOWLEDGMENT

The Ministry of Education, Sports, Science and Culture in Japan, Grant-in-Aid for Scientific Research, No. 11650181, funded this study.

REFERENCES

1. Ho, C.H. & Y.C. TAI. 1996. Review: MEMS and its applications for flow control. Trans. ASME J. Fluid Eng. **118**: 437–447.
2. LISSAMAN, P.B.S. 1983. Low-reynolds-number airfoil. Annu. Rev. Fluid Mech. **15**: 223–239.
3. SUNADA, S., et al. 1997. Airfoil section characteristics at a low reynolds number. Trans. ASME J. Fluid Eng. **119**: 129–135.
4. FUCHIWAKI, M., et al. 1999. Flow patterns behind pitching airfoil and unsteady fluid forces. ASME Fluid Engineering Division Summer Meeting, FEDSM99-7286.
5. SATO, J. & Y. SUNADA. 1995. Experimental research on blunt trailing-edge airfoil sections at low Reynolds numbers. AIAA J. **33**(11): 2001–2005.
6. FUCHIWAKI, M. & K. TANAKA. 2000. Arrangement and dynamic behavior of vortices from a pitching airfoil. JSME Int. J. Series B **43**(3): 443–448.
7. LAI, J.C.S. & M.F. PLATZER. 1999. Jet characteristics of a plunging airfoil. AIAA J. **37**(12): 1529–1537.
8. JONES, K.D., et al. 1998. Experimental and computational investigation of the Knoller–Betz effect. AIAA J. **36**(7): 1240–1246.
9. PETER, F. 1988. Propulsive vortical signature of plunging and pitching airfoil. AIAA J. **26**(7): 881–883.
10. JOSEPH, C.L. & M.F. PLATZER. 1998. The characteristics of a pitching airfoil at zero free-stream velocity. ASME, Fluid Engineering Division Summer Meeting, FEDSM98-4946.

Imaging the Transient Boundary Layer on a Free Rotating Disc

BRANIMIR MATIJAŠEVIĆ, ZVONIMIR GUZOVIĆ, AND VINKO MARTINIS

*Faculty of Mechanical Engineering and Naval Architecture,
University of Zagreb, Zagreb, Croatia*

ABSTRACT: This report presents a visual study of the transition process of the laminar boundary layer (BL) in a turbulent BL on a free rotating disc. The imaging is based on an experimental investigation that aimed to analyze the structure of the BL by relating it to the ratio between turbulent energy and vortex energy, the critical and the transient Reynolds numbers (*Re*), the vortex numbers and their dependence on Re, and on the distance from the rotating disc.

KEYWORDS: transients; Reynolds number; rotating disc; turbulence; vortex energy

INTRODUCTION

Comprehensive investigations of the thermal and dynamic characteristics were performed (e.g., heat transfer between disc and fluid and vibrations of the disc) during the realization of an experimental device.[1] A disc of aluminium alloy with a radius of 0.323 m ensured an approximately isothermal temperature field. By means of a highly sensitive ceramic microphone as a measuring probe, the periodical time changes of the characteristic physical scalar quantity—the static pressure in the BL—were recorded. The probe for static pressure was designed and manufactured so that it could maximally preserve the fundamental characteristics of the flow field. Measurements of the periodic change of static pressure in the BL were performed at about 320 points of the *r-z* (radial–axial) planes. The measurements yielded the critical Reynolds number (*Re*), $Re_c = 289$, and the transient *Re*, $Re_T = 590$. These results matched very well reported results,[2–4] obtained by using a hot wire, and could be used for the validation of numerical calculations.[5,6]

PROCESSING AND ANALYSIS OF RECORDED SIGNALS

The measured signals were translated from a time domain to a frequency domain by means of a fast Fourier transformation (FFT) analyzer. Basic and reference signals were recorded. The basic signal included the periodic phenomenon, parasitic background noise, and the dynamic characteristic of the probe, recorded in the

---

Address for correspondence: Branimir Matijašević, Faculty of Mechanical Engineering and Naval Architecture, University of Zagreb, I. Lučića 5, 10000 Zagreb, Croatia.
branimir.matijasevic@fsb.hr

boundary layer. The reference signal consisted only of the dynamic characteristic of the probe and the background noise. These were recorded sufficiently far outside of the boundary layer, where the periodic phenomenon disappeared. By dividing the basic signal by the reference signal, we obtained a signal that consisted only of the periodic phenomenon, without the parasitic noise and the dynamic characteristic of the probe.

FIGURE 1 shows typical examples of the power spectra obtained from the measurements. The power spectra characteristic for the interval $1 \leq Re/Re_c < 1.314$, is

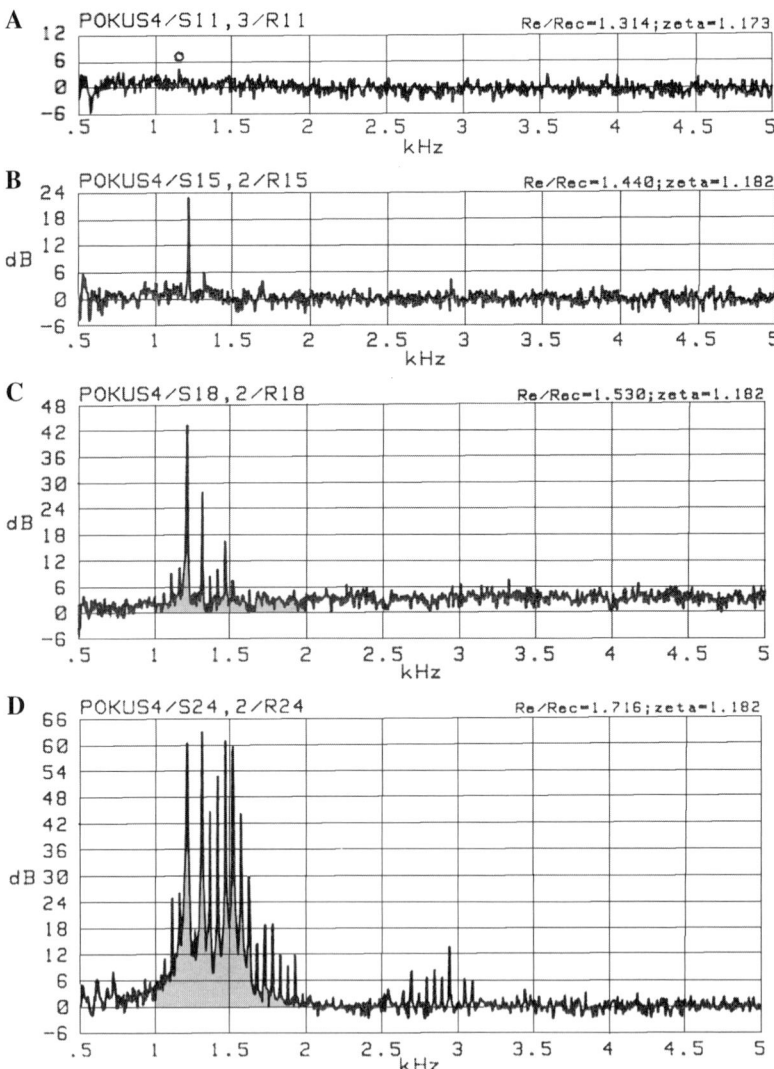

FIGURE 1. Typical examples of the power spectra obtained by measurement.

presented in FIGURE 1A, where only the wavy form of flow is shown. This region is characterized by a different number of periodic pressure disturbances (in the diagram this depends on the distance from the disc), which indicates a three-dimensional heterogeneous development of waves. A similar conclusion follows from the measurements of Wilkinson.[2]

FIGURE 1B shows a typical power spectrum in the interval $1.314 \leq Re/Re_c < 1.410$. An expressive peak, which corresponds to 24 stable vortices without turbulent noise, can be seen. FIGURE 1C shows the power spectrum at the beginning of the region, which corresponds to the interval $1.410 \leq Re/Re_c < 2.044$. Except for an insignificant peak in the region close to the dominant number of 24 vortices, peaks can be seen at higher frequencies, corresponding to between 24 and 29 vortices. This is a consequence of a periodic decay of a particular vortex with turbulent spots (see FIG. 5, below). The occurrence of duplication of all vortices is noted towards the end of the transient region of the boundary layer. This confirms the power spectrum in FIGURE 1D. Double frequencies between 2.5 and 3.5 kHz are observed, apart from a *peak comb* of basic frequencies ranging between 1.25 and 1.75 kHz. The occurrence of significant turbulent noise in the region of basic disturbances is also obvious, resulting from the transition into the turbulent flow at the periphery of all basic vortices.

IMAGE OF PERIODIC PRESSURE DISTURBANCES

An initial notion of the transition process from the laminar BL into the turbulent BL on a free rotating disc can be based on the visualization,[1] represented by FIGURE 2,

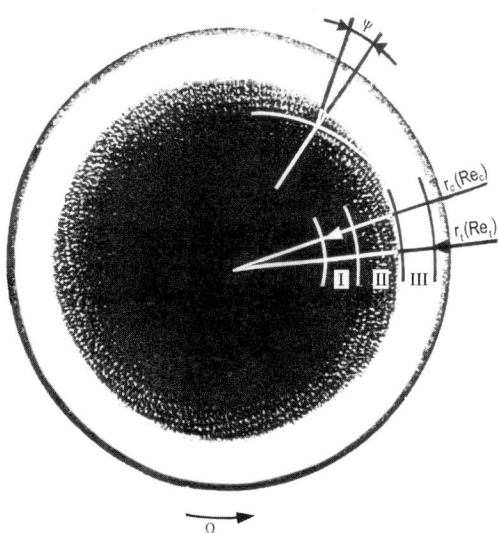

FIGURE 2. Visualization of vortices in the BL of the free rotating disc[1] with critical and transient Reynolds numbers marked.

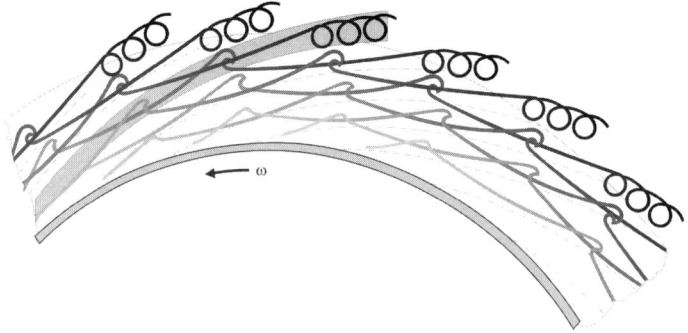

FIGURE 3. Image of flow during the first stage—transition from wavy to vortex form.

which clearly shows the stable vortices region. The Clarkson's visualization[7] shows the detail of the transition from the basic number of vortices to the duplicated number. The increased area of white spots at the transition from the second into the third region in FIGURE 2 is the result of powder deposition under conditions of significant turbulent characteristics that still preserve the vortex cores.

Analysis of the number of periodic pressure disturbances, that is, the vortex number, during one disc rotation shows that the process of development and breakdown of vortices in the BL occurs in three characteristic stages. In the first stage (see FIGURE 3), which follows the stable laminar BL, there are between 15 and 19 periodic disturbances, depending on Re. The power spectra of the signals show that there is no turbulent noise during the first stage.

In the second stage (see FIGURE 4) there are 24 stable spiral vortices. The relative amplitude of disturbance does not increase with an increase in Re. Similarly, the distance from the disc at which the periodic disturbances were registered does not change. The power spectra of the signals from the second stage of transition show no turbulence noise. At the end of the second stage some of the spiral vortices are periodically duplicated (see FIGURE 5), and newly generated vortices do not necessarily have the same dimensions. There are 24–26 periodic disturbances. The

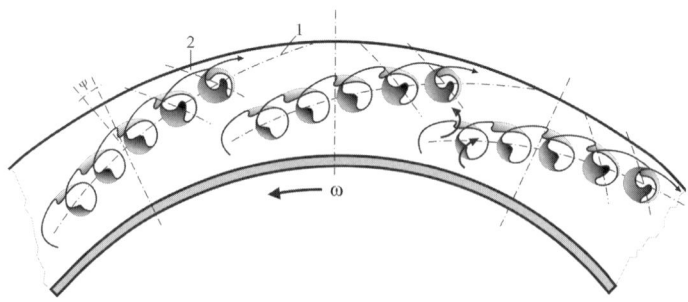

FIGURE 4. Image of flow during the second stage—formation of secondary vortices: 1, the axis of the basic vortex; 2, the axis of the secondary vortex.

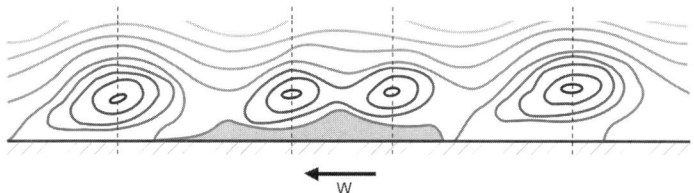

FIGURE 5. Image of duplication of the basic vortex.

duplication is accompanied by the occurrence of turbulent noise, which disappears from the signal by expulsion from disc. The frequency peak 24 remains in the signal.

In the third stage of the transition, the flow becomes non–stationary, which is manifested in periodic breakdown of the vortices that existed during the second stage, so that besides 26 vortices, 29 and 30 now appear. The non–stationary characteristic of the vortex number results in the relative motion of their axis against the disc, which is manifested by modulation of the phase and amplitude of the signal. The time durations of the various numbers of vortices are not equal and change with an increase in Re: longer duration of a larger number of vortices corresponds to a larger Re. The accompanying occurrence of the vortex breakdown processes is an increase in the turbulent part of the energy, which is manifested by an increase in turbulent noise in the signal power spectrum. The share of turbulent noise on the same radius decreases with the distance from the disc. The increase of the turbulent part of the energy and a decrease of vortex energy when Re approaches Re_T suggests successive *hulling* of the vortex periphery. As the outward distance from the disc toward the outside border of the BL increases, there are again 24 periodic disturbances dominating during one disc rotation, as a result of the wave shape of fluid flow that does not participate in the vortex flow at all.

RESULTS

A large number of power spectra of signals, recorded at points in the r–z plane of the BL, yielded images of the transition process from the laminar into the turbulent BL on the free rotating disc. The flow in the first transition stage has a wavy form (FIG. 3). The number of waves, the wave amplitudes and the intermittence factor during one disc rotation, increase with increasing Re. This was also shown in the graphic interpretation of measured results,[2] and is now once again validated by the present measurements. By increasing Re, the peak of the wave is successively swaddled up in the direction of the axis of a future vortex, until the crest assumes the shape of a hairpin. At the end of the first stage, the broadening peak of the wave is broken off and it twists into a vortex, the axis of which has a logarithmic spiral shape. The first stage is characterized by a successive increase of the relative amplitude (measured normal to the disc), and a considerably larger increase in the absolute amplitude (measured in the direction of the lengthened peak of the wave). Entering the second stage (FIG. 4), the number of spiral vortices remains stable. However, as Re increases, additional deformations of stream lines appear. Adjacent to the disc they are twisted in the direction of the vortex axis and disc periphery. At the outside border of the BL they are also twisted in the direction of vortex axis, but in the direction of the disc axis. In the middle of the second stage these deformations become expressed, and assume the shape of a hairpin, and are further twisted into a secondary vortex, whose existence during the third stage is very short. The formation of secondary vortices has also been registered in experiments using visualization.[3,7] The secondary vortex is broken down into entirely turbulent flow. The core of the vortex is still preserved, with considerably smaller dimensions. A sudden decrease in the dimension of the remaining vortex core starts when Re approaches its transient value Re_T. Changes of flow characteristics at the end of the third stage, when even the last forms of periodic flow disappear, occur at a small increase in Re, and the flow is very stormy. The transient Re value depends primarily on disturbances that affect the remainder of the undisturbed flow from the turbulent part of the BL and less on the activity of disturbances outside the BL.

SUMMARY

The work presented here is based on experimental measurements that allow insight into less visible occurrences during the transition from laminar into turbulent flow inside the BL on a free rotating disc.

REFERENCES

1. MATIJAŠEVIĆ, B. 1989. Analysis of Phenomena in Transient Boundary Layer of Rotating Disc (in Croatian). Ph.D. Thesis. Faculty of Mechanical Engineering and Naval Architecture, University of Zagreb, Croatia, Zagreb.
2. WILKINSON, S.P. & M.R. MALIK. 1985. Stability experiments in flow over a rotating disc. AIAA J. **23:** 588–595.
3. KOBAYASHI, R., Y. KOHAMA & C. TAKAMADATE. 1980. Spiral vortices in boundary layer transition regime on a rotating disc. Acta Mechanica J. **35:** 71–72.
4. MALIK, M.R., S.P. WILKINSON & S.A. ORSZAG. 1980. Instability and transition in rotating disc flow. AIAA J. **18:** 1485–1489.
5. DODIA, B., D.J. DOORLY & F.T. SMITH. 1997. Computational modeling of unsteady boundary layer processes. AIAA Paper 97-1834.
6. LIU, S.Z. & H.M. TSAI. 1998. Simulation of boundary layer transition with modified k-omega model. AIAA Paper 98-0340.
7. CLARKSON, M.H., S.C. CHIN & P. SHACTER. 1980. Flow visualization of inflexion instabilities on a rotating disc. AIAA Paper 279.

Effects of a Splitter Plate in the Near Wake of a Divergent Trailing Edge

L. NEAU,[a] J. PRUVOST,[a] O. RODRIGUEZ,[a] AND TA PHUOC LOC[a,b]

[a]*Office National d'Etudes et de Recherches Aérospatiales (ONERA), Lille Cedex, France*
[b]*Laboratoire d'Informatique pour la Mécanique et les Sciences de l'Ingénieur (LIMSI), Orsay Cedex, France*

ABSTRACT: This paper discusses experimental research aimed at defining a base drag reduction device based on splitter plates. The unsteady three-dimensional nature of the flow is studied by Schlieren photography, particle image velocimetry, and measurement of the total unsteady lift and drag loads.

KEYWORDS: base flow; wakes; unsteady flow; drag reduction; Schlieren photography; vortex; PIV

INTRODUCTION

Wave drag at transonic speeds can be reduced on wings with supercritical airfoils by thickening the trailing edge.[1] The gain in drag is, however, partly offset by an increase in base drag due to the thicker trailing edge. It should, therefore, be possible to improve operation of the basic design by developing base drag reduction systems.[2,3] This paper discusses experimental research aimed at defining such a system, by investigating the structure of the near wake of a divergent trailing edge with and without a splitter plate designed to reduce base drag. The structure of such flows is still not well known, especially the nature of the essentially unsteady exchanges at the origin of drag reduction.

One of the aims of the research was to construct a comparison database for direct numerical simulation (DNS) and large eddy simulation (LES). The experimental database, therefore, had to include data on the unsteady and three-dimensional (3D) features of a flow whose boundary conditions were steady and two-dimensional. The unsteadiness of the flow; that is, the structure of the instantaneous velocity field, was investigated using particle image velocimetry (PIV) measurements. The 3D character of the flow was determined using a two-directional Schlieren photography technique. The loads applied were diagnosed using a two-component (lift and drag) balance. The experimental work was conducted in a specially equipped water tunnel, assimilating the basic trailing edge to a blunt base. This paper discusses experimental results that appear to demonstrate that, although the strongly 3D nature of the spanwise flow is an important pseudo-random phenomenon, the wake, especially the near wake, consists, as expected, of large structures with a deterministic character whose simulation requires numerical techniques such as DNS and LES.

Address for correspondence: L. Neau, Office National d'Etudes et de Recherches Aérospatiales (ONERA), 5, Bd Paul Painlevé, 59045 Lille Cedex, France.
laurent.neau@imf-lille.fr

EXPERIMENTAL SETUP AND METHODOLOGY

The test section of the water tunnel was a $300 \times 300\,\text{mm}^2$ square section with a length of approximately 1 m. The plenum chamber was equipped with a partition in the middle protruding into the test section. The partition separated the chamber into two identical sections, each supplied by its own pump. This system generated two flows in the test section with different velocities, U_1 and U_2, on either side of the partition. Since the purpose of the work was to study the local flow on the trailing edge, the model had to reproduce shear conditions that were representative of those existing on a wing. The trailing edge was, therefore, assimilated to a blunt base with a thickness $H = 14\,\text{mm}$. This value was a tradeoff, allowing relatively strong loads to be applied, which required H to have a fairly high value, but ensuring that the ratio δ/H of the boundary layer thickness to the trailing edge thickness was of the same order of magnitude as on a wing. Shear was defined by the ratio $U_2/U_1 = 0.9$, where U_1 and U_2 were the assumed velocities on the upper and lower surfaces, respectively. The end of the base was removable to allow for installation of various trailing edge configurations, divergent or not, equipped or not with splitter plates designed to reduce base drag. The side walls of the test section were equipped with windows allowing visualization of the flow over the full length of the test section and in both directions normal to the far-field velocity. The Reynolds number $Re = U_1 H/\nu$ was below 20,000.

Schlieren Visualization in Water

The Schlieren method allows visualization of gradients in the refraction index. It is, therefore, very widely used to investigate compressible flows whose refraction index varies with the density. This technique can also be used in hydrodynamics, taking advantage of the fact that the refraction index of liquids varies very rapidly with the temperature.[4] This characteristic was used in the present study. The bases were equipped with flush-mounted heating elements located just upstream of the separation lines on the upper and lower surfaces. As they came in contact with the heating elements, the fluid particles generated temperature gradients. Their agglomeration in large structures downstream of the separation allowed visualization of the wake. Since the Schlieren method is an integration technique, it tends to smooth small structures and privilege large structures, which are reinforced by spanwise integration. Projection of these structures on a plane, therefore, gives the impression that they are bidirectional, although this is not necessarily true. Their 3D character is manifested by filming in the direction normal to the span. The two images were filmed simultaneously, allowing them to be correlated (see FIGURE 1).

Particle Image Velocimetry

PIV is a nonintrusive laser velocimetry technique that gives the instantaneous velocity field in a plane of the flow. When the velocity field is obtained by cross-correlation,[5] the PIV images are separated into interrogation windows. The area S of these windows in the object plane determines the size of the smallest vortex structures that can be resolved. The value of S depends on the window size, in pixels, and the magnification used. Consequently, several views are necessary when investigating

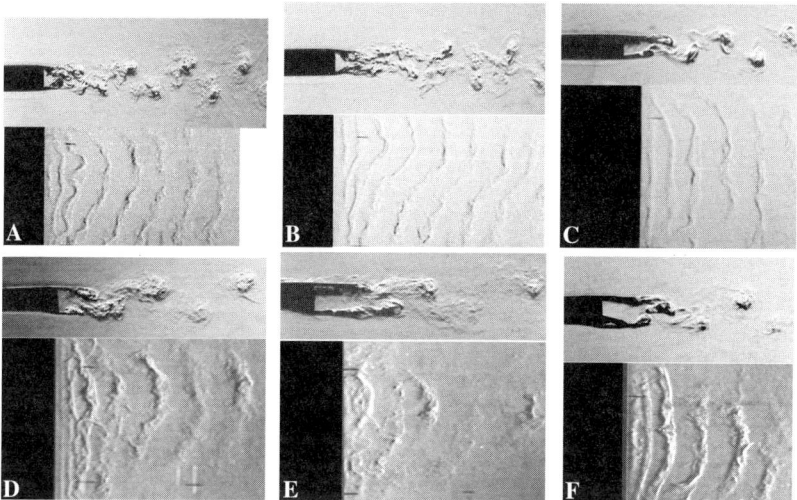

FIGURE 1. Schlieren photography for (**A**) and (**B**) the blunt base; (**C**) for the blunt base with horizontal splitter plate; (**D**) for the divergent base; (**E**) for the divergent base with horizontal splitter plate; (**F**) for a divergent base with inclined splitter plate.

flows, such as base flows, that include many spatial scales. A first view, corresponding to a value of S equal to $4 \times 4\,\text{mm}^2$, was made to resolve the large vortex structures evolving in the far wake. A second view, corresponding to $S = 1 \times 1\,\text{mm}^2$, was made in order to resolve the microstructures in the near wake of the base. However, it is doubtful if this value of S is small enough to characterize the smallest scales present in the flow. These structures may be smaller and, therefore, may not appear in the instantaneous velocity fields. This consideration is important when attempting to determine the spatial fields to be modeled and those to be directly solved in an LES calculation. Consequently, we used a different analysis technique than the cross-correlation, which is based on optical flow.[6]

Measurements of Applied Loads

A two-component (lift and drag) balance was included in the model. This balance measures the loads applied to the moving part of the model freed from any link with the fixed part and, in particular, the wires of the heating elements whose stiffness risked biasing the deformation of the strain gauges subjected to aerodynamic loads. The lift measurement resulted from the pressure difference between the lower surface and the upper surface and was straightforward. However, it was necessary to exercise caution in interpreting the drag measurements, since the measured load resulted from the pressure difference between the base and the cavity separating the weighed part from the fixed part. In order to be able to compare drag measurements for the various trailing edge configurations, it was necessary for the cavity pressure to remain constant if velocities U_1 and U_2 were constant. This condition was satisfied by affixing a mylar strip 20 micrometers thick to the lower surface of the fixed part to cover the cavity without applying any friction to the moving part. This strip was applied to the lower surface by local pressure. The pressure inside the cavity was, therefore, equal to the pressure outside the boundary layer on the upper surface.

RESULTS AND DISCUSSION

FIGURE 1A shows the Schlieren visualizations of the wake of the blunt base. It can be seen that the vortices of the vortex street in the direction parallel to the span are relatively well defined in the far wake. However, in the near wake, they appear to be deformed or in the process of agglomerating. This phenomenon was explained by examining the image made in the direction normal to the span. It can be seen that shedding of the vortex structures is not very well correlated along the span. The vortices are deformed and their projection parallel to the span spreads the image of the structures. In the far field, these vortices tend to recover a straighter form and the wake then appears more structured, a sign that this decorrelation of vortex shedding varies more or less randomly over time. It can lead to rather surprising spanwise flow visualizations when part of a vortex aligns itself with part of the previous vortex (FIG. 1B). This phenomenon is not specific to the set up, since it has also been observed in Schlieren visualizations made at subsonic speed for a blunt base[7] of Mach 0.66 (see FIGURE 2A and B). It appears that the wake of a 2D obstacle consists of large structures with a deterministic character, but that these structures can have very pronounced 3D forms. This phenomenon is not specific to base flows; Prasad and Williamson investigated it for a cylinder.[8,9] It is verified that the same phenomenon is observed when the base includes a splitter plate (FIG. 1C). It may, however, be considered that the flows remain locally two-dimensional since the wavelength of the spanwise deformation is large compared with the dimensions of

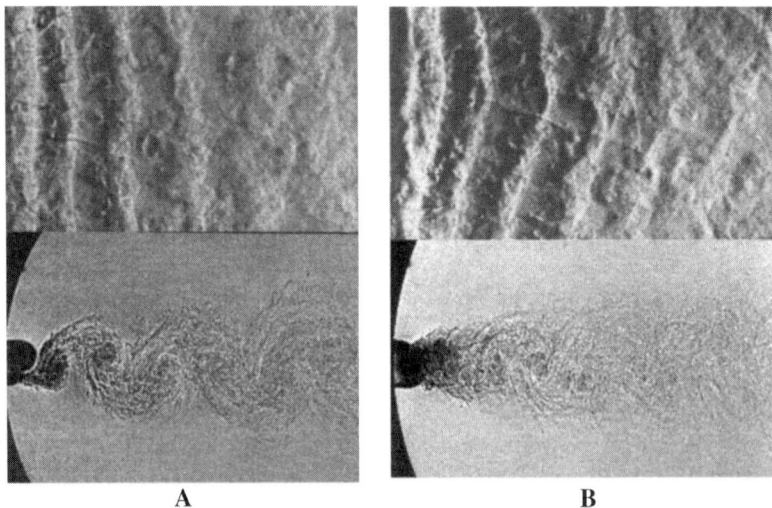

FIGURE 2. Schlieren photograph for a blunt base flow at subsonic speed, $M = 0.66$.

the large structures observed parallel to the span. This is especially true in the near wake. It confirms the advantage of PIV for investigating the instantaneous velocity field. However, it is important to determine whether this 3D character of the flow, which appears rather randomly, is predominant over time or is a low probability phenomenon that occurs more or less accidentally. This phenomenon can have major consequences on the magnitude of the unsteady loads applied, in particular drag, since these loads are directly related to the vortex shedding structure in the recirculation region. From a theoretical standpoint, the importance of the effects related to decorrelation must not be neglected. Although it is commonly considered that the difference between the drag on a cylinder given by 2D LES simulation and by 3D simulation is effectively due to a 3D effect, this effect is often ascribed exclusively to the presence of ribs that contribute to 3D turbulence by stretching small longitudinal vortices. It is, therefore, essential to determine precisely how spanwise decorrelation and the presence of ribs impact the loads applied (it is noted that these two phenomena are clearly visible in FIG. 2). This task is especially arduous because neither of the two phenomena can be taken into account by a PIV plane. This shows how important it is to measure the instantaneous loads. Furthermore, it should be noted that the lateral boundary layers probably play a role in decorrelation of the

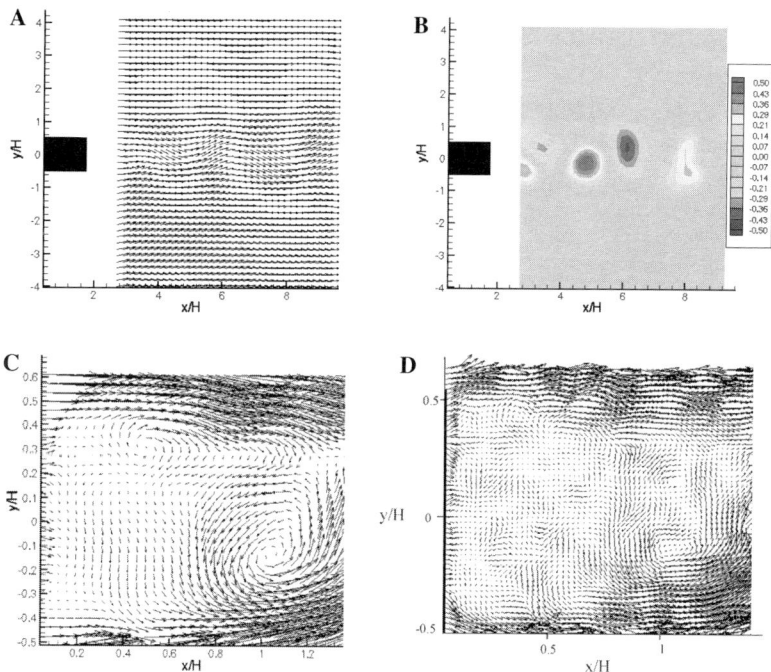

FIGURE 3. The blunt base configuration, $Re = 11{,}200$. Large view: (**A**) instantaneous velocity field, (**B**) vorticity map. Smaller view: (**C**) velocity field by intercorrelation,[5] (**D**) velocity field by optical flow.[6]

spanwise vorticity structures. Prasad and Williamson[8,9] related this spanwise decorrelation to the boundary conditions applied to the two ends of a cylinder.

The PIV velocity fields corresponding to the large view show the large vorticity structures that form the Von Karman vortex street (see FIGURE 3A and B). However, these velocity fields give no information about the near wake of the base. This is because the resolution of these PIV fields is not sufficient to show the vortex microstructures shed from the lower and upper surface separation lines and forming the large vortex structures of the wake by agglomeration in the recirculation region. Viewing smaller scales gives better access to the microstructures—that is, the structure of the mixing regions that result from the separation lines. The presence of small vortices is detected (FIG. 3C), but the velocity field obtained for the wake can obviously not be correlated with a velocity field corresponding to a larger view (FIG. 3A) since the images are not recorded at the same time. The best way of characterizing the unsteady nature of the flow would be to have a series of images separated by a sufficiently small time interval to be able to see development of the vortices over a complete cycle of the vortex street. Unfortunately, the laser operated at 5 Hz, which was too slow for such measurements. However, the velocity fields obtained can be used as a basis of comparison with the DNS and LES numeric calculation. This comparison will have to be limited to a plane, without being able to draw any conclusions

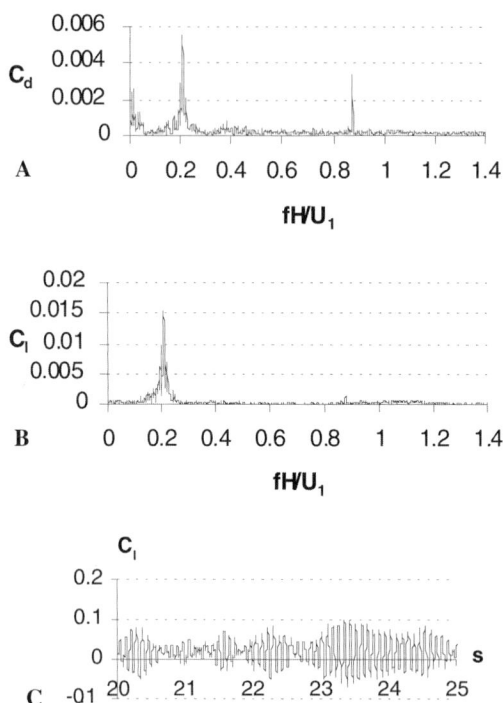

FIGURE 4. (A) C_d spectrum, (B) C_l spectrum, and (C) signal from the balance.

about the spanwise structure of the flow. To validate the spanwise flow structure by calculation, it will then be necessary to take the total loads into consideration and compare them with the measurements supplied by the balance. Furthermore, comparison of the velocity fields resulting from analysis by cross-correlation and by optical flow shows large differences. To improve the clarity of FIGURE 3D, the scales are different from those in FIGURE 3C. This means that at a given point, no direct comparison can be made between the two analysis techniques from the two figures based on the magnitude of the velocity vector. However, the field obtained by optical flow clearly shows the presence of microstructures that appear to have been filtered by cross-correlation.

It can be seen on the drag spectrum (see FIGURE 4A) and lift spectrum (FIG. 4B) obtained from the balance for the blunt base that the unsteady component is dominated by a spectral line whose Strouhal number fH/U_1 is about 0.2, which indicates that the unsteady loads fluctuate at a frequency f equal to that of the vortex street (the spectral line corresponding to $fH/U_1 \approx 0.9$ is of electrical origin and is due to the acquisition system). A systematic analysis of the signals versus the time should allow the identification of states for which the loads have maximum and minimum values, corresponding to 2D and 3D vortex structures (FIG. 4C). Work continues to conclude this analysis and classify the different systems according to their effect on drag.

CONCLUSIONS

Complementary experimental techniques were used to investigate the 3D and unsteady characteristics of the near wake that developed downstream of a 2D base, with and without a drag reduction system, for comparison with numerical simulation by DNS and LES.[10,11] Two-directional Schlieren visualizations show a spanwise decorrelation of the vorticity structures, but the flow can still be considered locally 2D. This property justifies the use of PIV to determine the instantaneous velocity field in the near wake. However, the resolution of this velocimetric technique used in conjunction with a cross-correlation analysis technique is not sufficient to account, simultaneously, for all the vorticity structures on all scales. It should be possible to improve the resolution by analyzing the images using optical flow. Since PIV cannot take into account the 3D character of the flow, validation by numerical simulation will have to be carried out using total load measurements aimed at attempting to determine the respective contributions of spanwise decorrelation and the presence of ribs to the loads, which was not possible experimentally. Detailed analysis of the signals from the balance over time and in the Fourier plane need to be carried out to attempt to identify the influence of decorrelation on the applied loads. The ideal, which is an ultimate objective, would be to correlate the two-directional visualizations, the PIV fields and the unsteady loads, making the measurements simultaneously.

REFERENCES

1. HENNE, P.A. & I.R.D. GREGG. 1991. New airfoil design concept. J. Aircraft **28**(5): 289–299.
2. RODRIGUEZ, O. 1998. Amélioration du concept de bord de fuite divergent en transsonique par réduction de la traînée de culot, Rapport technique ONERA.
3. RODRIGUEZ, O., J. PRUVOST & J.F. LE ROY. 1996. Etude préliminaire de dispositifs destinés à améliorer les performances d'un profil équipé d'un bord de fuite épais, Contrat SPAé-Etude IMFL No. 94328.
4. RODRIGUEZ, O. 1991. Base drag reduction by control of the three-dimensional unsteady vortical structures, Exp. Fluids **11**: 218–226.
5. MONNIER, J.C., G. CROISIER & A. GILLIOT. 1999. Characterisation of the EUROPIV nozzle by PIV, using a CCD recording device. *In* Particle Image Velocimetry: Progress Towards Industrial Application. Kluwer Academic Publishers. 226–233.
6. QUENOT, G.M., J. PAKLEZA & T.A. KOWALEWSKI. 1998. Particle image velocimetry using optical flow for image analysis, 8th Int. Symp Flow Visualization.
7. RODRIGUEZ, O. 2000. Etude des propriétés instationnaires de sillages proches, Thèse d'Etat.
8. PRASAD, A. & C.H.K. WILLIAMSON. 1997. The instability of the shear layer separating from a bluff body. J. Fluid Mech. **333**: 375–402.
9. PRASAD, A. & C.H.K. WILLIAMSON. 1997. 3D effects in turbulent bluff-body wakes. J. Fluid Mech. **343**: 235–265.
10. MARY, I. & P. SAGAUT. 2001. Large eddy simulation of flow around an airfoil. AIAA Paper 2559.
11. LARDAT, R. 1997. Simulations numériques d'écoulements externes instationnaires décollés autour d'une aile avec des modéles de sous-maille, Thèse de doctorat de l'université Paris 6.

Start-up Behavior of Viscoelastic Fluid Flow Near a Capillary Entry

MASATAKA SHIRAKASHI,[a] TSUTOMU TAKAHASHI,[a]
ATSUSHI WATANABE,[b] AND YOHSUKE ARUGA[a]

[a]*Department of Mechanical Engineering, Nagaoka University of Technology, Nagaoka, Niigata, Japan*

[b]*Toray Engineering Co. Ltd., Nihonbashi, Tokyo, Japan*

ABSTRACT: The effect of elasticity on the capillary entry flow development process was investigated by applying a laser doppler velocimeter (LDV) and flow visualization. The fluid filling the test section was kept at rest under a pressure of compressed air or a natural head and then a valve below the capillary was quickly opened to start the flow. Although the development of the capillary entry flow field of a Newtonian fluid is virtually quasi-steady, it takes a considerable time for a viscoelastic PAA-solution to attain its terminal regime. The transient behavior of the flow rate of the PAA-solution is attributed to the gradual development process of the flow field upstream from the entry.

KEYWORDS: viscoelastic fluid; capillary entry; start-up; laser doppler; LDV; PAA solution

INTRODUCTION

Fluid elasticity can have a dominant effect on the flow near a die entry, or in an abruptly contracting pipe flow, because the fluid elements are elongated unsteadily even when the flow field is itself steady. Due to its practical importance, many studies have been carried out on viscoelastic fluid flows in a contracting channel, mainly in steady flows.[1–5] This investigation is focused on the effect of elasticity on the development process of a capillary entry flow, where the influence of the side walls of the large upstream channel is negligible.

EXPERIMENTAL

Channel Geometry

The test section channel was composed of a transparent square cross section reservoir and a capillary attached flush to the bottom of the reservoir. The reservoir and the capillary had a common vertical axis. The dimension of the reservoir was $100\,\text{mm} \times 100\,\text{mm} \times 170\,\text{mm}$ and the capillary had a diameter $d = 2\,\text{mm}$ and a length $L = 157\,\text{mm}$. Since the contraction ratio was virtually infinite, the influence of the side walls of the upstream reservoir was negligible. The coordinate system had its

---

Address for correspondence: Masataka Shirakashi, Department of Mechanical Engineering, Nagaoka University of Technology, 1603-1 Kamitomioka, Nagaoka 940-2188, Japan.

origin at the center of the capillary entry, its z-axis along the capillary axis vertically downward, and a horizontal x–y plane.

Test Fluids

An aqueous solution of 0.2 wt% polyacrylamide (PAA) was used as the test fluid. A rice-syrup/water mixture was used as a Newtonian fluid for comparison. A cone-plate type rheometer was applied to obtain their material functions in the steady simple shear flow. The shear viscosity η and the first normal stress difference N_1 of the PAA-solution thus obtained are well represented by the Denn model:

$$\eta = K_1 \cdot \dot{\gamma}^{n-1}$$
$$N_1 = K_2 \cdot \dot{\gamma}^m \tag{1}$$

with $K_1 = 0.371$ (Pa·sec^n) and $n = 0.471$ in the range of shear rates $\dot{\gamma} = 1$–10^3 (1/sec). $K_2 = 1.08$ (Pa·sec^n) and $n = 0.764$ in the range of $\dot{\gamma} = 10$–10^3 (1/sec).

The generalized Reynolds number Re^* is defined

$$Re^* = \frac{\rho v^{2-n} d^n}{8^{n-1}\{(3n+1)/(4n)\}^n K_1} \tag{2}$$

for the developed capillary flow, where the characteristic velocity v is the terminal flow rate Q_∞ divided by the capillary cross sectional area and ρ is the density of the fluid. The maximum value of Re^* in this experiment is less than 90. The characteristic time λ and the Weissenberg number We, based on the Denn-model, are estimated from[1]

$$\lambda = \frac{N_1}{2\tau\dot{\gamma}}$$
$$We = \frac{\lambda v}{d} \tag{3}$$

where $\dot{\gamma}$ is the shear rate at the capillary wall for the fully developed power-law fluid capillary flow at the flow rate Q_∞.

Experimental Procedure, Measurement, and Flow Visualization

In an experimental run, the fluid filling the test section to the depth of 140 mm in the reservoir was kept at rest for a period under pressure of compressed air or natural head. The valve below the capillary was quickly opened at time $t = 0$ to start the flow. The change in the liquid level was negligible and the driving pressure was kept constant during a test run.

The tracer-particle and light-sheet technique was applied to visualize the flow. A two-component LDV was used for velocity measurements in the x–z plane. The flow rate Q and the pressure drop through the capillary were measured simultaneously. When the driving pressure was high and, hence, the terminal flow rate was high, say $Re^* > 90$, the flow field of the PAA-solution became unstable and the velocity continued to fluctuate even when the flow rate attained a constant terminal value. The experiments reported here were carried out with a driving pressure lower than this criterion.

RESULTS AND DISCUSSION

Transient Behavior of the Newtonian Fluid

The Newtonian rice-syrup/water mixture flow pattern is shown in FIGURE 1A, compared with the velocity vectors obtained by the LDV measurement. The exposure time for the photograph was from $t = 0$ to 9 sec. The start-up behavior of the axial and the radial velocity components, v_z and v_r, on the center line upstream from the capillary entry are plotted in FIGURE 1B. The velocity v_z rose in a stepwise manner just after the start-up valve was opened and attained its terminal value at $t < 0.5$ sec. The velocity vectors in FIGURE 1A were obtained from the terminal steady values obtained from LDV measurements, such as seen in FIGURE 1B. Simultaneous measurement of the flow rate Q shows that Q also increases in a stepwise manner and the transient time t_{tr}, the time for Q to attain its terminal value, is less than 1.0 sec in the experimental range of Re used here. The results shown in these figures suggest that the development of the flow field of a Newtonian fluid upstream the capillary entry is virtually quasisteady at Re values in the range of this experiment; that is, the flow pattern with radial streamlines shown in FIGURE 1A is established immediately after the flow is started and the absolute values of velocities increase proportionally with the flow rate Q.

Transient Behavior of the PAA-Solution

The normalized PAA-solution flow rate is plotted against the non-dimensional time t/λ in FIGURE 2A for three values of We. The transient time t_{tr} for the PAA-solution is quite long (e.g., $t_{tr} = 8$ sec at $We = 4.1$) compared with that of the Newtonian fluid. A very large overshoot of the flow rate is observed and the maximum value of Q/Q_∞ is seen to increase with We in the experimental range of this work. However, the non-dimensional transient time t_{tr}/λ is almost independent of We.

The axial and the radial velocity components, v_z and v_r, on the center line upstream from the capillary entry at $We = 4.1$ is shown in FIGURE 2B. The measurement was

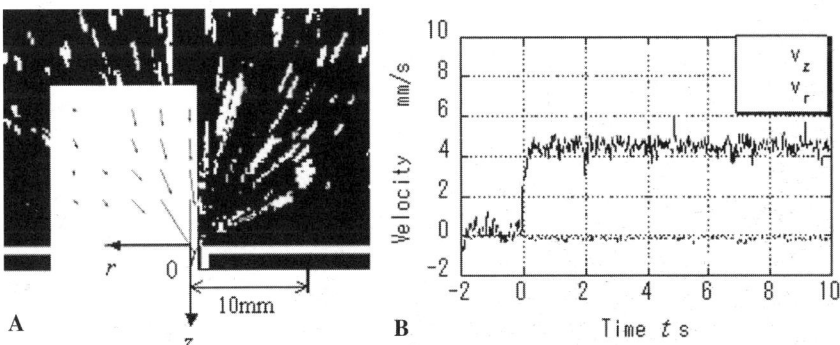

FIGURE 1. Results from a Newtonian fluid (rice-syrup/water mixture). **(A)** Flow pattern: $Re = 0.8$, exposure time $t = 0$–9 sec. **(B)** Transient behavior of velocity: $Re = 3.6$, $x = y = 0$, $z/d = -5$.

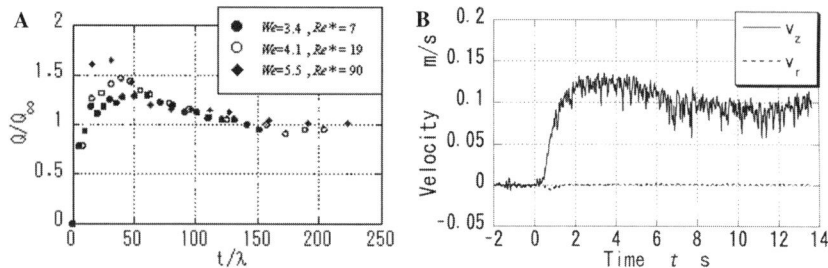

FIGURE 2. Transient behavior of the PAA-solution. (**A**) Normalized flow rate versus non-dimensional time. (**B**) Transient behavior of velocity. $We = 4.1$, $x = y = 0$, $z/d = -5$.

carried out simultaneously with Q for $We = 4.1$ in FIGURE 2A. The delayed development and the overshoot behavior are also observed in v_z corresponding to Q. However, the relative value of maximum overshoot is considerably lower than that of Q, meaning that the pattern of the streamline changes during the transient period.

The photographs in FIGURE 3 were taken with exposure times in the transient period to see the development of flow pattern for $We = 4.1$, when the time t at the maximum flow rate, t_{max}, was 3 sec and $t_{tr} = 8$ sec. The flow pattern just after the start of flow is similar to that of the Newtonian fluid, and the vortex surrounding the flowing region grows gradually until the terminal flow pattern is established.

The development of the velocity vector field corresponding to the photographs in FIGURE 3 is shown in FIGURE 4. The distribution of velocity v_z along the channel center line, $[v_z]_{r=0}$, was also obtained from the same LDV data as those used for

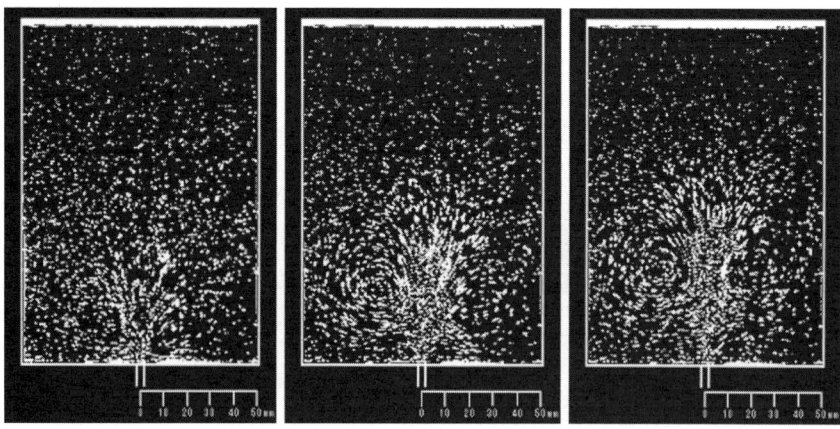

FIGURE 3. Flow pattern of the PAA-solution upstream from the entry, $We = 4.1$. (**A**) $t = 0–3$ sec, (**B**) $t = 3–6$ sec, (**C**) $t = 6–9$ sec.

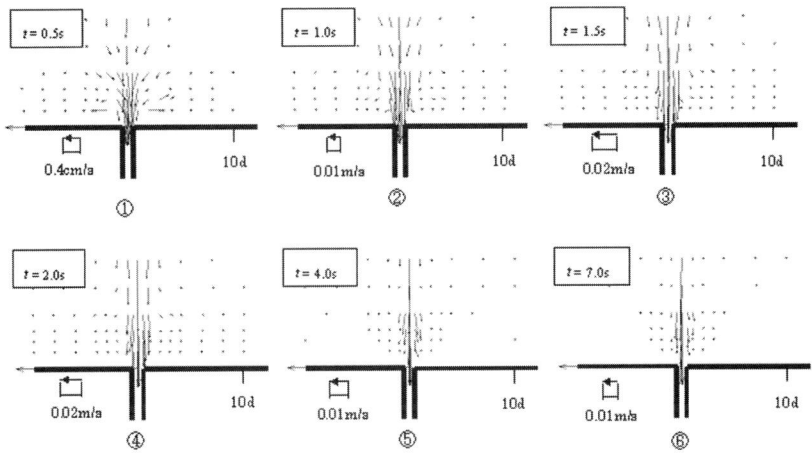

FIGURE 4. Development of velocity vector field, $We = 4.1$.

FIGURE 4. The results and the $[v_z]_{r=0}$–z plots thus obtained show that the region of z/d on the center line, where v_z has a considerable value, extends much higher than that of the Newtonian fluid,[6] and grows upward with time. Consequently, the angle of the flow region surrounded by the vortex, wherein the fluid is flowing toward the capillary entry, decreases with time.

The $[v_z]_{r=0}$–z plots obtained from LDV measurements confirm that the elongation rate at the die entry is high at the beginning of the flow, and decreases gradually with time t. Hence, the entry pressure loss associated with the elongation stress due to elasticity, is considered to decrease with time. The developing process of the flow field seen in FIGURE 4 causes the long transient time of the flow rate. However, this mechanism cannot give an explanation that is compatible with the flow rate overshoot, since the elongation rate at the die entry seems to decrease monotonously.

SUMMARY

Although the developing process of the capillary entry flow field of a Newtonian fluid is virtually stepwise and quasisteady, it takes a considerable time for the PAA-solution for a viscoelastic fluid to attain its terminal regime. The transient time for the PAA-solution normalized by the fluid relaxation time is independent of the Weissenberg number, We, but the flow rate overshoot ratio is seen to increase with We.

The flow pattern upstream from the entry determines the elongation rate, which governs the elongation stress at the die entry, and the entry pressure loss in turn. Hence, the delayed transient behavior of the flow rate is attributed to the gradual development process of the flow field upstream from the entry. Additional studies, including the start-up behavior of the pressure drop through the capillary, are needed to clarify the mechanism of flow rate overshoot.

REFERENCES

1. BOGER, D.V. & K. WALTERS. 1993. *In* Rheological Phenomena in Focus. 6–9. Elsevier, Amsterdam.
2. BOGER, D.V. 1987. Viscoelastic flow through contraction. Ann. Rev. Fluid Mech. **19:** 157–182.
3. LARSON, R.G. 1992. Instabilities in viscoelastic flow. Rheologica Act **31:** 213–263.
4. RAMAMURTHY, A.V. & J.C.H. MCADAM. 1980. Velocity measurement in the die entry region at a capillary rheometer. J. Rheology **24:** 167–188.
5. MCGLASHAN, S.A. & M.E. MACKAY. 1999. Comparison of entry flow techniques for measuring elongation flow properties. J. Non-Newtonian Fluid Mech. **85:** 213–227.
6. WATANABE, H., *et al.* 1998. Study of the flow of dilute polymer solution through a small orifice and of the measurement of elongational stress. Nihon Reoroji Gakkaishi **26:** 21–26.

Development and Validation of an X-ray Tomograph for Two-Phase Flow

ERIC HERVIEU,[a] EMMANUEL JOUET,[a] AND LAURENT DESBAT[b]

[a]*Commissariat à l'Energie Atomique, DEN/DTP/SETEX/LTAC, Grenoble, France*
[b]*TIMC-IMAG, UMR CNRS 5525, Faculté de Médecine, La Tronche, France*

> ABSTRACT: This paper describes the development and validation of a high spatial resolution X-ray tomograph designed for the investigation of air–water two-phase flow. The device hardware mainly comprises a 60 keV X-ray source, a detector, and an accurate mechanical bench. Our study concentrated on accurate quantification with emphasis on the reconstruction procedure. As is well known, absorption gradients induce reconstruction artifacts when using standard algorithms based on uniform regularization. In the particular case of two-phase flow in a pipe, this leads to poor measurement accuracy in the vicinity of the walls. To overcome such effects, improved algorithms were developed during this study that involve spatially adaptive regularization methods. Preliminary calibration performed on static phantoms clearly exhibited the benefits of the advanced reconstruction algorithms. A validation procedure was carried out on an air–water bubble column, equipped with an optical probe, which could be translated in order to explore the 80 mm × 80 mm square cross section. Comparisons of local void fraction measurements were performed pixel by pixel. They demonstrate the accuracy improvement induced by the advanced reconstruction algorithms.
>
> KEYWORDS: two-phase flow; X-ray tomography

INTRODUCTION

Modeling three-dimensional (3D) two-phase flow for industrial purposes is at present mainly based on averaged formulations, such as the two-fluid model. This kind of approach will prevail, especially for industrial applications, until replaced by finer models. Only this approach is compatible with a foreseeable computer performance levels in the medium term and is capable of handling all flow patterns. However, modeling efforts are still necessary, particularly concerning unstructured flows that are essential in many industrial problems at the boundary between strongly coupled bubbly flow and separated flow. In the field of nuclear safety, the coolant distribution in the primary system is often ruled by this transition. Although time averaging suppresses the need for detailed knowledge of the interface position, the complexity of the phenomena involved dictates maintaining a certain empirical part in the models of the future. The need for local experimental data requires instrumentation developments that must be closely coordinated with modeling advances.

Address for correspondence: Eric Hervieu, Ph.D., DTP/SETEX/LTAC, 17 Rue des Martyrs, 38 054 Grenoble Cedex 9, France.
  eric.hervieu@cea.fr

Experimental validation of models of a new generation will rely upon our ability to measure, at least, local phase velocities, interfacial area concentration, and void fraction. The 3D nature of unstructured flow prohibits using local measuring techniques, which are usually limited to one-dimensional flow. For this particular reason, non intrusive imaging techniques, such as X-ray tomography, are of great interest, especially for investigating domains containing internal structures and obstacles, such as rod bundles.

X-ray absorption has been widely used to obtain time- and space-averaged void fraction in various two-phase systems. However, the present work concerns the development of a local measuring technique, with a high spatial resolution, leading to a *quantitative* value of the void fraction in flows exhibiting *large amplitude* fluctuations. This goal differs from the medical point of view, where X-ray tomography is used to provide *qualitative* information on the localization of *static* objects; for example, tissues and organs. As a consequence, the present work is related to the way we solve the problems arising from the application of the original medical technique to the specific case of two-phase flow.

Recent studies relate to the application of X-ray and γ-ray tomography to investigate two-phase flow,[1,2] even at high frame rate,[3] we, however, emphasize reconstruction artifacts or local void fraction accuracy.

THE MINI-TOMIX EXPERIMENT

The Mini-Tomix experiment is a versatile tomographic bench, consisting of an X-ray generator, two detectors, an associated acquisition device, and a moving mechanism. A schematic representation is illustrated in FIGURE 1. The X-ray source is a Bruker/Siemens two-window tube with a copper anticathode. It is supplied with a 40 kV high voltage and 5 mA intensity that can be adjusted by using a microcomputer connected through a serial connection. Both windows are equipped with a nickel filter, a 0.5 mm diameter collimator, and a 0.1 mm-thick lead filter to reduce beam hardening effects. The detectors are classical quartz and silica NaI scintillation crystals modified to measure the integral value of the photonic flow. The first (PM1 in FIG. 1), measures the intensity of the beam crossing the test-section and the two-

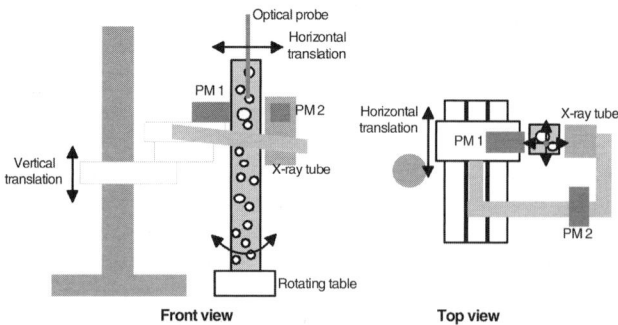

FIGURE 1. Schematic representation of the Mini-Tomix tomographic bench.

phase flow. The second (PM2 in FIG. 1), measures the intensity of the reference beam. The ratio of these two measurements eliminates eventual photonic fluctuations of the X-ray source. The X-ray tube and the detectors are fixed together on the same mechanical structure, which can be translated horizontally and vertically using two step motors (cf., FIG. 1). The resolution of these motions is 1 µm.

The "target" to be scanned (static phantom or test-section) was fixed on a rotating table, moved by a step motor. The angular resolution was 0.01°. The detector outputs were connected with an A/D acquisition board, and the sampled signals were averaged over several seconds. The sampling frequency and the number of samples were imposed according to the characteristic time scale of two-phase flow, in order to guarantee a converged average with a ±0.1% relative accuracy. A microcomputer (master) controlled the overall tomographic acquisition process and a fully automated LabView program had been developed to scan the target by achieving a looped procedure including translation and rotation controls and attenuation measurements.

The test-section was a plexiglas square vertical column, 1.2 m long with an internal square cross-section $80 \times 80 \, mm^2$, and a wall thickness of 10 mm. The column was filled with water and air was injected at its bottom through a drilled plate. Deflectors could be inserted inside the column, thus directing the flow and inducing non-uniform bubble repartition. An optical probe plunged inside the pipe measured the local void fraction 5 mm above the X-ray beam crossing the flow. This optical probe was fixed on a motorized two-axis translation mechanism. A second microcomputer (slave) controlled the translation step motors and the optical probe RBI acquisition device. The LabView program achieved the automated exploration of the whole cross-section of the bubble column.

RECONSTRUCTION ALGORITHMS

Choice of Reconstruction Method

All numerical developments were made with the TIMC-IMAG tools of the Institute Albert Bonniot. The numerical inversion methods frequently used in medical or industrial tomography can be classified in two categories: (1) analytical methods, the reconstructed attenuation map is given by the analytical expression of the Radon transform, which is inverted and *then* discretized;[4-7] and (2) algebraic methods, the attenuation map is the solution of a linear system, derived from the Radon transform, which is discretized *before* being inverted.[8] Comparison of their capabilities, performance, and calculation costs led to the choice of the algebraic reconstruction technique (ART) due to its flexibility concerning the projection geometry. Its ability to handle arbitrary beam positions and orientations guaranteed that the algorithm improvements developed with a parallel-beam geometry on the Mini-Tomix would be easily adapted to the fan-beam geometry of the future Tomix tomograph.

Comparison of Various ART Algorithms

Four ART algorithms were compared, in their standard or regularized versions: (1) least squares QR algorithm (LSQR),[9,10] (2) Chen's algorithm (RRLS),[9,10] (3) minimal residual algorithm (RM3),[11] and (4) conjugate gradient least squares

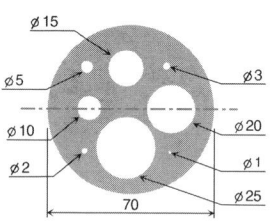
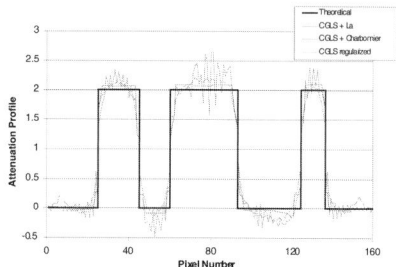

FIGURE 2. Left, geometry of the static phantom. **Right**, comparison between CGLS uniform and adaptative regularization procedures.

algorithm (CGLS).[10,12,13] Due to oscillations and artifacts exhibited by the reconstructions obtained with these algorithms in their standard versions, a regularization process was included in the linear system inversion, for all the algorithms tested. The regularization parameter was automatically estimated by a generalized cross-validation calculation.[14,15]

The behavior and capabilities of these four algorithms were compared by first reconstructing simulated data corresponding to the static phantom sketched in FIGURE 2(left). This is a Plexiglas cylinder drilled with holes of various sizes, the smallest being 1 mm in diameter. First, we defined the sampling geometry (position and orientation of the projections). Then, following the Beer–Lambert law, we calculated the attenuation of an infinitely thin and monochromatic X-ray beam, crossing the phantom at each of the chosen positions and orientations. This set of simulated data was used to reconstruct the attenuation map, which was compared with the image of the phantom. The global reconstruction error was quantitatively evaluated by the norm of the difference (pixel-by-pixel) between the original and reconstructed maps.

The corresponding global errors are given in TABLE 1, for the four selected algorithms, in their standard and regularized versions (127×81 measurements, 160×160 reconstructed map).

Since ART is based on the iterative resolution of a linear system, another important criterion in the choice of a reconstruction method is the calculation cost; that is, the CPU required by the tested algorithms in their regularized version. The convergence efficiency can be estimated by studying how fast the residue decreases before the linear system resolution converges. The study revealed that the performance of the regularized CGLS algorithm was better.

TABLE 1. Global error between original and reconstructed attenuation maps, using the various algorithms

	LSQR	CGLS	RM3	RRLS
Standard version	0.461815	0.461815	0.42438	0.461815
Regularized version	0.378442	0.362071	0.406905	0.378491

The last criterion that must be taken into account when comparing algorithms concerns their sensitivity to the input data noise level. To check this behavior, we superimposed white noise on the simulated attenuation data set, and studied the evolution of the global reconstruction error when increasing the relative amplitude of the added random fluctuations. At a 35% noise ratio, CGLS revealed a global error that was 32% above the value calculated without noise, whereas the other algorithms exhibited reconstruction errors that were more than 37% higher.

Similar comparisons of the same kind were performed on other simulated data sets. They revealed the same trend. The regularized CGLS algorithm was consequently preferred due to its higher accuracy, higher convergence efficiency, its lower CPU cost, and lower noise sensitivity.

Reconstruction Enhancement in Attenuation Gradient Zones

We now consider the attenuation profile of the above phantom, across a horizontal diameter, as illustrated in FIGURE 2 (left, horizontal dotted line). FIGURE 2 (right) compares the original profile and the reconstructed values obtained with a regularized CGLS algorithm under the same conditions as previously described (127×81 measurements, 160×160 reconstructed map). On the red profile, note the smoothing effect of the uniform regularization process, which damps the slope of the reconstructed profile in the vicinity of large attenuation gradients. This leads to poor reconstruction accuracy in areas where neighboring pixels exhibit large attenuation differences. In the case of two-phase flow, the consequence is that the void fraction is inaccurately determined near the walls. To overcome this difficulty, various solutions and algorithm enhancements were tested and implemented in CGLS. The first class of solutions concerns adaptive regularization methods. The idea is to restrict the regularization process outside of the areas exhibiting large attenuation gradients, these areas being automatically determined or *a priori* determined. Solutions of the second kind consist in "injecting" into the resolution process the *a priori* knowledge of the region of interest, where the reconstruction itself is restricted.

In the field of adaptive reconstruction methods, La[10] developed an algorithm that used two regularization parameter values. The small one was applied on pixels in which a gradient was detected—that is, when the difference between this pixel and one of its neighbors was greater than a chosen scale parameter. On all the other pixels, a large value was applied, that value obtained from the automatic cross-validation calculation performed in the uniform regularized CGLS algorithm. This adaptive method, developed in the field of medical tomography, required the imposition of a small regularization parameter and a scale parameter that forms a threshold between weak and strong regularization. An exhaustive parametric analysis was performed to determine the optimal value of these parameters. As shown by the profile in FIGURE 2 (right), this method induced accuracy enhancements in the gradient areas, but also induced large oscillations outside of these areas.

A simplified version of La's algorithm was implemented, in which the area of weak regularization was *a priori* imposed instead of being automatically determined. Using the above phantom, this procedure did not provide better results than those obtained with the uniformly regularized CGLS algorithm.

Employing the same idea of a weaker regularization in the gradient areas, Charbonnier[16] developed another adaptive method. The regularization parameter was given by a decreasing function that depends on the first derivative of the pixel values. The regularization parameter was about 1 where the first derivative was near zero, and tended toward zero (i.e., no regularization) when the first derivative was large. The threshold between weak and strong regularization was determined with a chosen scale parameter. Several tests provided an acceptable scale parameter and the possibility to implement an automatic determination of this parameter. The curve in FIGURE 2(right) reveals sharper gradients, and more flat attenuation profiles outside of the gradient areas. However, the difference between the reconstructed attenuation profile and the theoretical profile in the vicinity of the fronts implies a great error on the reconstructed void fraction.

THE ALGORITHM USED

Since the test-section geometry and position were precisely determined, we could use this *a priori* known information in the reconstruction process. The test section was scanned once when empty and once again with bubbly flow. The data sets were subtracted in order to remove the effect of the pipe walls. During the reconstruction process, the pixels outside of the flow area were set to 0 and no regularization was applied to them. The pixels of the reconstruction of simulated data of the test section full of water had the expected theoretical value.

MEASUREMENT AND RECONSTRUCTION OF TWO-PHASE FLOW

We performed measurements on the Mini-Tomix experiment for various injected air flow rates ranging between 5 and 50 Nl/min. The sampling geometry, that always consisted of 109 rotations and 70 projections, was used to reconstruct a 70×70 map.

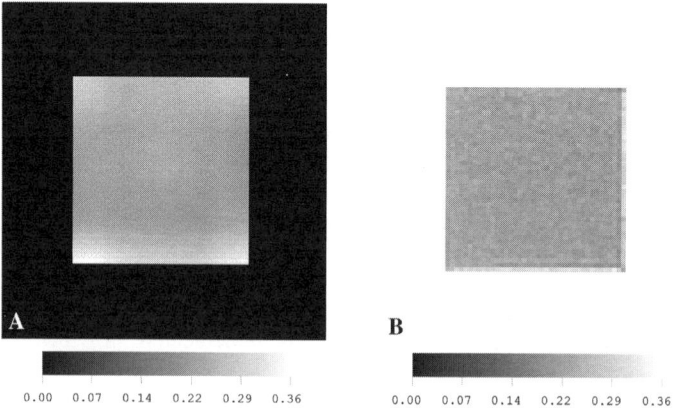

FIGURE 3. Uniform void fraction distribution in a square pipe, measured by (**A**) X-ray tomography and (**B**) optical probe.

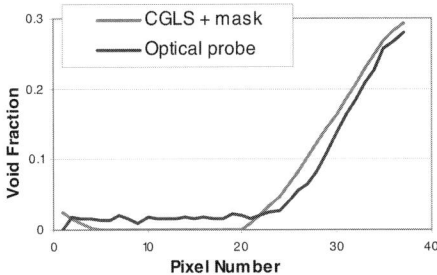

FIGURE 4. Non-uniform void fraction distribution, measured by X-ray tomography and optical probe.

Thus, the flow cross section had an extent of 40 × 40 pixels, which exactly matched the grid explored by the optical probe.

Uniform Void Fraction Distribution

FIGURE 3 presents the void fraction maps obtained with the X-ray tomograph (FIG. 3A) and the optical probe (FIG. 3B), for a 14 Nl/min air flow rate. A pixel-by-pixel comparison revealed that the void fraction discrepancy was about 9% (absolute value). This value was the same as the measurement error, a direct consequence of the instability of the detectors when submitted to an attenuation front.

Non-Uniform Void Fraction Distribution

FIGURE 4 presents the void fraction sections obtained with X-ray tomograph and the optical probe, for a 5.3 Nl/min air flow rate. The same sampling and reconstructing geometries was used. However, in order to generate a non-uniform bubble distribution in the cross-section, a deflector was inserted into the pipe. The difference between the two measurements was smaller than 10%, the same as the measurement error. Measurements of bundle flows were also obtained.

SUMMARY

A small-scale parallel beam bench was constructed to perform analytic studies, and an exhaustive investigation concerning the reconstruction algorithms was carried out. ART methods were preferred because of their easy adaptation from parallel-beam to fan-beam geometry. Since standard algorithms led to a poor reconstructed accuracy near the walls, emphasis was given to various adaptive regularization techniques. Reconstructions from simulated data clearly demonstrated their benefit. At the same time, when used with real data, their enhanced capabilities were demonstrated by the good agreement between X-ray tomography and optical probe measurements. However, it was shown that the accuracy could be significantly improved by using more efficient detectors. An industrial and large-scale fan beam X-ray tomographic bench is under development in CEA/Grenoble, aimed at fulfilling the increasing need for local measurements required by the new generation of local models.

ACKNOWLEDGMENTS

CEA, IPSN, EDF, and Framatome ANP financially support this study. TIMC-IMAG takes part in this study by its advice and reconstruction programs.

REFERENCES

1. CHAOUKI, J. & F. LARACHI. 1997. Non Invasive Monitoring of Multiphase Flows. Elsevier, Amsterdam.
2. GRASSLER, T. & K.E. WIRTH. 1999. X-ray computer tomography—potential and limitation for the measurement of local solids distribution in circulating fluidized beds. 1st World Congress on Industrial Process Tomography, Buxton, England.
3. HORI, K., et al. 1997. Advanced high-speed X-ray CT scanner foe measurement and visualization of multiphase flow. OECD/CSNI Specialist Meeting an Advanced Instrumentation and Measurement Techniques, Santa Barbara, CA.
4. HERMAN, G.T. 1979. Image Reconstruction from Projections: Implementation and Applications. 32. Springer-Verlag, Berlin, Heidelberg, New York.
5. NATTERER, F. 1985. Fourier reconstruction in tomography. Numerische Mathematik **47:** 343–353.
6. NATTERER, F. 1986. The Mathematics of Computerized Tomography. John Wiley and Sons.
7. LEWITT, R.M. 1983. Reconstruction algorithms: transform methods. Proc. IEEE **71**(3): 390–408.
8. BROOKS, R.A. & G. DI CHIRO. 1976. Principles of computer assisted tomography (CAT) in radiographic and radioisotopic imaging. Phys. Med. Biol. **21**(5): 689–732.
9. PAIGE, C.C. & M.A. SAUNDERS. 1982. LSQR: an algorithm for sparse linear equations and sparse least squares. ACM Trans. Math. Software **8**(1): 43–71.
10. LA, V. 1997. Correction d'atténuation en géométrie conique avec mesures de transmission en tomographie d'émission mono-photonique. Ph.D. Thesis, Inst. National Polytechnique de Grenoble, France.
11. AXELSSON, O. 1980. Conjugate gradient type methods for unsymmetric and inconsistent systems of linear equations. Linear Algebra Appl. **29:** 1–16.
12. BARRETT, R., et al. 1993. Templates for the solution of linear systems: building blocks for iterative methods. SIAM J. Appl. Math.
13. RIDDELL, C., et al. 1995. The approximate inverse and conjugate gradient: non-symmetrical algorithms for fast attenuation correction in SPECT. Phys. Med. Biol. **40:** 269–281.
14. GIRARD, D. 1989. A fast "Monte-Carlo cross validation" procedure for large least squares problems with noisy data. Numerische Mathematik **56:** 1–23.
15. GOLUB, G.H., M. HEATH & G. WAHBA. 1979. Generalized cross-validation as a method for choosing a good ridge parameter. Technometrics **21:** 215–224.
16. CHARBONNIER, P. 1994. Reconstruction d'image: régularisation avec prise en compte des discontinuités. Ph.D. Thesis, Université de Nice-Sophia Antipolis, France.

Visualization of Turbulent Wedges under Favorable Pressure Gradients Using Shear-Sensitive and Temperature-Sensitive Liquid Crystals

TZE-PEI CHONG,[a] SHAN ZHONG,[a] AND HOWARD P. HODSON[b]

[a]*School of Engineering, University of Manchester, Manchester, United Kingdom*

[b]*Whittle Laboratory, Department of Engineering, Cambridge University, Cambridge, United Kingdom*

ABSTRACT: Turbulent wedges induced by a three-dimensional surface roughness placed on a flat plate were studied using both shear sensitive and temperature sensitive liquid crystals, respectively denoted by SSLC and TSLC. The experiments were carried out at a free-stream velocity of 28 m/sec at three different favorable pressure gradients. The purpose of this investigation was to examine the spreading angles of the turbulent wedges, as indicated by their associated surface shear stresses and heat transfer characteristics, and to obtain more insight about the behavior of transitional momentum and thermal boundary layers when a streamwise pressure gradient exists. It was shown that under a zero pressure gradient the spreading angles indicated by the two types of liquid crystals are the same, but the difference increases as the level of the favorable pressure gradient increases. The result from the present study is important for modelling the transition of thermal boundary layers over gas turbine blades.

KEYWORDS: turbulent wedge; shear sensitive liquid crystals; temperature sensitive liquid crystals; boundary layer; pressure gradient; transition

INTRODUCTION

Turbine blades in turbomachines usually operate at a temperature close to the melting point of the material they are made of. The Reynolds number (Re) of these blades is often such that a substantial portion of their surface is covered by transitional flow.[1] Since laminar to turbulent transition is associated with a rapid spatial increase in heat transfer rate, our ability to predict the behavior of transitional boundary layers is vital in the design of effective blading.

Emmons[2] was the first to discover that boundary layer transition first occurred in localized regions known as *turbulent spots*. These turbulent spots formed randomly in time and space and grew as they propagated downstream until they merged with each other and covered the entire flow field. Using the Emmons spot hypothesis, and

Address for correspondence: T.P. Chong, Ph.D., School of Engineering, Goldstein Research Laboratory, University of Manchester, Manchester M13 9PL, UK.
 chongtzepei@hotmail.com

knowing the spot propagation rate and spreading angle, one can predict the boundary layer intermittency distribution effectively, hence the transition process.

The transition that is initiated via formation of a *turbulent wedge* by a three-dimensional surface roughness element has been studied for some time.[3–6] The turbulent wedge can be interpreted as being caused by a continual generation of turbulent spots. Therefore, a turbulent wedge provides a simplified case to study the growth of turbulent spots as they propagate downstream in a laminar boundary layer, and the transition mechanism for this process.

As reported by Blair,[7] the transition zone length, indicated by the velocity profile shape factor when a favorable pressure gradient exists, is generally smaller than that indicated by the wall heat transfer. This difference increases when the free-stream turbulence intensity is low. Such a discrepancy questions the conventional *Re* analogy, which relates skin friction to heat transfer and has been widely adopted by turbomachine designers. Although efforts have been made to study the various effects of pressure gradient on the momentum and thermal boundary layers, this phenomenon is still not fully understood. In this paper, two types of liquid crystals that respond to wall shear stress and surface temperature, respectively, were used to visualize the development of turbulent wedges at three different levels of favorable pressure gradients. The purpose of this work is to investigate the effect of pressure gradients on the momentum and heat transfer processes that occur in turbulent wedges, and thus, obtain further insight about the difference in the behavior of transitional momentum and thermal boundary layers. The result from the present study is expected to have an important implication for the transition modelling of thermal boundary layers over gas turbine blades.

EXPERIMENTAL FACILITY AND EQUIPMENT

Experimental Setup

The experiment was conducted in the Farnborough wind tunnel at the Goldstein Laboratory of the Manchester School of Engineering. The wind tunnel has a test section of 460 mm × 207 mm and a maximum speed of 28 m/sec. The turbulence intensity of the wind tunnel is about 0.3%. The top and two side walls of the test section are made of glass to allow optical access. A perspex plate, 6 mm thick and 625 mm long, with a 1:3 super-elliptical leading edge was mounted horizontally across the entire width of the test section.

For the tests with the SSLC, the test surface was first coated with water-based matt black paint in order to improve the color spectrum displayed by the crystals. The black paint is free from chemical contaminants, thus prolonging the life of the liquid crystals. Two 1 KW halogen lights were found to give desirable illumination of the liquid crystals when projected from downstream of the test plate at each side of the test section. In the present experiments, the liquid crystal color changes produced by the turbulent wedge were recorded at an angle of 54.5° to the horizontal plane through the top glass window from downstream.

To track the thermal footprint of the turbulent wedge, a heated plate consisting of multiple layers, as shown in FIGURE 1, was constructed. The test surface was heated uniformly by a 0.1-mm thick metallized film, which was connected to a DC power

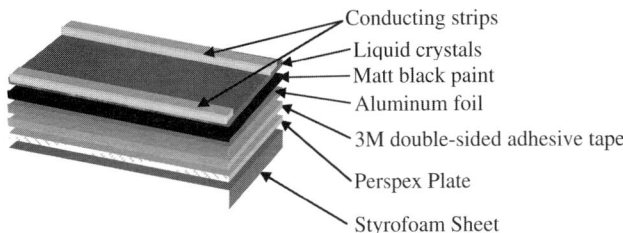

FIGURE 1. Schematic diagram of test plate used with temperature sensitive liquid crystals (TSLC).

supply through metal foil strips adhered to the edges of the surface. A 14-mm thick Styrofoam sheet was glued to the backside of the perspex plate to reduce the heat loss by conduction. Two 60 W fluorescent lights were clamped at each side of the test section to provide a uniform illumination to the crystal coating. The optimum viewing position was found to be normal to the test surface.

Favorable streamwise pressure gradients were achieved by attaching a flat perspex sheet at various angles to the ceiling of the test section. At the maximum operating velocity, the two wedge angles used in the present experiments produced accelerating flows with constant pressure gradient parameter $K = 0.24 \times 10^{-6}$ (K_1) and 1.02×10^{-6} (K_2), respectively. Here

$$K = \frac{\nu}{U^2} \frac{dU}{dx},$$

where ν and U are the kinematic viscosity and free-stream velocity, respectively. The corresponding velocity and K distributions for these two cases are shown in FIGURE 2.

To produce a turbulent wedge, a sphere with a diameter of 1.1 mm was glued on the test surface at a distance of 187 mm from the leading edge. According to Hall,[8] a critical Re value, Re_d, of 500 has to be reached in order to initiate a turbulent wedge immediately downstream of the roughness. Here, $Re_d = ud/\nu$ and u is the velocity in the undisturbed boundary layer at the height of the roughness d. In the present tests, Re_d was estimated to be 1,833 for the zero pressure gradient boundary layer. It may be plausible to assume that Re values for the other two favorable pressure gradient

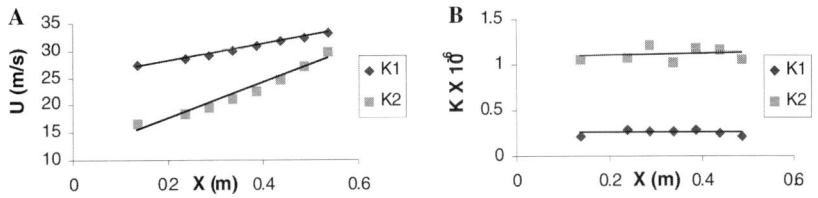

FIGURE 2. Velocity and K distribution for two favorable pressure gradients. **(A)** Velocity distribution for K_1 and K_2. **(B)** Pressure gradient distribution for K_1 and K_2.

cases were also sufficiently large, since according to Stüper[9] the presence of favorable pressure gradient does not greatly affect the "sensitivity" of the laminar boundary layer to disturbances.

Liquid Crystal Techniques

Shear sensitive liquid crystals (SSLC) (chiral-nematic) are capable of responding to various levels of shear stress by color changes. As shear stress increases, the color of the liquid crystals coating shifts from red through the visible spectrum to violet. Since, under the same Re value, the shear stress underneath a turbulent boundary layer is considerably higher than that in its laminar counterpart, the color variations displayed by the crystals can be used to differentiate a turbulent region from a laminar background, thus providing the information about the growth of a turbulent wedge. To apply the liquid crystals to the test surface, the slurry was first heated to about 50°C to reach its clearing point, then brushed on the surface using an artistic paintbrush and ensuring uniform coating. The thickness of the layer was estimated to be 15 µm, although this value was subjected to some bias because a certain amount of slurry deposited on the paintbrush.

Temperature sensitive (TSLC) (encapsulated cholesteric) liquid crystal shows variations of color from red through the visible spectrum to blue as temperature increases. Since a turbulent boundary layer exhibits high heat transfer rates, the surface temperature is lower than its laminar counterpart. Hence, this type of liquid crystal can be used to visualize the thermal footprint of a turbulent wedge. Zhong et al.[10] used this to visualize the growth and development of artificially created spots in a water tunnel experiment. The crystal used here has an active color bandwidth of 5°C, showing red at 22°C and blue at 27°C. The slurry and resin were mixed with an equal amount of distilled water and sprayed using an artistic airbrush to the test surface. The thickness of the coating was estimated to be about 30 µm.

RESULTS AND DISCUSSION

All the color information was recorded using a digital photographic camera. FIGURES 3 and 4 show the color variation of both shear sensitive and temperature sensitive liquid crystals at zero, mild, and strong favorable pressure gradients, respectively. The green color (displayed as grey in FIG. 3) represents the high shear region

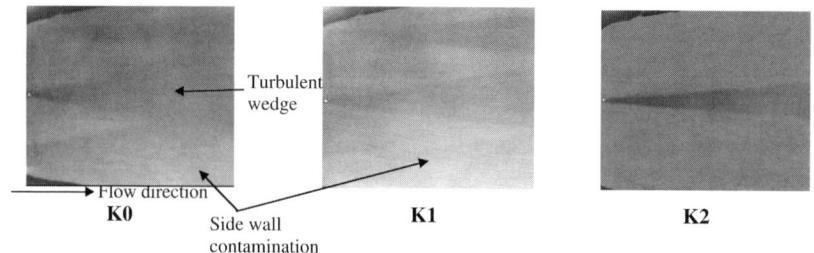

FIGURE 3. Turbulent wedges by SSLC.

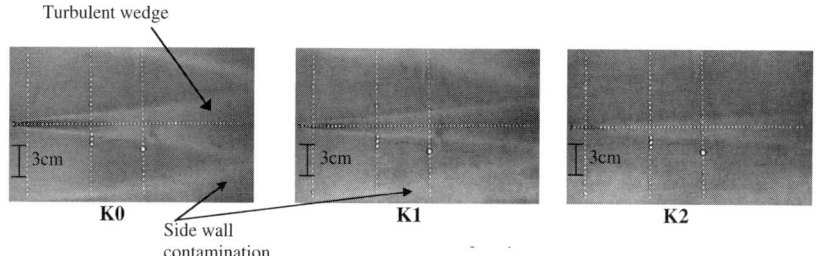

FIGURE 4. Turbulent wedges by TSLC.

making the turbulent wedge clearly visible. A similar color variation was also observed in FIGURE 4 for the temperature sensitive liquid crystals, which showed red and blue in the turbulent and laminar regions, respectively. It can be seen that the spreading angle of the wedge decreases as the level of favorable pressure increases, consistent with results from other experimental methods.[11,12] The side-wall contamination zones on each side of the plate are also clearly visible. The side-wall contamination zone is smaller at the mild pressure gradient in comparison to the zero pressure gradient case and at the strong favorable pressure gradient it almost completely disappears.

It should be noted that all of the images from shear sensitive liquid crystals were taken at an oblique angle downstream to the surface roughness, thus the lateral spreading angle obtained directly from the original images is considerably larger than its actual value. Therefore, a perspective transformation of the images is necessary before the results can be compared with those from the temperature liquid crystals images, which were taken with the camera normal to the test surface. The methodology of the perspective transformation can be found elsewhere.[13]

In order to determine the lateral spreading angles of the turbulent wedge more accurately, a program was written to convert the red, green, blue (RGB) values of the SSLC images to *intensity*. The boundary of the turbulent wedge was then determined by taking the intersection point of the nearly constant *intensity* (laminar region) to the gradual decrease of *intensity* (turbulent region) as sketched in FIGURE 5. This process was repeated at several streamwise locations on both sides of the turbulent wedge. Finally, the wedge spreading angle was taken as half the angle included between two boundaries that depict the envelope of the wedge. The same detection algorithm also applies to TSLC, except that the distribution of heat transfer coefficient (h) of the turbulent wedge was investigated instead. A summary of the half-spreading angles for all of the test cases is presented in TABLE 1.

It was found that the spreading angle indicated by the shear sensitive liquid crystals was nearly the same as that obtained by the temperature sensitive liquid crystals in the zero pressure gradient case. However, at a mild favorable pressure gradient K_1, the spreading angle of the shear stress footprint was about 20% larger than that of the thermal footprint of the turbulent wedges. The difference is more pronounced, 29%, at the strong pressure gradient K_2. Since a larger spreading angle corresponds to a faster spanwise growth of the turbulent wedge or turbulent spots, thus, to a shorter transition zone length, the result from the present study is consistent with Blair:[7]

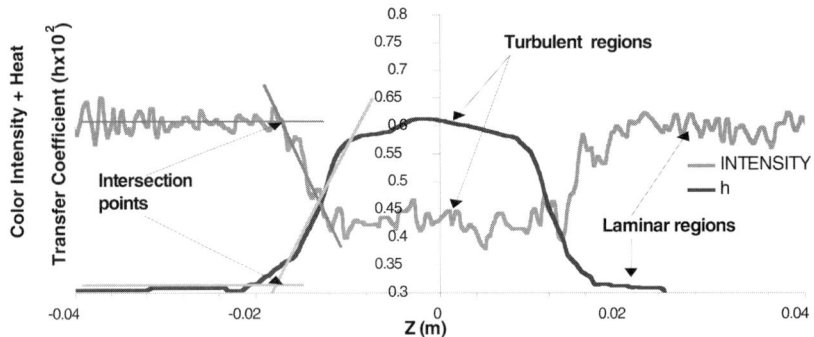

FIGURE 5. Distributions of INTENSITY (SSLC) and heat transfer coefficient, h (TSLC) across turbulent wedge, 100 mm downstream of the three-dimensional roughness.

that is, in a transitional flow subject to a favorable pressure gradient, the momentum boundary layer establishes the fully turbulent flow state faster than the thermal boundary layer. This result also suggests that, when turbulent spot based models are employed to model a transitional boundary layer, different empirical relations have to be used to estimate the spot growth rate in order to obtain the correct transition zone lengths for the momentum and thermal boundary layers.

In the present experiment, turbulent wedges were found to spread at an average angle of 6.4° for the zero pressure gradients in the momentum boundary layer. This result is consistent with that of Zhong et al.[10] who obtained a 6.5° angle for turbulent spots in zero pressure gradient flow using temperature sensitive liquid crystals. However, this value is smaller than the widely reported[3,11,12] values of 10°. There are two explanations for this discrepancy. The first is that the spot has a three-dimensional shape and, thus, a smaller contact area with the wall than the main body in the boundary layer—that is, some form of an "overhang" exists.[4] It is also worth noting that all of those who obtained a larger spreading angle used a hot wire as the key experimental apparatus for their measurement. Because of the physical size of the hot wire, the measurement had to be carried out at a certain distance away from the wall. The other possible explanation for the small spreading angle is that the liquid crystals only respond to the fully turbulent region in the turbulent wedge. The intermittent region, which is outside the turbulent core, cannot be detected due to its fluctuating behavior. By coincidence, Gad-El-Hak et al.,[4] who coated a dye of one color on their three-dimensional roughness element and seeped a dye of another color onto a part of the span of the plate in their turbulent wedge visualization in a

TABLE 1. Summary of half-spreading angles indicated by SSLC and TSLC, respectively

	K_0	K_1	K_2
SSLC (°)	6.4	5.5	3.4
TSLC (°)	6.3	4.4	2.4
Difference (%)	1.6	20	29

towing tank, obtained half-spreading angle of 6 ± 0.5° in the total turbulent region and 10 ± 0.5° in the outer region of the wing tips. These two explanations are currently being investigated using both hot wires and an array of surface mounted hot films. The results will be reported elsewhere.

SUMMARY

The purpose of this investigation was to examine the spreading angles of the turbulent wedges, as indicated by associated surface shear stresses and heat transfer characteristics, respectively, and obtain insight into the difference in the behavior of transitional momentum and thermal boundary layers when a streamwise pressure gradient exists.

It was found that under a zero pressure gradient the spreading angles indicated by the two types of liquid crystals are similar, but the difference increases as the level of favorable pressure gradient increases. Since a larger spreading angle corresponds to a faster spanwise growth of turbulent wedge or turbulent spots and, thus, to a shorter transition zone length, the result from the present study implies that the momentum boundary layer will reach a fully turbulent state more quickly than the thermal boundary layer under a favorable pressure gradient. This finding has an important implication for modelling the transitional boundary layers in gas turbine blades. It suggests that when the turbulent spot based models are employed, different empirical relations have to be used to estimate the spot growth rate in order to obtain the correct transition zone length for the momentum and thermal boundary layers.

REFERENCES

1. MAYLE, R.E. 1991. The role of laminar-turbulent transition in gas turbine engines. ASME 91-GT-261.
2. EMMONS, H.W. 1951. The laminar-turbulent transition in a boundary layer—Part 1. J Aerospace Sci. **18**(7): 490–498.
3. SCHUBAUER, G.B. & P.S. KLEBANOFF. 1955. Contributions on the mechanism of boundary layer transition. NACA TN-3489.
4. GAD-EL-HAL, M., R.F. BLACKWELDER & J.J. RILEY. 1981. On the growth of turbulent regions in laminar boundary layer. J. Fluid Mech. **110**: 73–95.
5. PETRIE, H.L., P.J. MORRIS, A.R. BAJWA & D.C. VINCENT. 1993. Transition induced by fixed and freely convecting spherical particles in laminar boundary layers. Report No. AD-A267757; TR-93-07, Penn. State University, Applied Research Lab.
6. CLARK, J.P., P.J. MAGARI & T.V. JONES. 1993. On the distribution of the heat transfer coefficient in turbulent and "transitional" wedges. Int. J. Heat Fluid Flow **14**(3): 217–222.
7. BLAIR, M.F. 1982. Influence of freestream turbulence on boundary layer transition in favourable pressure gradients. ASME J. Eng. Power **104**: 743–750.
8. HALL, G.R. 1967. Interaction of the wake from bluff bodies with an initially laminar boundary layer. AIAA **5**(8): 1386–1392.
9. STÜPER, J. 1949. The influence of surface irregularities on transition with various pressure gradients. Division of Aeronautics, Australia, Report A59, Melbourne.
10. ZHONG, S., C. KITTICHAIKAN, H.P. HODSON & P.T. IRELAND. 2000. Visualisation of turbulent spots under the influence of adverse pressure gradients. Exp. Fluids **28**: 385–393.

11. SANKARAN, R., A.J. CHAMBERS & R.A. ANTONIA. 1986. The influence of favourable pressure gradient on the growth of turbulent spot. 9th Australian Fluid Mechanics Conf. Auckland.
12. GOSTELOW, J.P., N. MELWANI & G.J. WALKER. 1996. Effects of streamwise pressure gradient on turbulent spot development. ASME J. Turbomachin. **118:** 737–743.
13. GONZÁLEZ, R.C. 1993. Digital Image Processing. Addison-Wesley, Wokingham.

Visualization and Measurement of Multiphase Flow in Porous Media Using Light Transmission and Synchrotron X-Rays

CHRISTOPHE J.G. DARNAULT,[a,b] DAVID A. DICARLO,[c] TIM W.J. BAUTERS,[a] TAMMO S. STEENHUIS,[a] J.-YVES PARLANGE,[a] CARLO D. MONTEMAGNO,[d,e] AND PHILIPPE BAVEYE[f]

[a]*Department of Biological and Environmental Engineering, Riley-Robb Hall, Cornell University, Ithaca, New York, USA*

[b]*Malcolm Pirnie, Inc., Environmental & Water Resources Engineering, Newport News, Virginia, USA*

[c]*USDA/ARS National Sedimentation Laboratory, Oxford, Mississippi, USA*

[d]*Department of Biomedical Engineering, University of California Los Angeles, Los Angeles, California, USA*

[e]*Department of Mechanical and Aerospace Engineering, University of California Los Angeles, Los Angeles, California, USA*

[f]*Laboratory of Geoenvironmental Science and Engineering, Bradfield Hall, Cornell University, Ithaca, New York, USA*

ABSTRACT: Non-aqueous phase liquids enter the vadose zone as a result of spills or leaking underground storage facilities, thus contaminating groundwater resources. Measuring the contaminant concentrations is important in assessing the risk to human health and the environment and to develop effective remediation. This research presents the development and application of the light transmission method (LTM) for three-phase flow systems, aimed at investigating unstable fingered flow in a soil–air–oil–water system. The LTM uses the hue and intensity of light transmitted through a slab chamber to measure fluid content, since total liquid content is a function of both hue and light intensity. Evaluation of the LTM is obtained by comparing experiments with LTM and synchrotron X-rays. The LTM captures the spatial resolution of the fluid contents and can provide new insights into rapidly changing, two-phase and three-phase flow systems. Application of the LTM as a visualization technique for environmental and physical phenomena is noted. Visualization by LTM of groundwater remediation by surfactants as well as visualization of model cluster growth and fractal dimensions was also explored.

KEYWORDS: multiphase flow; preferential flow; fingered flow; light transmission method (LTM); synchrotron X-rays; surfactants; Eden cluster; diffusion; aggregation; non-aqueous phase liquids

Address for correspondence: Christophe J.G. Darnault, Ph.D., Environmental & Water Resources Engineering, Malcolm Pirnie, Inc., 701 Town Center Dr., Newport News, VA 23606, USA.

cdarnault@pirnie.com

BACKGROUND

Multiphase flow and transport phenomena in the unsaturated and saturated zones of the subsurface environment are the focus of numerous research efforts. One of the less understood transient flow phenomena is unstable fingering. Fingering decreases the fluid retention time in the vadose zone, thus increasing groundwater contamination. Consequently, there is a need for fast, non-destructive, and accurate measurements of transient three-fluid phase flow in porous media.

Transient visualizations with no direct measurement of fluid content have been made in Hele-Shaw cells with smooth walls or with imprints of porous media on glass. Currently, very few methods exist that allow rapid determination of fluid contents in three-phase, non-aqueous phase liquids (NAPL)–air–water systems. Most of the methods that allow determination of fluid contents in these three-phase systems involve some form of radiation. Synchrotron X-rays allow accurate and fast measurements of fluid contents in transient flow fields, in any soil type, but can measure only a small section of the flow field at one time due to the small beam size of 1 mm by 8 mm.[1] The LTM is a non-destructive method that allows visualization and measurement of fluid content in transient air–oil–water flow occurring in sandy porous media, over the entire flow field, with a time resolution of tenths of seconds.[2]

EXPERIMENTAL

For fluid measurements in porous media, LTM involves placing a two-dimensional chamber in front of an uniform light source and recording the transmitted light.[2] Experiments were performed at a constant temperature of 20°C. A light source composed of a bank of 24 fluorescent, high-frequency light bulbs, located in front of a white background, was used. The transmitted light was recorded with a Sony color video camera employing three half-inch charge couple device (CCD) images, each having a total of 250,000 effective picture elements. The camera was located 1 m in front of the chamber with constant settings (zoom 0.95 m and aperture f5.6). The images were stored on a Hi8 video cassette in red, green, and blue (RGB) format. Recorded images were converted from RGB to hue, saturation, and intensity (HIS) format and analyzed using a Power Macintosh equipped with a video digitizer (RasterOps 24 XLT from RoasterOps Corporation, 1992) and scientific image processing software (IPLab Spectrum V3.00 software from Signal Analytics Corporation). The advantage of the HSI format is that it treats color roughly the same way that humans perceive and interpret color.

LTM Calibration

A calibration method was developed to measure water, oil, and air content in porous media as a function of hue, intensity, and porosity with LTM.[2] This calibration involved a two-dimensional chamber consisting of cells filled with porous media and known fluid ratios of oil, water, and air, allowing the relationship for water, oil, and air content to be obtained as functions of hue, intensity, and porosity.

The oil used was Soltrol 220, a transparent isoparaffine solvent. The distilled water was dyed blue with $CuSO_4$ at a concentration of 28%. The calibration chamber was 62 cm high, 52 cm wide, and 1 cm thick. It was partitioned into 24 cells each 3 cm high by 23 cm wide. The cell walls were made of 1 cm thick Hyzod polycarbonate sheet. The cells were packed to a porosity of 38% with 20/30-sieve size industrial-quartz silica sand. The sand packed cells were then filled with all possible fluid combinations. The fluid content of each of the three fluids were varied by $0.076 \text{cm}^3/\text{cm}^3$ increments. Average hue and intensity values were plotted versus the corresponding water and liquid contents to obtain the calibration curves; fluid contents were expressed by a set of equations,[2] in which water content was uniquely related to hue and oil content was a function of hue and intensity.

LTM Validation

The accuracy of the LTM in determining water, oil, and air content was compared with synchrotron X-rays in static experiments. X-ray measurements of the fluid content in the soil–air–oil–water systems were performed at the F-2 beam line in the Cornell high-energy synchrotron source (CHESS), using dual energy attenuation measurements.[1] These consisted of an initial white X-ray beam that was reflected off a double-bounce Si(220) monochrometer producing a beam of two distinct energies (20 and 40 Kev) that entered the experimental hutch. For this experiment the slab was filled with 20/30 quartz silica sand. Oil (Soltrol 220) and water were injected from the bottom in a specific sequence and the slab was then drained accordingly to achieve three layers and two transition zones: starting at the bottom, there was a 5 cm layer of water-saturated sand, a 5 cm transition zone, a 20 cm layer of oil-saturated sand, a 15 cm transition zone, and a 5 cm layer of air-saturated sand.

Water Fingering Experiments

The transient flow experiment consisting of unstable fingered flow in a porous media of quartz silica sand was performed in a two-dimensional slab chamber filled with quartz silica 20/30 sand and saturated with a combination of oil–air–system to meet the initial experimental conditions. To form a finger, water was applied as a point source through a needle, 1 cm above the sand surface, at a rate of 3 ml/min. The oil level was kept constant via overflow tubing located 1 cm below the oil in the chamber. Vertical profiles were analyzed for fluid content, at the center of the finger, as the finger progressed into the various sand–oil–air saturated phases and interfaces.

Surfactants Experiments

An unstable flow field of pure water and surfactant solutions were generated into oil-saturated 12/20 sand within a two-dimensional slab chamber. The oil used was Soltrol 220. The water or surfactant solutions were rained through a needle attached to a cam driver that reciprocated. These experiments were carried out with 1% surfactant solution of Alfonic 810-4.5 from Vista Chemical Company (a nonionic ethoxylated alcohol) and Neodol R from Shell Chemical Company (an anionic ethoxylated alcohol). Pure water and surfactant solutions were added uniformly at flow rates of 4.5 ml/min (pure water) and 4.8 ml/min (surfactant solutions) to the oil-saturated 12-20 sand soil. Fingered flow characteristics were visualized and analyzed.

Model Cluster Growth and Fractal Dimensions Experiments

Horizontal slabs, plastic or glass chambers, were filled with either hydrophilic or hydrophobic media, and saturated with either water or oil (Soltrol 220). Water was injected in oil-saturated media, and oil was injected in water-saturated media, in either hydrophilic or hydrophobic media.

Two kinds of water repellent beads were used separately for the plastic slab: 2 mm square by 1 mm thick plastic beads (Lexan Polycarbonate Resin) and 1 mm diameter glass beads (Jaygo Inc) coated with 2X988B silicone lubricant (Dem-Kote). The water wetting beads were 1 mm diameter glass beads (Jaygo Inc).

The fluids were clear water and oil, and coloring them by USDA Blue Dye #1 (water) and Sudan IV (oil) enhanced the contrast between the oil and water phases. The primary (saturating) fluid remained clear in each experiment whereas the secondary (injected) fluid was colored. The primary fluid was introduced at the bottom of the slab, placed in vertical position, using a pump. The secondary fluid was injected at the center point source from the subsurface when the slab was put in horizontal position. Injection was performed by a pump, at a rate of 1 ml/min. Overflow from the primary fluid was drained off through side holes in the slab connected to tubes that opened to the atmosphere 61 cm above the slab surface. A light shining on the slab was used to enhance visualization of the flow phenomena occurring in the slab. The progression of each experiment was recorded by a photocamera, set in a fixed location, about 60 cm above the slab horizontal surface.

APPLICATIONS AND RESULTS

Validation of LTM by Synchrotron X-Rays

LTM based water, oil, and air content were compared with corresponding synchrotron X-rays data for a static oil–water–air system (see FIGURE 1). In general, the fluid contents compare well; the water content is nearly identical for both methods. The oil content shows the same trend for both methods. An image of the static experiment setup is shown in FIGURE 1 A. The various oil, air, and water contents are clearly distinguishable. From the bottom up, different colors identify water, water and oil, oil, oil and air, and air. Visualization of hue (FIG. 1 B) and intensity (FIG. 1 C) allows the qualitative and quantitative characterization of water content and liquid content, respectively, as well as their spatial distribution.

Fluid content profiles for the fully formed fingers in various soil–oil–air–water systems are presented in FIGURE 2, in (A) RGB format, (B) hue image, and (C) intensity image. As expected from previous experiments,[1] and as can be clearly observed in FIGURE 2D, the tip of the finger is the wettest and the rest is much dryer. In all fingering experiments, the finger does not expand once it is formed due to hysteresis in the soil constitutive relationships. The finger core is on the drainage branch, whereas the surrounding fluid is on the imbibing branch, making it possible to have a high water content (drainage branch) and low water content (imbibing branch) at the same matrix potential. The LTM measurements, in conjunction with simultaneous matrix potential measurements, can be used to determine the constitutive relationship for fingered flow in porous media.

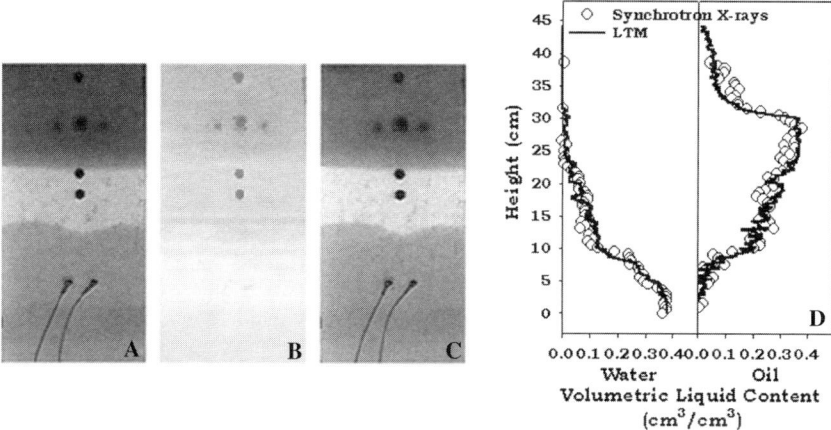

FIGURE 1. Visualization of a static experiment in soil–air–oil–water systems, using **(A)** RGB system, **(B)** hue image, and **(C)** intensity image. **(D)** Vertical fluid content profile from the drainage experiment using LTM and synchrotron X-rays.

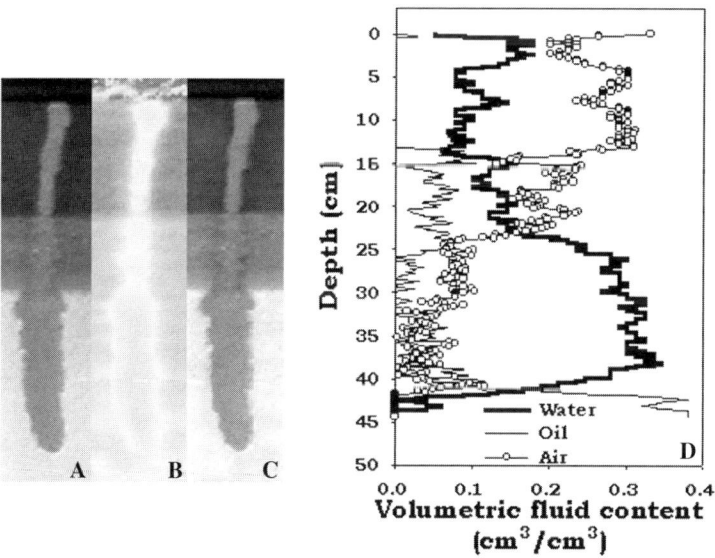

FIGURE 2. Visualization by LTM of water fingering phenomena in soil-air-oil system using **(A)** RGB system, **(B)** hue system, and **(C)** intensity system. **(D)** Vertical fluid content profile of unstable fingered in a three-phases flow system: water finger in soil–air–oil system.

Visualization of Groundwater Remediation by Surfactant Solutions

Surfactants applied for the groundwater cleanup have shown significant fingered flow behavior. Groundwater remediation was visualized with LTM and simulated as rainfall application of water or surfactant to oil-saturated porous media. FIGURE 3 visualizes the fingered flow resulting from water (FIG. 3A), Alfonic surfactant solution (FIG. 3B), and Neodol (FIG. 3C) in oil-saturated porous media. More fingers occurred with surfactant solutions than with water, because of the lower interfacial surface tension between surfactant solution and oil (Alfonic–oil, 5.91 dynes/cm; Neodol–oil, 1.94 dynes/cm) than between water and oil (32.84 dynes/cm).

When Alfonic and Neodol surfactant solutions were applied to the oil-saturated porous media, a uniform wetting front was initially formed. In the upper part of the layer, the surfactant solution crossed the coarse layer at many points, each generating a small finger. In each of these small fingers, the finger tip was the region of highest water content (dark color). Many of these fingers then merged to form larger, stronger and faster moving fingers that continued to move downward, therefore reducing the numbers of fingers. The finger tips formed the zone of the highest water content inside the finger cores. These merging fingers persisted in time and formed "core" areas that conducted most of the flow through the chamber.[3]

Surfactants show a significant potential for enhancing the remediation of contaminated groundwater, either by solubilization of contaminants in surfactant micelles or by mobilizing residual liquids by reducing capillary forces trapping liquid droplets in the aquifer porous medium. The concentration of surfactants at which micelles begins to form is called the critical micelle concentration (CMC). There are three locations within a micelle where contaminants may be solubilized: the hydrophobic core of the micelle, the polar surface of the micelle, and the transition region between the surface of the micelle and the core of the micelle. The capillary forces that cause trapping of a non-aqueous liquid depend primarily on the interfacial tension between oil and water. When the interfacial tension is greatly reduced, remediation is improved. Such low interfacial tensions are found only in the presence of a middle-phase microemulsion.

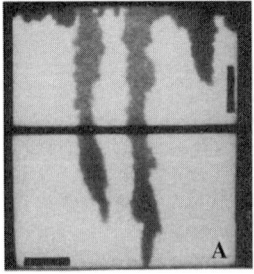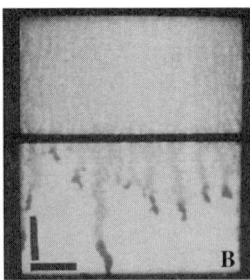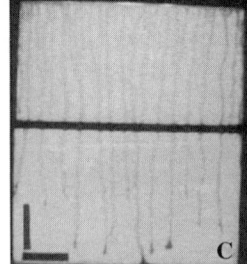

FIGURE 3. Visualization by LTM of the fingering flow phenomena in oil-saturated porous media resulting from **(A)** water application, **(B)** Alfonic surfactant solution application, and **(C)** Neodol surfactant application.

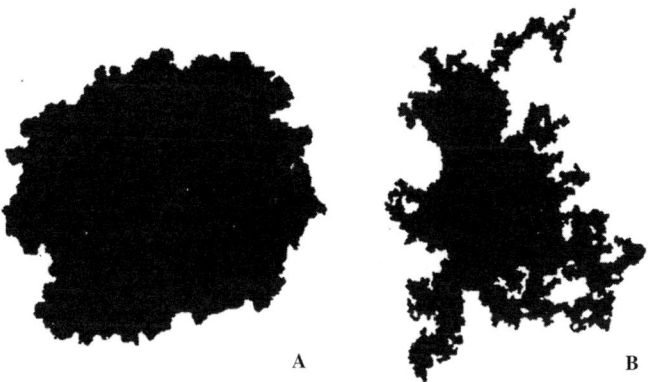

FIGURE 4. Visualization by LTM of **(A)** Eden model cluster (wetting fluid displacing a non-wetting fluid in saturated porous medium) and **(B)** diffusion limited aggregation (DLA) model cluster (non-wetting fluid displacing a wetting fluid in saturated porous medium).

Visualization of Model Cluster Growth and Fractal Dimensions

Many growth processes in nature are governed by their spatial distribution.[4,5] In the Eden model cluster growth is simulated by placing particles at randomly chosen sites on the perimeter of the cluster. This leads to a dense cluster with a few holes. In the diffusion limited aggregation (DLA) model the cluster grows by the addition of particles that diffuse from randomly chosen points outside of the cluster. Analogies between DLA and flow in porous media have been pointed out, including viscous fingering in porous media.

Two clusters that result from stable or unstable flow phenomenon, Eden cluster (blob) and viscous fingering (see FIGURE 4), were simulated and observed in a horizontal slab chamber using the LTM principle. FIGURE 4A can be described as an Eden model cluster. It was obtained as a wetting fluid displaced a non-wetting fluid in saturated porous medium (oil into water in a hydrophobic porous medium environment). A cluster growth model was introduced by Eden in 1961 to simulate the growth of tumors. In addition to its biological applications, this model has relevance to many other types of growth phenomena with stable or marginally stable interfaces. When growing an Eden cluster, one of the empty sites next to the aggregate (perimeter sites) is chosen randomly, and is added to the cluster. FIGURE 6B can be defined as a DLA model cluster. The phenomena of viscous fingering takes place when a non-wetting fluid is injected in a porous medium saturated with a wetting fluid (oil into water in a hydrophilic porous medium environment).

CONCLUSIONS

The LTM is a noninvasive nondestructive method that permits visualization and measurement of multiphase flow, such as in transient air–oil–water flows occurring in sandy porous media. The LTM has been found to yield reliable fluid content in steady state, air–oil–water flow fields, as verified by comparing the results obtained with synchrotron X-rays. In transient measurements the LTM is capable of representing the fast changing fluid content with high temporal resolution. The advantage of the LTM is that it records the fluid contents of the entire flow in less than 0.1 sec, so that we are able to visualize and analyze, qualitatively and quantitatively, fingering phenomena in soil–air–oil–water systems. Qualitative data, such as visualization of the phenomena, and quantitative data, such as the dimensions, velocity, and fluid content of the flow, are of primary importance in understanding flow patterns in soil–air–oil–water systems. Additionally, the LTM is able to provide data to validate one- and two-dimensional computer codes for transient air, oil, and water flow, and to develop models of three-phase flow phenomena. Applications of the LTM in visualization of simulated of groundwater remediation by using surfactants, and to visualize physical phenomena, such as model cluster growth and fractal dimensions, were also demonstrated.

ACKNOWLEDGMENTS

This research was performed at the Department of Biological and Environmental Engineering, Cornell University, and was sponsored by the US Air Force Office of Scientific Research, under Grant/Contract Number F49620-94-1-0291.

REFERENCES

1. DiCarlo, D.A., et al. 1997. High speed measurements of three-phase flow using synchrotron X-rays. Water Resour. Res. **33:** 569–576.
2. Darnault, C.J.G., et al. 2001. Measurement of fluid contents by light transmission in transient three-phase oil–water–air systems in sand. Water Resour. Res. **37:** 1859–1868.
3. DiCarlo, D.A., et al. 1999. Surfactant induced changes in gravity fingering of water through a light oil. J. Contam. Hydrol. **41:** 317–334.
4. Vicsek, T. 1992. Fractal growth phenomena. World Scientific. Singapore.
5. Baveye, P., et al. 1998. Influence of image resolution and thresholding on the apparent mass fractal characteristics of preferential flow patterns in field soils. Water Resour. Res. **34:** 2783–2796.

Assessment of Physiologic and Pathologic Radiative Heat Dissipation Using Dynamic Infrared Imaging

MICHAEL ANBAR

Department of Physiology and Biophysics, School of Medicine and Biomedical Sciences, University at Buffalo (SUNY), Buffalo, New York, USA

> ABSTRACT: This paper reviews the mechanism and assessment of regulated radiative heat dissipation, involving the circulatory system and the skin. It describes the quantitative assessment of skin temperature modulation. The main regulating process, which can be quantitatively monitored by fast and sensitive dynamic infrared imaging, involves autonomic nervous control of cutaneous and subcutaneous perfusion. This control is significantly affected by a variety of local or systemic pathologic conditions, including cancer and certain neuropathies. A potential clinical application that objectively assesses local attenuation of temperature modulation in the presence of breast cancer is described in some detail. Systemic aberrations in skin temperature modulation can be clinically useful also in neurology. It can be used also in psychology and psychiatry to evaluate transient effects of mental stress on the autonomic nervous system.
>
> KEYWORDS: radiation; heat dissipation; dynamic imaging; infrared; perfusion; skin temperature; breast cancer; nitric oxide; neuropathy; mental stress; computerized diagnosis

INTRODUCTION

In addition to its well-known function as carrier of oxygen and nutrients to bodily tissues, blood is the primary physiologic heat exchange fluid in vertebrates. The removal of heat produced by oxidation of nutrients is more important than removal of carbon dioxide or other metabolites from live tissues. Effective heat removal is more important because, unlike the effect of buildup of metabolites that provides a negative feedback, warming up of tissues results in a strong *positive* feedback, due to the high activation energy of most metabolic processes. This might lead to a catastrophic overheating of tissues, associated with irreversibly denatured proteins and tissue death. Consequences of substantially impaired heat removal can, therefore, be dramatic and immediate. However, excessive heat removal, especially from the viscera to the skin, may also result in an unwarranted outcome. It would result in slowing down of bodily metabolism, and under extreme conditions positive-feedback cooling may lead to malignant hypothermia and again to eventual death.

Address for correspondence: Michael Anbar, Ph.D., Department of Physiology and Biophysics, School of Medicine and Biomedical Sciences, University at Buffalo (SUNY), Buffalo, NY 14214, USA.
amara@adelphia.net

Metabolic heat is dissipated primarily by radiative loss from the skin to the environment, with blood acting as the heat transfer fluid.[1] To avoid both hyper- and hypothermia, tissue perfusion and dissipation of heat by cutaneous and subcutaneous perfusion must be effectively regulated. The term "perfusion" is preferred over "blood flow" because heat dissipation by the skin is essentially independent of the direction blood flows in the local vasculature. This is particularly true of capillary beds in heat generating tissues, such as the liver or brain, or in skin—the main heat-dissipating organ. Biochemically generated heat can also be transferred by fluids other than blood. For instance, heat generated by the neurons of the retina is transferred by convection in the vitreous fluid of the eye and dissipated directly by the cornea.[2] The term "biochemical" is used here because, in addition to metabolism, also ion transport into and out of cells, neurons in particular, is associated with absorption and release of heat when ions undergo dehydration and hydration.

Any regulation of temperature is associated with its modulation. Such modulation can be driven by a large variety of regulatory mechanisms. Quantitative evaluation of such modulation, in terms of frequency and amplitude, can help to elucidate its mechanism. Non-contact temperature measurement of blackbody radiation emitted by tissues is an ideal tool to quantitative study of physiologic thermal regulation from the cellular level up to the level of the whole organism. This paper focuses on radiative heat dissipated by the skin, which is the most important process in the systemic thermal regulation in man.

PHYSIOLOGIC TEMPERATURE MODULATION

Physiologic temperature modulation can be observed at two levels—at the microscopic cellular level, where it is driven by local metabolic or neuronal activities, and at the systemic level, as part of thermoregulatory process aimed at optimizing physiologic activities of the whole organism. At the cellular level, heat generation in all tissues is modulated by the positive feedback associated with the activation energy of biochemical processes. Since most cellular metabolic processes are exothermic, the heat generated by them enhances the metabolic rate. On the other hand, most enzymatic processes have a narrow optimal temperature range of activity, above which the kinetic activation energy becomes negative. This provides negative feedback that generally offsets potential run-away cellular hyperthermia. Another modulation of cellular heat output is caused by neuronal stimulation. Neuromuscular contraction and relaxation is followed by corresponding changes in metabolic activity, associated with modulation of thermal output. Activity of neurons by itself involves temperature output modulation, as we have recently demonstrated in small groups of neurons in retinas of extracted eyes.

At the level of the whole organism, cellular temperature modulation is averaged out by heat exchange with blood. The thermal regulation of the organism boils down to temperature regulation of blood—the systemic heat exchanging fluid. As stated, we focus here on systemic temperature regulation through radiative dissipation of blood heat by the skin, its measurement, its driving mechanism, and on some of its clinical applications.

MODULATION OF SKIN TEMPERATURE

Pulse rate and cardiac output provide a rough regulation of perfusion and consequently of heat dissipation of blood heat by the skin. Heart rate may increase in response to external heat stress. However, the main regulatory process of perfusion and heat transfer is by control of vascular tone, that is, control of the lumen of blood vessels. Blood vessels, both arteries and veins, dilate and contract under control of the autonomic nervous system. Local vasodilatation takes place following excretion of acetylcholine by nerve cells embedded in the walls of blood vessels. Another neuronal mechanism, involving epinephrine, accounts for vasoconstriction. Vasodilatation involves acetylcholine-induced influx of calcium ions into certain endothelial cells, resulting in enhanced enzymatic production of nitric oxide (NO) as a secondary chemical messenger. NO induces muscle relaxation and inhibits muscle contraction downstream in a sequence of biochemical steps that are outside the scope of this review paper. Certain pathologic processes can affect locally or systemically the NO-dependent neuronal vasodilatory process. This pathologic effect can be quite important from a clinical standpoint.

To achieve effectively controlled radiative heat dissipation, cutaneous and subcutaneous perfusion is modulated by autonomic nervous impulses induced by central or by peripheral temperature neuro-sensors. This modulation is manifested by a corresponding modulation of blackbody radiative heat output. At normal human skin temperature, this output maximizes at about 12 mµ. Under physiologic conditions, the tight control of bodily heat dissipation results in temperature modulation of skin of the order of millidegrees Kelvin (mK) within the range from 10 to 50 mK, depending of the specific skin area. Combination of anatomic complexity of blood vessels, staggered propagation of autonomic nervous signals, interference between propagating and reflected wave fronts of blood flow, and sonic hydrodynamic effects produced by cardiogenic pulse waves, result in a highly complex FFT spectrum of temperature modulation. This is measurable in the range from 0.01 to 10 Hz.[3–5]

Monitoring this modulation requires high sensitivity and fast response rate of radiative heat measuring devises. It is not surprising, therefore, that before the recent advent of modern infrared detectors, such as the focal plane arrays (FPA) of quantum-well infrared photodetectors (QWIP),[3] this physiologic behavior of human skin or other tissues has not been observed. Assessment of skin temperature modulation involves accumulation of thousands (e.g., 2,048 or 4,096) of sequential infrared images over short periods, followed by fast Fourier transformation (FFT) of the time series of each pixel or of the average temperature of small groups (4, 9, or 16) of pixels.[4,5] For instance, in many of our studies we accumulated 2,048 infrared 256×256 pixel images at a rate of 100 images per second. These yielded FFT power spectra of 1024 discrete equally spaced frequencies between 0 and 24.4 Hz. The frequency resolution could be increased by collecting more images in an experiment, and the frequency range could be doubled by increasing the acquisition rate to 200 images per second. The technique described here is generally referred to as dynamic area telethermometry (DAT),[1] sometimes also referred to as dynamic infrared imaging (DIRI). The former name is more general, since it also includes other temperature measuring techniques, such as magnetic resonance imaging (MRI) or microwave thermometry.

In addition to the observed temperature modulation, there is also modulation of spatial thermal homogeneity (STH) (thermal spottiness) of skin—that is, modulation of temperature variance over small areas (less than 1 cm^2) of skin or other tissue.[6] STH modulation of skin, which generally ranges up to 5 mK, manifests modulation of saturation of the cutaneous capillary bed. STH modulation of skin, which provides another regulatory mechanism of radiative heat dissipation by skin, manifests primarily neuronically controlled capillary shunting, and its modulation is strongly related to modulation of vascular tone in larger vessels.[3]

Locally enhanced modulation at specific frequencies, distinctive for arteries and veins, can be observed over superficial large blood vessels. Closer observation can identify the propagation of blood pulses in them and quantify their propagation velocity.[7]

PATHOLOGIC ABERRATIONS OF TEMPERATURE AND STH MODULATION

There are two kinds of aberrations of skin temperature dynamics: (1) Local aberrations due to a local lesion—occlusion of vessels, aneurysm, inflammation (due to infection, injury, or local autoimmune response), and cancer. In all these cases localized attenuation of skin temperature modulation is expected.[1] (2) Systemic abnormalities in skin temperature regulation, which may be due a neurologic disorder or to mental stress.[1] Both types of aberration can be detected by DIRI, which can be used for clinical diagnosis and management.[1] We discuss first the use of localized attenuation of skin temperature modulation, emphasizing the use of DAT in objective detection of breast cancer.

Colored bitmaps can display the spatial distribution of amplitude modulation at specific frequencies over areas of interest.[8] Localized clusters of subareas of attenuated modulation can be visualized[8] or algorithmically detected and displayed.[4,5,9] The latter approach is substantially more reliable since it avoids subjective human judgement. Identification of subareas with attenuated modulation can be useful in vascular surgery to detect anatomic or functional abnormalities in superficial vessels. Vascular occlusion, aneurysm, or interference with regulatory neuronal signals of deeper vessels result in attenuated modulation at specific frequencies over extended regions of affected skin. Vascular occlusions and aneurysms are manifested primarily at frequencies driven by cardiogenic pulses. Neuronal aberrations are manifested by attenuated modulation at other, generally higher, frequencies. These aberrations include nerve irritation and nerve blocks, as well as interference with the function of intravascular chemical messengers involved in control of vascular tone.

Visual inspection of color-coded bitmaps can identify inflamed joints as subareas with attenuated modulation. Local nervous disorders can result in poorer control of vascular tone, hence, again in subareas of attenuated modulation. Unlike classical static thermal imaging, which monitor spatial temperature distribution, dynamic imaging can help to more readily assess the severity of the disorder and the efficacy of treatment. Visual inspection of modulation amplitude maps can also be used to assess the efficacy and extent of nerve blocks, which inhibit autonomic regulation of vascular tone and result in over-perfused (thus hyperthermic) regions of strongly attenuated amplitudes at specific frequencies. Also advanced cancerous lesions can be detected by visual inspection of modulation bitmaps as attenuated regions,[8] which can indicate local hyperperfusion due to pathologic, poorly innervated hypervascularity, in addition to the local vasodilatation by cancer-generated NO. However, this is not true of less advanced cancerous lesions.

Observation of attenuated or enhanced modulation of a subcutaneous point source or line source (e.g., a blood vessel) will show enhanced edge blurring, depending on its depth, due to the isotropic propagation of heat. This effect is frequency dependent because of selective attenuation of the high frequency components by the dispersive heat impedance of tissue. Spatial distribution of abnormal attenuation as a function of frequency could, therefore, be used to determine the subcutaneous depth of pathological lesions. This approach may be useful to localize aneurysms or cancerous lesions.

OBJECTIVE DETECTION OF CANCEROUS LESIONS

Most interesting, from a clinical standpoint, are situations where NO, produced outside of blood vessels under pathologic conditions, diffuses into the vasculature and saturates the endothelial NO-receptors. These then become non-responsive to endothelial neuronal NO signals, resulting in uncontrolled vasodilatation and subsequent attenuation of neuronally driven modulation of vascular tone.[10,11] Pathologic extravascular production of NO occurs in two groups of common disorders—joint inflammation[12] and cancer.[4,5,8–11]

Significant differences in the spatial distribution of attenuated areas between cancerous breasts and breasts with benign lesions could be observed at many distinct frequencies but at no single frequency was the difference significant enough to be diagnostically useful. Consequently, visual inspection of colored bitmaps, which display the spatial distribution of amplitudes at a single frequency, or over a limited range of frequencies, was found to be of limited diagnostic value.

Computation procedures, which are much more sensitive than visual inspection, have overcome that limitation.[4,5,9] These have been used to differentiate between cancerous and non-cancerous breasts. Comparing the extent of clustering of subareas of attenuated amplitudes at each discrete FFT frequency between 2 and 10 Hz, one finds highly significant differences at specific frequencies between groups of cancerous and non-cancerous breasts.[5,9] Various algorithms have been developed to identify and characterize clusters of attenuated subareas on the skin of cancerous breasts,[4,5,9] and even more effective algorithms have been recently developed.

FIGURE 1 (top part) shows the frequency spectra of the means of first cluster ratio (FCR), a newly developed *clustering tendency* function, for 27 cancerous breast and 77 breasts with benign lesions, all excised and pathologically examined. These means are distinct at many frequencies. Clearly FCR(cancerous) is greater than FCR(non-cancerous) at most frequencies, and the difference is generally significant. The middle part of FIGURE 1 shows, using the classic Student's t-test, the level of significance of the difference between the two means at each frequency in terms of $1 - p$ ($1 - p$ is the specificity).[13] The lower part of FIGURE 1 shows the diagnostic sensitivity of FCR at $1 - p = 0.95$ (95% specificity) at each frequency. Comparing the clustering of an

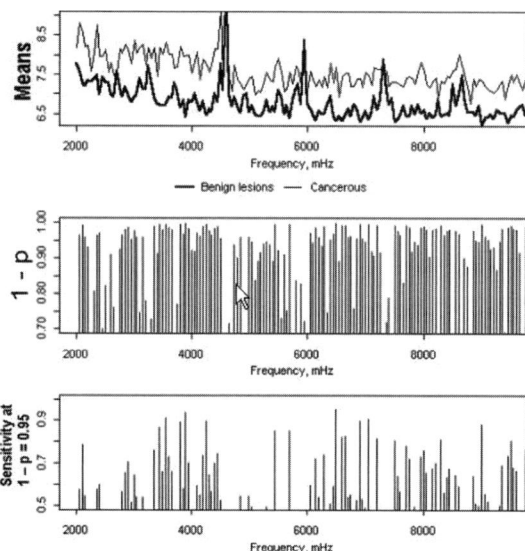

FIGURE 1. Mean CPR of cancerous and non-cancerous breasts at various frequencies; significance of difference between the means, and sensitivity of differentiation between the groups at 95% specificity.

unknown case to the means of non-cancerous and cancerous breasts at all frequencies with $1 - p > 0.95$, yields a diagnostic parameter z-value difference (ZVD),[9] which measures the level of pathology of the given case. Using this objective, fully computerized methodology, diagnostic sensitivity and specificity of more than 95% has been obtained, although for a limited number of clinical cases. FIGURE 2 compares the ZVD values for 30 cancerous breasts, 87 breasts with benign lesions and 14 breasts

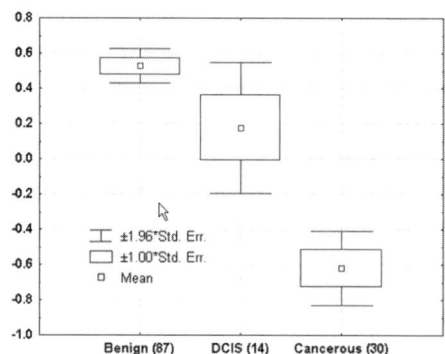

FIGURE 2. ZVD values of breasts with benign lesions, ductal carcinoma *in situ*, and invasive breast cancer.

with ductal carcinoma *in situ* (DCIS). The results clearly demonstrate the potential usefulness of this methodology in the diagnosis of breast cancer.

SYSTEMIC ABERRATION OF SKIN TEMPERATURE MODULATION

Several systemic neurologic disorders, including Raynaud's disease, complex regional pain syndrome I (RSD), lupus erythematosus, Guillain-Barre syndrome, diabetic neuropathy, and vasoconstrictive hypertension, as well as metabolic disorders such as hypo- and hyperthyroidism, diabetes mellitus (metabolic), and paramenopausal syndrome (hot flushes) that affect regulation of bodily temperature, are highly likely to be differentially diagnosed by DAT.[1,14] Another group of applications of DIRI involves detection of changes in autonomic function induced by anxiety. DAT can be used as a non-contact lie-detector to assess sympathetic response under mental stress,[1,15] replacing less reliable polygraphy. Unlike the latter, a totally objective DIRI test can also be applied to non-cooperative subjects. Moreover, the test can be absolutely objective since a computer can administer the verbal or visual stimulation, as it can the assessment of the level of abnormality in autonomic function. The same technique can be used to assess drug-induced impairment of mental function, and provide biofeedback in the treatment of phobias and other mental anxieties.[1,15]

SUMMARY

Skin temperature is modulated by oscillations of perfusion that regulate body temperature. This regulation is mediated by the autonomic nervous system. Pathologies that affect this neuronal regulation of vascular tone, including joint inflammation and cancer, can be detected by aberrations in the spatial distribution of modulation amplitudes at specific frequencies. Visualization of such aberrations is not always sufficient to be clinically useful. Computerized assessment of abnormal modulation can yield objective quantitative evaluation of pathology with higher sensitivity and specificity than is achievable by visual diagnostic imaging, which does require subjective human expertise. This methodology has been shown to be potentially useful in the detection of cancerous lesions, breast cancer in particular. DAT can also be used to assess systemic aberrations in the autonomic nervous system associated with neuropathies, metabolic disorders and mental stress.

REFERENCES

1. ANBAR, M. 1994. Quantitative Dynamic Telethermometry in Medical Diagnosis and Management. CRC Press, Boca Raton.
2. HAVERLY, R.F. & M. ANBAR. 1994. A telethermometric study of the age dependence of corneal temperature. Biomed. Thermol. **14**(4): 10–26.
3. ANBAR, M., *et al.* 1997. Fast dynamic area telethermometry (DAT) of the human forearm with a Ga/As quantum well infrared focal plane array camera. Eur. J. Thermol. **7**: 105–118.

4. ANBAR, M., C.A. BROWN & L. MILESCU. 1999. Objective identification of cancerous breasts by dynamic area telethermometry (DAT). Thermol. Intl. **9:** 137–143.
5. ANBAR, M., *et al.* 2001. Objective detection of breast cancer by DAT—an update. Thermol. Intl. **11:** 11–18.
6. ANBAR, M. & R.F. HAVERLY. 1994. Local "micro" variance in temperature distribution evaluated by digital thermography. Biomed. Thermol. **13**(4): 173–187.
7. ANBAR, M., *et al.* 1997. Study of skin hemodynamics with fast dynamic area telethermometry (DAT). Proc. 19th IEEE EMBS International Conference 644–648.
8. ANBAR, M., *et al.* 2000. The potential of dynamic area telethermometry in assessing breast cancer. IEEE Eng. Med. Biol. Magazine **19**(3): 58–62.
9. ANBAR, M., *et al.* 2001. Detection of cancerous breasts by dynamic area telethermometry (DAT), IEEE Eng. Med. Biol. Magazine **20**(5): 80–91.
10. ANBAR, M. 1994. Hyperthermia of the cancerous breast—analysis of mechanism. Cancer Lett. **84:** 23–29.
11. ANBAR, M. 1995. Mechanism of hyperthermia of the cancerous breast. Biomed. Thermol. **15**(2): 135–139.
12. ANBAR, M. & B.M. GRATT. 1998. Nitric oxide in the physiopathology of temporomandibular pain. J. Oral Maxillofacial Surg. **56:** 872–882.
13. FEINSTEIN, A.R. 1977. Sample size and the other side of 'statistical significance'. *In* Clinical Biostatistics, Chapter 22. C.V. Mosby Co., St. Louis.
14. ANBAR, M. 1998. Clinical thermal imaging today—shifting from phenomenological thermography to pathophysiologically based thermal imaging. IEEE Eng. Med. Biol. Magazine **17**(4): 25–33.
15. ANBAR, M. 1998. Telethermometric psychological evaluation in skin perfusion induced by the autonomic nervous system. U.S. Patent No. 5,771,261.

Molecular Motion and Cardiac Muscle Motor Dynamics

AMIR LANDESBERG,[a] YOLANDA LANDESBERG,[a] SAMUEL SIDEMAN,[a] AND HENK E.D.J. TER KEURS[b]

[a]*Department of Biomedical Engineering, Technion, Israel Institute of Technology, Haifa, Israel*

[b]*University of Calgary, Health Care Sciences, Alberta, Canada*

ABSTRACT: This paper reports on the kinetics of molecular motion in cardiac (and skeletal) muscles, studied by image analysis of the motility assay of an isolated actin filament sliding over isolated myosin heads that perform as linear motors. Image analysis allows us to study the dynamics of the crossbridges (Xbs) formed by the interactions between the actin molecule and the myosin heads, and to explore the rate kinetics of the interactions between the two. A most significant result pertains to the identification and characterization of two kinds of kinetics involved in Xb dynamics. The analysis of the acquired images suggests that Xb dynamics is determined by two distinctive kinetic mechanisms: fast physical kinetics that relates to the actin–myosin Xb attachment and detachment cycle and a process that is orders of magnitude *slower* in biochemical kinetics that relates to the reactions of nucleotide binding and dissociation.

KEYWORDS: myosin–actin; crossbridge; motility assay; molecular motors; filaments; sliding velocity

INTRODUCTION

Cardiac muscle contraction and mechanical function is generated by the sarcomere, the intracellular contractile element, and is determined by the interaction between actin–myosin filaments and the creation of Xbs. The Xbs are linear motor units that convert biochemical to mechanical energy, generate force, and contract the sarcomere, hence the muscle, by filament sliding and sarcomere shortening. Theoretical[1,2] and experimental (tissue and whole heart)[3–5] studies suggest that intracellular control of the mechanical activity in the cardiac muscle is based on Xb cycling between a *weak*, non-generating force, state and a *strong*, force generating, state.[6] It is important to note that the head of the myosin is 19 nm long and 5 nm thick, much smaller than any human-made nanomotor. The isolated myosin heads create a unitary force of about 2 pN and a stroke step of 5 nm. Each cubic mm of muscle tissue contains 40×10^{12} motor units. It is obviously interesting to know how the muscle regulates this huge number of motor units and what are the mechanisms that yield the outstanding efficiency of about 70% of the muscle motors.

Address for correspondence: A. Landesberg, M.D., Ph.D., Department of Biomedical Engineering, Technion, Israel Institute of Technology, Haifa, Israel.
amir@biomed.technion.ac.il

This study continues the search for the basic control mechanisms underlying Xb cycling, and the relationship between the physical processes of force and work generation and the chemical kinetics of energy (ATP) consumption.[7–10] Deciphering the regulation of biochemical to mechanical energy conversion by the Xb is a crucial step in understanding the excitation–contraction coupling in the muscle and is of great importance in normal and pathologic conditions.

Our previous studies reveal that the contractile filament function in the sarcomere at the sub-cellular organelle level is modulated by two main feedback mechanisms that regulate XBs recruitment[11] and dictate the linear relationship between the energy consumed (ATP hydrolysis), and the mechanical energy.[12,14] The two intracellular feedback mechanisms conveniently and convincingly explain the regulation of energy conversion in the cardiac muscle.[2,11,13]

The *cooperativity mechanism* provides a closed-loop feedback mechanism, whereby the loading condition determines the rate of energy consumption.[11,13] Xb cycling is modulated by the calcium feedback cooperativity mechanism that affects the calcium–troponin–tropomyosin interactions.[11] An increase in the flow resistance (afterload) increases the affinity of troponin for calcium, which hastens the rate of Xb recruitment and ATP hydrolysis, and vise versa. Sarcomere shortening decreases the number of strong Xbs, thus decreasing the affinity of troponin for calcium by the cooperativity mechanism.[11]

Mechanical feedback is an inherent property of the myosin motor unit, whereby the filament sliding velocity determines the rate of Xb turnover (weakening) from the strong, force generating, state to the weak, non-force generating, state.[12–14] This feedback provides the linear relationship between the amount of ATP hydrolysis (i.e., the energy consumption) and the mechanical energy,[12,14] including the external work, energy dissipation as heat due to the viscous properties of the Xbs,[15] and the pseudopotential energy.[11,12]

These studies[12,13] suggest that Xbs dynamics is determined by two distinctive kinetics: a slow kinetics that relates to XB cycling between the weak and strong biochemical states and a fast kinetics that relates to the myosin head interactions (attachment/detachment) with the actin binding sites. Moreover, since XB turnover from the weak to the strong state requires hydrolysis of one ATP and the strong XB can perform several stroke steps (interaction) until it returns back to the weak state, these studies suggest that the XBs perform multiple attachment and detachment cycles per single ATP hydrolysis. The motility assay technique allows study of the biophysics of the isolated molecular motor, and to measure the interactions between isolated actin and myosin molecules: that is, the displacement (nm) and force (pN) generated by the XB motor unit.[9,16–18] Laser tweezers are used to measure the unitary (isometric) force per XB and the XB stroke step size. Motility assay and laser tweezers permit the study of the molecular basis for XB function and malfunction and exploration of the pathogenesis of various heart diseases. Current image analysis of the filament trajectory in the motility assay studies is based on the analysis of the center of filament mass velocity. The present study aims to measure precisely the sliding velocity of each segment of the filament and to use this data in estimating the rate kinetics of actin and myosin interactions, at the isolated molecules level. The data is used to test the hypothesis that XB dynamics is determined by two distinct kinetics.

EXPERIMENTAL

Actin and myosin were obtained by the methods of Pardee and Spudich.[19] The myosin was purified by a method developed by Shiverrick et al.[20] and the actin was purified as described by Zol and Potter.[21] The purified actin filaments were fluorescently labeled with tetramethyl rhodamin isothiocyanate phalloidin, as described by Warshaw et al.[10] Using nitrocellulose as adhesive, myosin head molecules were attached to a microscope coverslip. The flow cell was created by separating two glass coverslips by spacers made from thin coverslips. The microchamber volume was about 15 µl. The actin filaments were loaded into an ATP solution covering the myosin heads. The flurocently labeled[10] actin filaments were propelled by the myosin heads and were observed with an inverted epifluorescent microscope (excitation 520 nm and emission 589 nm). The fluorescent images of actin motion were obtained using an image intensifier CCD camera.

IMAGE ANALYSIS

Tailor-made analytic computer software allowed us to recognize the actin filaments in the myosin motility assay with hitherto unattainable detail, and to analyze the velocity of the actin filament segment[26] using the following steps: (1) object identification in each frame; (2) recognition of the same filaments in successive frames, based on spatial correlation algorithm; (3) detection of filament backbone, the center line through the filament, using dedicated algorithms for background correction and for overcoming uneven light emissions along the filament; (4) identification of the object trajectory (see FIGURE 1); (5) calculating filament length at the various frames; and (6) partitioning the filament into equidistant segments and calculating the trajectory velocities and acceleration in the two-dimensional plane of each segment.

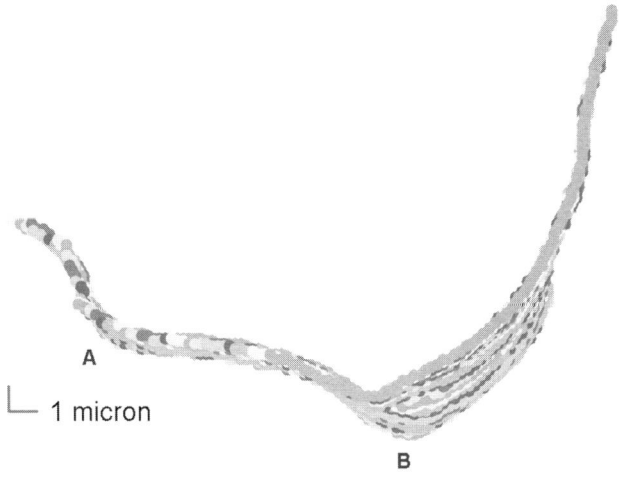

FIGURE 1. Trajectory of an actin filament during one second.

Image analysis of the actin trajectory allows us to determine the rate of myosin detachment and attachment to the sliding actin filament at its maximal unloaded velocity and to evaluate the relationship between the rates of Xb cycling between the strong and weak conformations and the kinetics of Xb attachment and detachment. Note that the relationship between the sliding velocity and the unitary force generated by the Xbs, which characterizes the viscoelastic property of the Xb and the physical interaction between actin and myosin, is outside the scope of this report.

RESULTS

The data obtained in this study provided information about Xb kinetics and the propulsion of the actin filaments. FIGURE 1 presents 33 successive frames taken during 1 sec. Image analysis suggests that, although the myosin heads provide energy for filament sliding, the direction of the actin filament trajectory is determined by the polarity of the actin filament. The actin filaments follow winding paths, with well-defined heads and tails, and well-defined deflection points. One side of the filament always remains ahead and the other forms the tail, although the myosin head are randomly scattered over the slide. The deflection points in FIGURE 1 remain stable for several hundreds of milliseconds. Measurement of the actin filament length at the various frames reviled that the lengths of the actin filaments remain constant during filament sliding, within the spatial resolution of the optical system. The length of the actin filament shown in FIGURE 1 was 23.32 ± 0.34 µm in the various frames. This allowed us to mark points along the filament, based on the distance from the edges of the filament, and to quantify the velocity of motion of each point along the actin filament. The speed of the center of gravity of each filament was 0–10 µm/sec. However, the speed of an individual marked point along the actin filament was 12–20 µm/sec. FIGURE 2 presents the velocities and the acceleration of the actin filament at a given instant. At a given point along the actin trajectory (point A in FIG. 1) the observed actin filament passed through an identical trajectory, within the optical resolution of the system. The reported values of the distance that myosin can propel

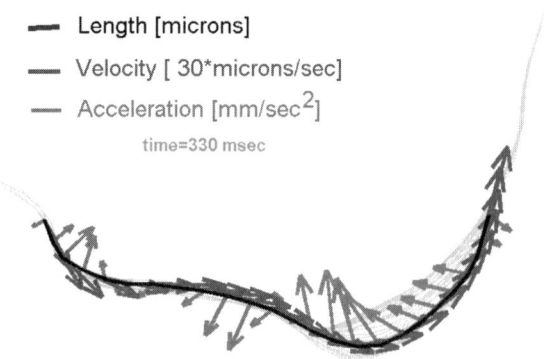

FIGURE 2. Velocity and acceleration of the actin filament at a given instant of time. (COLOR PLATE 5.)

the actin filaments per single stroke step range is between 4 and 10 nm.[8,28,29] If the distance that the myosin can propel the actin filaments is estimated at 10 nm per step and the actin filament velocity over that same point is 20 μm/sec, then the attached myosin head detaches within 0.5 msec.

DISCUSSION

Image analysis allows us to verify and/or void the various hypotheses and theories that purport to explain muscle function. The sliding velocity defines the minimal rate of XB attachment–detachment. The stroke step of the myosin head is determined by the structure of the actin filament. The actin filament is composed of double strand polymers of actin monomers. The size of single actin monomer is 5 nm, and indeed most of the studies suggest that the myosin head motion is composed of multiple steps of 5 nm each.[9,16–18] The classic Huxley theory of muscle contraction[28] suggests that the XB stoke steps are of the order of 6 to 10 nm. The classical theory of muscle function also suggests that the myosin head consumes one molecule of ATP per stroke step and that the rate of ATP consumption and XB attachment–detachment is low (order of 40 sec^{-1} at 25°C). The finding that the sliding velocity of a marked point along the actin filament is of the order of 20 μm/sec suggests that the myosin heads have to detach at a rate of at least 2,000 per second. The filament sliding velocity defines the minimal rate of Xb attachment–detachment. Consider for example a myosin head in the vicinity of the deflection point A. The Xb stroke step, that is, the maximal length that the Xb can move, is limited to 10 nm, according to Huxley's model.[28] Since the filament passes through the same trajectory and deflaction points, if the same myosin heads support the actin filament sliding at a given points along the trajectory, the myosin heads need to perform at least 2,000 strokes in steps of 10 nm per second to allow the actin filament to slide at 20 μm/sec.

The present data analysis supports our hypothesis that XB dynamics involves two distinctive kinetics: a fast kinetics (attachment–detachment) which relates to the physical aspects of Xb motion and a slow biochemical process of XB cycling between strong and weak conformations.

The Fast Kinetics

The data and the consequent analysis characterize the mechanical kinetics of the Xb attachment/detachment[28] phenomenon. The image analysis verifies that multiple steps of XB attachment/detachments occur per hydrolysis of a single ATP molecule.[7,9,30,31] The measured velocity of the actin filament propulsion suggests that the Xb physical detachment rate indeed exceeds 2,000/sec. Note that the fast kinetics of Xb attachment–detachment limits the maximal unloaded velocity. This fast kinetics is related to the kinetics of the force response to quick releases and to the fast kinetics observed in the measurements of the T_1–T_2 plots.

The Slow Kinetics

The slow mechanism of Xb cycling between the strong and weak conformations determines the time over which the Xb is in the strong state and can generate force.

This kinetics describes the rate of nucleotide binding and release and, hence, the rate of energy (ATP) consumption by the cycling XBs. The time during which the filament's deflection points are steady and active extends well above 200 msec (at 25°C). The filament passes though the same points as long as the Xbs there are in the strong state (causing deflection in the trajectory). Hence the time over which the filament passed over the same point can be used to evaluate the duty cycle of strong XBs. The estimated rate of Xb weakening is in the order of 5/sec. Thus, the kinetic of Xb cycling is two orders of magnitude slower than Xb attachment/detachment.

The development of specialized computer tools for image analysis greatly enhances our understanding of molecular motion and muscle contraction. Motility assay studies[7,10,14,16,18,24,30,32] focused on the unloaded sliding velocity (i.e., zero force) and on the generated force under isometric conditions (i.e., zero velocity). In these cases the product of force and velocity is zero and no work is generated, thus leaving the effect of the sliding velocity on the generated force obscure. The goal of our future studies is to evaluate the effect of the sliding filament velocity on the time during which the myosin is at the strong state, and to quantify the relationship between the slow and the fast kinetics.

SUMMARY

The data support the hypothesis that Xb dynamics is determined by two distinctive kinetic mechanisms: a fast physical one, which relates to Xb attachment–detachment, and a slow one, which relates to the biochemical reactions of nucleotide binding and dissociation. Image analysis suggests that the kinetics of attachment–detachment is two orders of magnitude faster ($2,000\,\text{sec}^{-1}$) than the kinetics of Xb cycling between the strong and the weak states.

Consistent with our isolated fiber studies,[27] the motility assay study validates our analytic description of Xb dynamics, it elucidates the underlying molecular mechanisms,[13,14] and may establish the basis for an analytic description of energy conversion in the cardiac muscle. This includes verification of multiple steps of Xb attachment/detachment per hydrolysis of a single ATP molecule.[2,9,13,30,31]

In general, the study enhances our understanding of the mechanisms of motion in the biological world. It links basic molecular studies to the biophysics of the whole muscle and cardiac contraction. Furthermore, it may be used to differentiate between normal and pathological hearts. Consequently, the study has significant implications for future understanding and diagnosis of cardiac disease and opens an important research avenue into molecular and genetic engineering.

ACKNOWLEDGMENTS

This study was sponsored by the MRC–Canada (H.E.D.J. ter Keurs & LA), Yigal-Allon Fellowship (LA) and the Technion foundation for the promotion of science (LA). The authors acknowledge with thanks the generous bequest of the late Larry and Horti Fairberg, to ATS, NY, which facilitated the presentation of this report.

REFERENCES

1. LANDESBERG, A. & S. SIDEMAN. 1994. Coupling calcium binding to troponin-C and crossbridge cycling in skinned cardiac cells. Am. J. Physiol. **266:** H1260–H1266.
2. LANDESBERG, A. & S. SIDEMAN. 1994. Mechanical regulation in the cardiac muscle by coupling calcium binding to troponin with crossbridge cycling. A dynamic model. Am. J. Physiol. **267:** H779–H795.
3. ALLEN, D.G. & J.C. KENTISH. 1985. The cellular basis of the length tension regulation in cardiac muscle. J. Mol. Cellular Biol. **17:** 821–840.
4. CHALOVICH, J.M. & E. EISENBERG. 1986. The effect of troponin–tropomyosin on the binding of heavy meromyosin to actin in the presence of ATP. J. Biol. Chem. **261:** 5088–5093.
5. BRENNER, B. & E. EISENBERG. 1986, Rate of force generation in muscle: correlation with actomyosin Atpase activity in solution. Proc. Natl. Acad. Sci. USA **83:** 3542–3546.
6. EISENBERG, E. & T.L. HILL. 1985. Muscle contraction and free energy transduction in biological system. Science **227:** 999–1006.
7. BRENNER, B. 1991. Rapid dissociation and reassociation of actomyosin during force generation: a new observed facet of crossbridge action in muscle. Proc. Natl. Acad. Sci. USA **88:** 10490–10494.
8. MOLLOY, J.E., J.E. BURNS, J. KENDRICK-JONES, et al. 1995. Movement and force production by single myosin head. Nature **378:** 209–212.
9. YANAGIDA, T., T. ARATA & F. OOSAWA. 1985. Sliding distance of actin filament induced by a myosin crossbridge during one ATP hydrolysis cycle. Nature **316:** 366–369.
10. WARSHAW, D.M., J.M. DESROSIERS, S.S. WORK & K.M. TRYBUS. 1990, Smooth muscle myosin crossbride intraction modulate actin filament sliding velocity *in vitro*. J. Cell Biol. **111:** 453–463.
11. LANDESBERG, A. & S. SIDEMAN. 1999. Regulation of energy consumption in cardiac muscle: analysis of isometric contractions. Am. J. Physiol. **45**(3): H998–H1011.
12. LANDESBERG, A. & S. SIDEMAN. 2000. Force–velocity relationship and biochemical to mechanical energy conversion by the sarcomere. Am. J. Physiol. **278**(4): H1274–H1284.
13. LANDESBERG, A. 1997. Intracellular mechanism in control of myocardial mechanics and energetics. *In* Advances in Experimental Medicine and Biology, Vol. 430. Analytical and Quantitative Cardiology: From Genetics to Function. S. Sideman & R. Beyar, Eds.: 75–87. Plenum Publishing, New York.
14. LANDESBERG, A. 1996. End systolic pressure–volume relation based on the intracellular control of contraction. Am. J. Physiol. **270:** H338–H349.
15. DE TOMBE, P.P. & H.E.D.J. TER KEURS. 1992. An internal viscous element limits unloaded velocity of sarcomere shortening in rat myocardium. J. Physiol. **454:** 619–642.
16. DUPUIS, D.E., W.H. GUILFORD, J. WU & D.M. WARSHAW. 1997. Actin filament mechanics in laser trap. J. Muscle Res. Cell Motility **18:** 17–30.
17. SIMMONS, R.M., J.T. FINER, S. CHU & J.A. SPUDICH. 1996. Quantitative measurements of force and displacement using an optical trap. Biophys. J. **70:** 1813–1822.
18. FINER, T.J., R.M. SIMMONS & J.A. SPUDICH. 1994. Single myosin molecule mechanics: piconewton force and nanometer steps. Nature **368:** 113–119.
19. PARDEE, J.D. & J.A. SPUDICH. 1982. Methods Enzymol, Vol. 85. Academic Press, NewYork.
20. SHIVERRICK, K.T., L.L. THOMAS & N.R. ALPERT. 1975. Purification of cardiac myoisn, application to hypertrophied mycardium. Biochim. Biophys. Acta **393:** 124–133.
21. ZOT, H.G. & J.F.D. POTTER. 1981. Purification of actin from cardiac muscle. Prep. Biochem. **11:** 381–395.
22. BLOCK, S.M. 1990. Optical tweezers: a new tool for biophysics. *In* Noninvasive Technology in Cell Biology. 375–402. Wiley-Liss Inc.
23. CHE, S. 1991, Laser manipulation of atoms and particles. Science **253:** 861–866.

24. ASHKIN, A., J.M. DZIEDZIC, J.M. BJORKHOLM & S. CHU. 1986. Observation of single beam gradient force optical trap for dielectric particles. Optics Lett. **11:** 288–290.
25. YANAGIDA, T., Y. HARADA & A. ISHIJIMA. 1993. Nano-manipulation of actomyosin molecular motor *in vitro*: a new working principle. Trends Biochem. Sci. **18:** 319–323.
26. LANDESBERG, A., L. LIVSHITZ & H.E.D.J. TER KEURS. 2000. The effect of sarcomere shortening velocity on force generation. Ann. Biomed. Engin. **28**(8): 968–978.
27. LANDESBERG, A. & H.E.D.J. TER KEURS. 1997. Regulation of force output by the sarcomere shortening in the rat cardiac trabeculae. American Heart Association, Orlando. (Abstr.)
28. HUXLEY, A.F. & R.M. SIMMONS. 1971. Proposed mechanism of force generation in striated muscle. Nature **233:** 533–538.
29. UVEDA, T.Q.P., H.M. WARRICK, S.J. KOM & J.A. SPUDICH. 1991. Quantized velocity at low myosin densities in an *in vitro* motility assay. Nature **352:** 307–311.
30. HARADA, Y., K. SAKURADA, T. AOKI, *et al.* 1990. Mechanochemical coupling in actomyosin energy transduction studied by *in vitro* movement assay. J. Mol. Biol. **216:** 49–68.
31. ISHIJIMA, A., Y. HARADA, H. KOJIMA, *et al.* 1994. Single molecule analysis of the actomyosin motor using nano manipulation. Biochem. Biophys. Res. Commun. **199:** 1057–1063.
32. GUILFOLD, W.H., D.E. DUPHIS, G. KENNEDY, *et al.* 1997. Smooth muscle and skeletal muscle myosin produce similar unitary forces and displacement in laser trap. Biophys. J. **72:** 1006–1021.

Modeling the Spreading Cortical Depression Wavefront

UGUR BAYSAL[a] AND JENS HAUEISEN[b]

[a]*Hacettepe University, Department of Electrical and Electronics Engineering, Beytepe, Ankara, Turkey*

[b]*Friedrich-Schiller University, Biomagnetic Center, Jena, Germany*

> ABSTRACT: Spreading cortical depression (SCD) is a wave of depolarization that spreads across the cortex at 2–5 mm/min and is followed by a 5–10 minute reduction in EEG activity. It is assumed that SCD is involved in the pathophysiology of migraine. We present a new model and a visualization technique for the spread of excitation on realistic brain surfaces. The usefulness of the technique is demonstrated on a rat brain that had been segmented from a three-dimensional magnetic resonance image data set. With the help of this methodology it is possible to create specific patient models to better understand the mechanisms of SCD.
>
> KEYWORDS: spreading cortical depression (SCD); brain surface; three-dimensional modeling; magnetic resonance imaging (MRI); magnetoencephalography (MEG); electroencephalography (EEG); neurons

BACKGROUND AND BASIC DEFINITIONS

Neurons are specialized cells in the nervous system of the body. The main input to the nervous system is via sensory transducer neurons, and the main output is through the triggering of muscle fiber contraction by motor neurons. All other neurons take their inputs from, and send their outputs to, groups of other neurons, transforming a number of inputs that represent the outputs of other neurons, into a single output signal. A group of interconnected neurons is called a *neural network*.

Neurons are filled with, and immersed in, an ion-rich fluid. The fluid inside the cell is separated from the fluid outside by the cell wall, or *membrane*, of the neuron. The membrane is only slightly permeable to ions, allowing the potentials inside and outside the cell to differ. A slow but continuous process pumps ions across the membrane to maintain the potential inside the cell at about $-70\,\mathrm{mV}$ relative to the potential outside. This is the *rest potential* of the cell. The potential of the fluid in which the neurons reside is usually assumed to be at zero potential, and intracellular potentials are usually measured relative to this extracellular fluid. Depolarization indicates a decrease in the membrane potential.

The output of a cell is through its *axon*, which provides inputs to other neurons by forming *synapses* with the target neuron membrane. A synapse is notionally the

---

Address for correspondence: Ugur Baysal, Department of Electrical and Electronics Engineering, Hacettepe University, Beytepe 06532 Ankara, Turkey.
ubaysal@hacettepe.edu.tr

gap between the impinging axon and the neuron membrane, although a synapse is often taken to describe the entire structure mediating an axon to neuron connection. Structures on and parameters concerning the axon side of the gap are called *presynaptic,* whereas those on the membrane side are referred to as *postsynaptic.* Synapses can form anywhere on the membrane of a neuron, either on the *soma* (the main body of the cell where the nucleus is located) or on the *dendrites* (the thinner extended membrane branches). Postsynaptic potential are thought to be the main basis for the electroencephalogram and magnetoencephalogram recorded outside the head.

The synaptic input alters the potential in that area of the cell, creating an intracellular potential gradient. This results in voltage changes and current flows being propagated throughout the cell, in the same way that pulses propagate through an electrical cable. The speed of propagation of voltage changes is a function of both the conductance of the cell fluid and the conductance and capacitance of the cell membrane. Given sufficient entry of positive charge at the synapses over a short enough period, the potential at the soma will rise above a *threshold potential,* usually about $-55\,mV$, triggering a series of currents that cause a sharp rise and fall in potential at the head of the neuron axon. This potential *spike* is very brief (less than 2 ms) and very high (peak potential about $+30\,mV$), and is propagated without decrement down the axon. When an axonal spike reaches a synapse it causes the release of a brief burst of neurotransmitter, thus triggering ion currents in the target neuron. The production of a voltage spike at the head of a neuron axon is referred to as the *firing* of the neuron. The spike in potential is felt in the soma, yet becomes heavily attenuated and spread out in time as it propagates back into the dendrites.[1]

Electroencephalography (EEG) and magnetoencephalography (MEG) measure the electrical activity of the brain. Due to the small signal amplitude of each neuron and to the distance between the site of recording and the site of the origin of the signal, both EEG and MEG are only able to pick up signals produced by simultaneous firing of large numbers of neurons.

The term cortical depression (CD), different from neural disorder "depression", means diminished cortical electrical activity. This diminishing electrical activity spreads like a wavefront along the surface of the cortex, hence the name spreading cortical depression (SCD).

SPREADING CORTICAL DEPRESSION

Leao[2] first discovered SCD in the EEG of a rabbit brain more than 50 years ago. He found that reduction of the normal electrical activity of rabbit neocortex could be induced by a variety of external stimuli[3] including KCl application.[4] Leao observed that the depressed activity begins in one hemisphere, at the stimulus site, and spreads slowly in all directions. He also observed vascular changes (pial arterial and venous dilatation) that occurred simultaneously with the onset of spreading electrical depression, and followed closely the pattern of that depression.[5] This led him and others to speculate about the possible involvement of SCD in the pathophysiology of migraine.[6,7]

Migraine is a moderate to severe headache that lasts from 4 to 72 hours; it is often accompanied by nausea and hypersensitivity to light and noise. Some patients experience sensory disturbances, like a so called *aura*, prior to the attack. The term aura indicates neurologic abnormalities, such as visual disturbances or partial loss of vision. Both clinical and experimental evidence suggests that the migraine aura results from a transient abnormality that originates locally in the cerebral cortex, and then propagates throughout this brain region. Understanding SCD would lead to a significant step forward in migraine research.

SCD is a relatively long lasting phenomenon, and baseline drift problems make it rather difficult to record it non-invasively by using electrodes on the scalp of the patient. Therefore, MEG recordings, which can be performed without the typical baseline drift problems of EEG, have the potential advantage of delivering new information about SCD. However, only little is known about the characteristics of MEG data in migraine patients.[3] Measurements in animals can provide a direct comparison between the invasive electrical recordings and simultaneously obtained non-invasive MEG data. Thus, in this study we have simulated the SCD on a realistic rat brain model.

MODELING AND ASSUMPTIONS

The wave of depolarization of the SCD is modeled using small dipoles representing small areas of pyramidal cell activities (see FIGURE 1). Each dipole represents a group of cells having a certain surface area. Dipole modeling has been used previously to describe SCD.[7–9] Previous studies mainly assumed simple symmetric geometries in order to obtain closed form analytic expressions. This study uses realistic cortical geometries and a numerical approach to simulate the electromagnetic field generated by the SCD wavefront. Moreover, the entire SCD wavefront circle is taken into account.

In order to implement the proposed algorithm, a three-dimensional MRI of a rat head was constructed from 256 slices of T1 weighted images. The brain surface and the inner skull surface were segmented using the biomagnetic inverse problem solver research software package CURRY®.[10] The brain surface was discretized with 1,804 triangles having 1.0 mm average edge length. The computation of the magnetic field utilized a one-compartment boundary element model constructed from the inner skull surface. The triangle side length for this surface was 1 mm to provide the required numerical accuracy.[11]

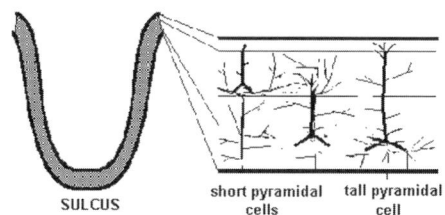

FIGURE 1. A simplified cross-sectional representation of a small part of cortex (sulcus) and neurons in the magnified portion.

It is assumed that the SCD starts at a point $P(x,y,z)$ on the cortical surface whose coordinates are obtained from the three-dimensional image of the brain. The wavefront then spreads as a circlet around the initial point P. The simulated wavefront is represented by discrete dipoles, where each dipole has an unit intensity, direction, and travel velocity. The dipoles are oriented perpendicular to the cortical surface.[8] The velocity is constant in strength and tangential to the cortical surface. As a dipole (hence the SCD wavefront) travels, it occasionally passes across an edge of two discrete surface elements (triangles). At these edges, the velocity vector of the dipole is rotated and the travel is continued. Since the SCD is accompanied by total depolarization and this depolarization period lasts on the order of minutes, it is also assumed that, when a part of the wavefront reaches an already depolarized region, this part no longer proceeds further from that point. This is called *collision*.

One step of travel is defined for all wavefront dipoles as the step that they have completed when their positional increment is equal to their velocity. The velocity, the number of steps, and the interdipole distance are all user defined. At each step, all dipoles are checked to see if any collision has occurred. If a collision has occurred, the latest dipole reaching the collision point is deleted. At the end of each step, the dipoles are redistributed along the perimeter of the wavefront, so that a uniform interdipole distance is preserved.

The simulation program was written in the Delphi language. It reads the ASCII surface data file output from CURRY and outputs another ASCII file having dipole positions and normals to be read into CURRY. The program flowchart is shown in FIGURE 2.

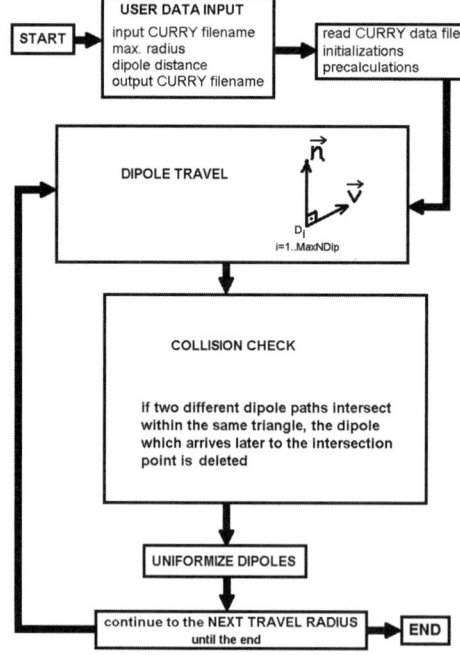

FIGURE 2. Flowchart of the program used in the modeling approach.

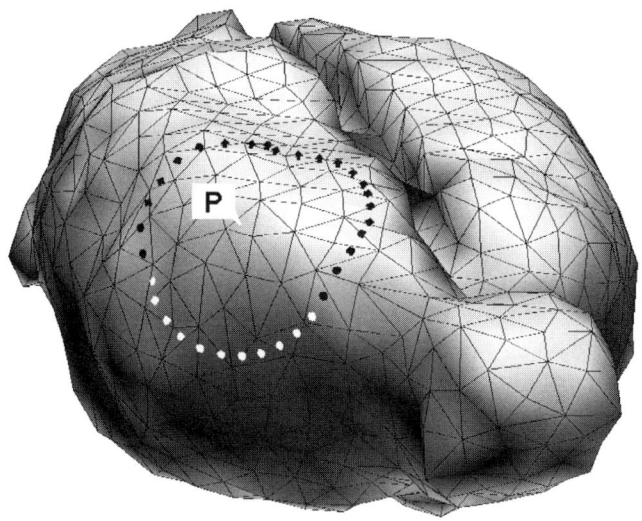

FIGURE 3. Dipoles simulating sources of a SCD wavefront on the cortical surface of the rat. The SCD starts from point P. The simulation is carried to a radius of 10 mm. The black or white dipoles representing the wavefront were assigned arbitrarily to improve visibility.

RESULTS

The results of the program computing the dipole distribution on the cortical surface are shown in FIGURE 3. Collisions of dipoles occur only rarely on the relatively flat cortical surface of the rat. Collisions are expected to occur very often on the complex cortical surface of the human brain.

The magnetic field was computed using the boundary element model described above on each sensor position of a 16 channel microSQUID biomagnetometer system.[12] FIGURE 4 shows the results of this simulation. With increasing radius the complexity of the field pattern increases.

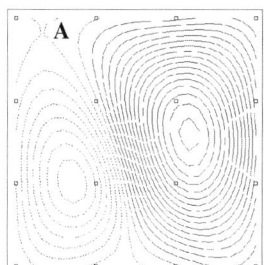 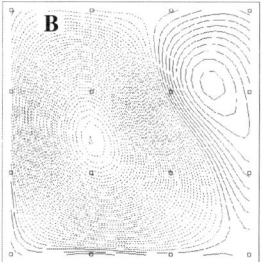 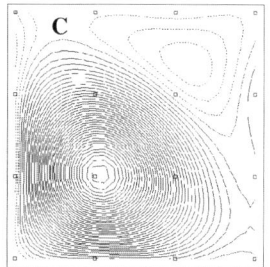

FIGURE 4. Simulated magnetic field maps for radii of **(A)** 10 mm, **(B)** 20 mm, and **(C)** 30 mm. The positive iso-field lines are indicated by *solid lines*, the negative ones by *dashed lines*.

SUMMARY

In this paper, simulations were used to visually represent spreading cortical depression (SCD) on a realistic brain model. Using this methodology, it is possible to create animal specific and patient specific SCD models to better understand the mechanism of SCD. The dipoles that are generated in this simulation were used to calculate the biomagnetic field distribution at the positions of the MEG sensors on the head of a rat. These simulated fields can be compared with future experimental measurements.

ACKNOWLEDGMENTS

The Turkish Scientific and Technical Research Council (TUBITAK) support Ugur Baysal's research activities. We thank Tobias Friedrich for fruitful discussions and help with Curry® File I/O routines. We thank Dr. Michael Eiselt for providing the rat MRI data.

REFERENCES

1. STUART, G.J. & B. SAKMANN. 1994. Active propagation of somatic action potentials into neocortical pyramidal cell dendrites. Nature **367:** 69–72.
2. LEAO, A.A.P. 1944. Spreading depression of activity in the cerebral cortex. J. Neurophysiol. **7:** 359–390.
3. BOWYER, S.M., et al. 1999. Neuromagnetic measurements of evoked and spontaneous migraine with aura. Cephalalgia **19:** 352–352.
4. EISELT, M., et al. 1999. KCl-induced spreading depression—simultaneous detection by 16-channel ECoG and MEG in rats. *In* Recent Advances in Biomagnetism. Yoshimoto, et al., Eds.: 353–356. Tohoku University Press, Sendai.
5. LEAO, A.A.P. 1944. Pial circulation and spreading depression of activity in the cerebral cortex. J. Neurophysiol. **7:** 391–463.
6. LEAO, A.A.P. & R.S. MORISON. 1945. Propagation of spreading cortical depression. J. Neurophysiol. **8:** 33–45.
7. MILNER, P.M. 1958. Note on a possible correspondence between the scotomas of migraine and spreading depression of Leao. Electroencephalogr. Clin. Neurophysiol. **10:** 705.
8. GARDNER-MEDWIN, A.R., et al. 1991. Magnetic fields associated with spreading depression in anaesthetised rabbits. Brain Res. **540:** 153–158.
9. TEPLEY, N. & R.S. WIJESINGHE. 1996. A dipole model for spreading cortical depression. Brain Topography **8:** 345–353.
10. NEUROSCAN CORP. (Sterling VA, USA). CURRY. <http://www.neuro.com/neuroscan/prod05.htm>
11. HAUEISEN, J., et al. 1997. The influence of boundary element discretization on the forward and inverse problem in electroencephalography und magnetoencephalography. Biomed. Tech. **42:** 240–248.
12. NOWAK, H., et al. 1999. A 16-channel SQUID-device for biomagnetic investigations of small objects. Med. Eng. Phys. **21:** 563–568.

Time-Varying Current Density Distributions in the Human Heart and Brain

JENS HAUEISEN,[a] MAREK ZIOLKOWSKI,[b] AND UWE LEDER[c]

[a]*Biomagnetic Center, Friedrich-Schiller University Jena, Jena, Germany*

[b]*KETiI, Electrical Engineering Department, Technical University Szczecin, Szczecin, Poland*

[c]*Clinic of Internal Medicine, Friedrich-Schiller University Jena, Jena, Germany*

> ABSTRACT: A new method for visualizing and postprocessing three-dimensional time varying vector fields is presented. This method is based on equivalent ellipsoids fitted to these fields. The new technique has been tested on vector fields representing current density reconstruction results based on biomagnetic data from a cardiac patient and a neurological patient. Multiple foci in the vector fields are extracted by multiple ellipsoids which are fitted in an iterative manner. The new method enables visualization of even very complex vector fields, as well as statistical postprocessing.
>
> KEYWORDS: magnetoencephalography; magnetocardiography; electroencephalography; electrocardiography

INTRODUCTION

Bioelectromagnetic measurements provide non-invasive information about the electrical activity within the human body. For example, magnetoencephalography (MEG) and electroencephalography (EEG) are used to assess electrical brain function, and magnetocardiography (MCG) and electrocardiography (ECG) provide information about the electrical heart activity. MEG and MCG are commonly measured without contact to the human body with the help of biomagnetometers using superconducting quantum interference devices (SQUIDs) as sensors.[1] EEG and ECG are measured with the help of electrodes on the body surface. Based on these measured data, current density reconstructions (CDRs) are used to assess cardiac activation[2] and brain function.[3] CDRs are vector fields where each vector represents the current density in a volume element or on a surface element. Often CDRs are very complex, containing several hundreds or thousands of single vectors; a data reduction method is needed to visualize these CDRs. Moreover, for statistical data analysis (e.g., in group studies) a method is needed that enables a comparison of time varying CDRs within groups of patients or volunteers. Previously, for example, a parameterization based on the visual inspection of up to eight subareas of the heart was applied.[4] These subareas have been classified manually into active or inactive

---

Address for correspondence: Jens Haueisen, Ph.D., Biomagnetic Center, Friedrich-Schiller University Jena, Philosophenweg 3, 07743 Jena, Germany.

haueisen@biomag.uni-jena.de

areas and statistics have been computed relating to the number of classified subareas. One clear disadvantage of this method is that it yields only a very rough statistical description of the location and extent of the activation maxima and minima.

In this paper, we expand a new technique, previously introduced for two-dimensional planes,[5] to three-dimensional time varying problems. This technique is based on the parameterization of CDRs with the help of equivalent ellipsoids. The usefulness of our new technique is demonstrated with the help of two examples. In the first example, data from a cardiac patient after myocardial infarction are analyzed. Here, the location and extent of the site of the infarction with respect to the origin and extent of so called late potentials that occur after the QRS complex, are of clinical interest. The second example illustrates the new technique by using data from a neurological patient. This patient suffers from migraine and data was obtained during an induced migraine related headache. The causes and mechanisms of migraine are not yet well understood. Thus, the investigation of this patient aims at a description of the electromagnetic phenomena related to migraine.

EQUIVALENT ELLIPSOID TECHNIQUE

An equivalent ellipsoid has been defined as a three-dimensional ellipsoidal object fitted to a current density distribution region in which the magnitude of the currents exceeds a certain threshold.[5] The threshold T_h used for marking the most important regions in the CDR is defined empirically by $T_h = 100\% \times (Q_{max} + Q_{mean})/2Q_{max}$, where Q_{max} is the maximum and Q_{mean} is the mean value of the dipole moment in the current distribution. Dipoles having a dipole moment greater than the threshold T_h are used to fit the equivalent ellipsoid. The equivalent ellipsoid is defined by its three orthogonal semi-axes in a local coordinate system. First, the center of gravity (COG) of the marked region is calculated. The position of the COG has been used as a new origin of the local coordinate system. The direction of the main axis is computed on the basis of the weighted longest distance (LD) from the COG according to:

$$d_{max} = \max_{i = 1, ..., N_T} M_i |\mathbf{P}_i - \mathbf{P}_{COG}|, \tag{1}$$

where M_i denotes the magnitude of the moment of ith current density vector and N_T is the number of current density vectors in the threshold region. In the next step, the local coordinate system is rotated so, that the main axis represents the z-axis. In the rotated coordinate system, the normalized position of the current density vector with the maximal distance from COG on the new x–y plane is used as the direction of the second semi-axis of the equivalent ellipsoid. The direction of the third axis has been determined by the cross product of the first two axis. For multifocal distributions, the basic algorithm is applied iteratively, including an additional parameter, the separation radius R_s. R_s selects a certain part of the threshold CDR to which the equivalent ellipsoid is fitted. Using a local COG and R_s, only the current density vectors within this restricted region are included in the fit of the equivalent ellipsoid. After the first equivalent ellipsoid fit, the second local maximum is searched in the rest of the threshold CDR. Based on the separation radius and the new position of the local maximum, the new local COG is computed, and the next equivalent ellipsoid is estimated. The procedure is repeated until the last current density vector in the threshold distribution is obtained or the user stops the process. The algorithm is applied to each time step separately in order to visualize time varying three-dimensional CDRs.

CARDIAC EXAMPLE

The measurements were made in a magnetically shielded room (AK3b, Vacuumschmelze, Hanau, Germany) at the Biomagnetic Center in Jena, Germany. The magnetic field was recorded with a twin Dewar biomagnetometer system (2 × 31 channels) with first order axial gradiometers (Philips, Hamburg, Germany).[6] We measured the magnetocardiogram of a 70-year-old patient who had non-sustained ventricular tachycardia that developed after anterior left ventricular (LV) myocardial infarction and apical aneurysm, recording 600 sec of signals at a sampling rate of 1,000 Hz. The last 40 msec of the bidirectional 30 Hz highpass filtered depolarization signal (late potentials, LP) were used for inverse computations. Late potentials outlasting ventricular depolarization reflect a slow and inhomogeneous conduction in scarred myocardium and slow conduction as a requisite condition for reentrant ventricular arrhythmias. A three-dimensional MRI data set of the chest of the patient was obtained using a GYROSCAN® ACS II machine (Philips). Cardiac gated transverse slices (T1 weighted, 5 mm) were measured and merged to a three-dimensional torso data set. A boundary element model consisting of the left and right lungs, as well as the outer torso surface, was applied for the magnetic field computations. The surface of the LV was segmented from the MRI data set and subsequently used to restrict the source space. The current density vectors were determined through a minimum norm least squares algorithm (L_2 norm) for all time points.[7] This method selects the source configuration with the minimum sum of the squared current magnitudes. Since currents close to the sensors evoke larger magnetic signals than currents further away from the sensors, a lead field normalization was applied. This method removes the bias of the L_2 norm towards the reconstruction of superficial sources.[7]

FIGURE 1 depicts the QRS signal and the bidirectionally high pass filtered signal. The time instants in FIGURE 1 indicate the times at which the source reconstruction

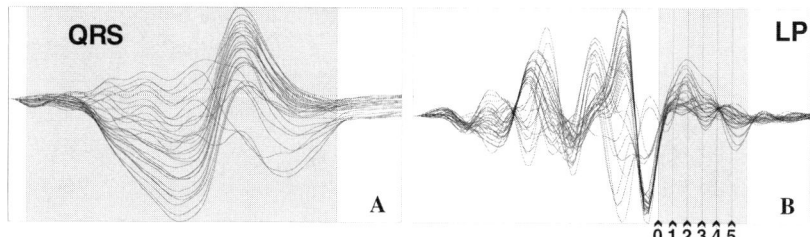

FIGURE 1. (**A**) Magnetic QRS. (**B**) Butterfly plot of the terminal 40 msec of the bidirectionally 30 Hz high pass filtered depolarization signal (LP).

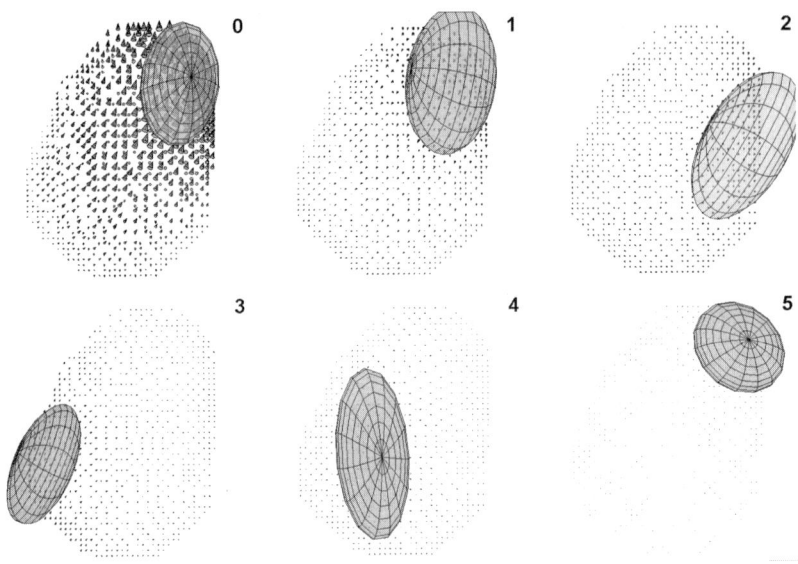

FIGURE 2. Equivalent ellipsoids found at various time steps in the LP interval, as indicated in FIGURE 1.

is presented in FIGURE 2. The results in FIGURE 2 show a mostly upper right (apical segment) activation maximum (position of ellipsoids). FIGURE 2-4 is close to a minimum in signal strength and, thus, the CDR vulnerable to the influence of noise. Therefore, the equivalent ellipsoid for this instant in time is not representative. The mainly apical position of the equivalent ellipsoids in FIGURE 2 agrees with the localization results obtained using the maximum CDR in FIGURE 3.

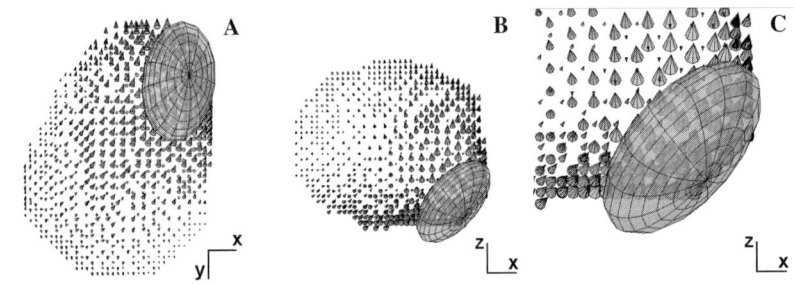

FIGURE 3. Equivalent ellipsoids fitted to the reconstructed current density distribution located on the surface of the LV for the maximum mode: **(A)** diaphragmal view, **(B)** apical view, and **(C)** magnified apical view. The threshold distribution is indicated by *light cones*.

NEUROLOGICAL EXAMPLE

As a part of a larger study we measured 31 channel DC MEG using the biomagnetometer described above for a patient with pharmacologic induced migraine (nitroglycerin spray). The biomagnetometer was positioned above the parietal-occipital cortex and one hour of continuous data with a sampling rate of 40Hz (0–15Hz bandwidth) was obtained. Artifact rejection was performed on the basis of the magnetic reference sensors, the electrooculogram (vertical and horizontal eye movements), and the electromyogram (musculus masseter and musculus trapecius). Large amplitude waves were identified in the raw data and used for minimum norm estimation, as described above. A three-dimensional MRI data set of the head was obtained, the inner skull surface was segmented and triangulated (triangle size 7mm),[8] and a one compartment boundary element model was constructed and applied for source reconstruction. The source model consisted of 23,564 current density vectors on the segmented brain surface (3.2mm average distance).

FIGURE 4A depicts the entire brain and the reconstructed current density distribution. The threshold CDR is given in FIGURE 4B and FIGURE 4C, where this multifocal CDR cannot be represented by a single ellipsoid, as given in FIGURE 4B. The set of ellipsoids in FIGURE 4C approximates the threshold CDR. The multiple ellipsoids indicate multiple sites of brain activation during this pharmacologically induced migraine headache. These multiple sites of activation support the theory of a spread of activation in migraine patients.[9]

SUMMARY

We have presented a new technique for post-processing CDRs that supports visualization and statistical analyses. Although we presented only examples from the field of biomagnetism, it is possible to apply this technique to other types of vector fields.

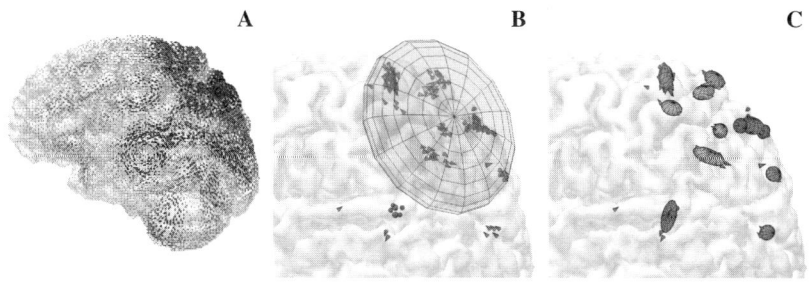

FIGURE 4. Results of the current density reconstruction for a patient with pharmacologically induced migraine: **(A)** side view, **(B)** and **(C)** enlarged side view. An equivalent ellipsoid with $R_s \to \infty$ **(B)** and ellipsoids with $R_s = 12$mm **(C)** fitted to the CDR.

ACKNOWLEDGMENT

This work was partly supported by the German Ministry of Science under project No 165V1271. We thank Christiane Karlowsky and Jacqueline Weiss for providing the measurement data on the neurological patient.

REFERENCES

1. ANDRÄ, W. & H. NOWAK, Eds. 1998. Magnetism in Medicine. Wiley-VCH, Berlin.
2. LEDER, U., J. HAUEISEN, M. HUCK & H. NOWAK. 1998. Non-invasive imaging of arrhytmogenic left-ventricular myocardium after infarction. Lancet **352:** 1825.
3. HÄMÄLÄINEN, M.S. & R.J. ILMONIEMI. 1994. Interpreting magnetic fields of the brain: minimum norm estimates. Med. Biol. Eng. Comput. **32:** 35–42.
4. NOWAK, H., U. LEDER, P. POHL, et al. 1999. Diagnosis of myocardial viability based on magnetocardiographic recordings. Biomed. Tech. **44**(Suppl. 2): 174–177.
5. ZIOLKOWSKI, M., J. HAUEISEN, H. NOWAK & H. BRAUER. 2000. Equivalent ellipsoid as an interpretation tool of extended current distributions in biomagnetic inverse problems, IEEE Trans. Mag. **34:** 1692–1695.
6. DÖSSEL, O., B. DAVID, M. FUCHS, et al. 1991. A modular approach to multichannel magnetometry. Clin. Phys. Physiol. Meas. **12**(Suppl. B): 75–79.
7. FUCHS, M., M. WAGNER, T. KÖHLER & H.A. WISCHMANN. 1999. Linear and nonlinear current density reconstructions. J. Clin. Neurophysiol. **16:** 267–295.
8. HAUEISEN, J., A. BÖTTNER, M. FUNKE, et al. 1997. The influence of boundary element discretization on the forward and inverse problem in electroencephalography and magnetoencephalography. Biomed. Tech. **42:** 240–248.
9. BOWYER, S.M., Y.C. OKADA, N. PAPUASHVILI, et al. 1999. Analysis of MEG signals of spreading cortical depression with propagation constrained to a rectangular cortical strip I. Lissencephalic rabbit model. Brain Res. **843:** 71–78.

Neutrophils Movement *in Vitro*

ANNA KORZYŃSKA

*Institute of Biocybernetics and Biomedical Engineering,
Polish Academy of Sciences, Warsaw, Poland*

> ABSTRACT: Ineffectiveness of the motility of neutrophils is observed in some immunodeficiency disorders. There is a need to find an adequate description and means to evaluate the activity of neutrophils. Early studies describe an individual cell behavior by the cell motility index vector with three components: penetration, directionality, and expansivenesses. Movement quantification has been improved by extending the description of cell activity to two stages of classification. In the first stage, one minute of each cell's behavior is classified into one of five types of cell behavior, using the k-NN method. In the second stage, the sample level that corresponds to the final classification, the Shapiro–Francio test has been used. The proposed new method allows us to describe differences between a normal child and a Chediak–Higashi syndrome patient.
>
> KEYWORDS: cell movement; cell motility; neutrophils; Chediak–Higashi syndrome

INTRODUCTION

Neutrophils granulocytes, also called polymorphonuclear leukocytes, are essential elements of the human immune system. Neutrophils demonstrate an active motility in response to inflammatory stimuli: they crawl towards the infected tissue and phagocyte or damage bacterial cells. If their movement becomes ineffective, a patient suffers from recurrent, severe, pyogenic infections.[1] Diagnosis of immunodeficiency disorders presently depends on indirect examination of neutrophils samples. Their activity is evaluated by using parameters that describe the population of cells in two static states. Clearly, this quantification entails a loss of information about individual cell behavior and the dynamics of the process.

The development of image analysis methods, combined with improvements and reduction in the costs of imaging technology (quality, resolution, and enlargement of microscopes), recording methods (low costs and high performance of video sensors), as well as power and speed of computers, have made the direct examination of individual cell behavior realizable in medical diagnostics.[2–7]

The method for describing single cell behavior in time using the cell motility index $CMI_{<\Delta t>}$ vector was proposed in previous studies.[8–10] This vector consists of three components (CMI_p, CMI_d, and CMI_e) that correspond with three aspects of activity: CMI_p quantifies the intensity of penetration and describes how compactly or densely the cell path covers a given area; CMI_d quantifies the directionality, which

---

Address for correspondence: Anna Korzyńska, Ph.D., Institute of Biocybernetics and Biomedical Engineering, Polish Academy of Sciences, ul. Trojdena 4, 02-109 Warsaw, Poland.
akorz@ibib.waw.pl

is a measure of how effectively the cell covers a distance, and how much its path differs from a straight line; and CMI_e quantifies the expansiveness, as a measure of the area on which the cell does its duty. This previous method[7] leads to the description of single cell behavior by a sequence of numbers. The numbers are the sums of values of all the three aspects calculated for Δt, the duration (minutes) of examination $[CMI_{\Delta t, i}]_{i=1,\ldots,q}$. Two numbers, the mean value of a time sequence and its standard deviation, are used to describe single cell activity during the entire examination period.[11]

The new approach proposed here involves classification of cell behavior during the interval Δt into one of the cell behavior classes in the three-dimensional space of certain features. Thus, single cell behavior becomes a time sequence of the class codes $[K_{\Delta t, i}]_{i=1,\ldots,q}$. Based on the distribution analysis of all cell behavior codes from one sample, a sample level classification is performed. The classification criteria consider the fraction of cells without any restriction on the movement ability in the sample.

CLASSES OF CELL BEHAVIOR

Three basic types of neutrophils behavior have been described by us:[7,9] chaotic, formal, and directed. FIGURE 1 shows examples of these types of behavior as sequences of an elliptic model of the cell contours, the construction based on microscopic images collected in two-second time increments.[11] All of them and some hybrid types of cell behavior have been observed in groups of normal subjects and in patients with cells showing ineffectiveness of movement. *Chaotic* movement is correlated with a high value of the penetration index and with an average or high value of the expansiveness. The *directed* movement is correlated with a high value of directionality and with an average or high value of expansiveness. *Formal* movement is correlated with a low value of expansiveness and with an average value of the penetration index or the directionality index.

Detailed examination of the experimental material showed that most of the neutrophils movement is more efficient in covering a field or a distance and is quicker and more expansive in normal subjects than in patients with a deficiency in the crawling mechanism. Some neutrophils move with a temporary restriction in their movement ability. All neutrophils from patients with immunodeficiency associated with dysfunction of movement abilities (e.g., Chediak–Higashi syndrome patients[1]) exhibit restricted movement abilities. The distributions of cell behavior codes differ in these two cases.

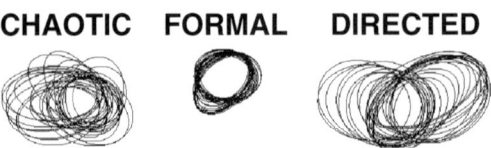

FIGURE 1. Basic types of neutrophils behavior.

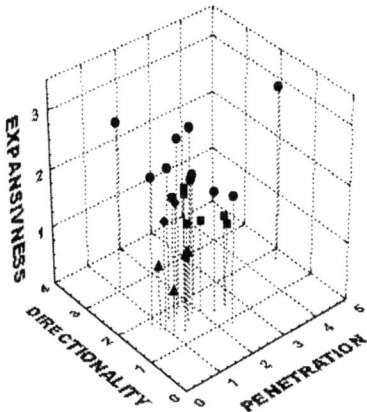

FIGURE 2. Classes of neutrophils behavior in three-dimensional space of penetration index, directionality, and expansiveness. A learning set of neutrophils one-minute behavior used in nearest neighbor classification method. ● Class I, ■ Class II, ◆ Class III, ▲ Class IV.

It is proposed to partition the three-dimensional space of features into five classes: Class I, cells showing expansiveness typical only of normal cells; Class II, cells showing average expansiveness and high penetration; Class III, cells with average expansiveness and a high directionality; Class IV, cells showing restricted movement ability; and Class V, for unclassified objects that do not correspond to any of the other classes. FIGURE 2 shows learning set of cell's behavior used in experimental study.[11] The nearest neighbor method (k-NN) is proposed to classify the examined cells' behavior according to the knowledge contained in the learning test.

The distribution of cell behavior codes was used as a background for the new method for final classification of the sample level. Because the distribution of neutrophils codes in normal subjects is normal, the Shapiro–Francio statistical test is proposed to examine if the distribution of cell behavior codes is different from a Gaussian distribution. The proposed methods of classification could easily be adjusted for a particular immunodeficiency, connected with the neutrophils movement ineffectiveness, given a learning set containing observations. For example, the distributions of the behavior of neutrophils, taken from a Chediak–Higashi syndrome patient, exhibit the characteristics of immunodeficiency bias.[11]

MATERIALS AND METHODS

To examine the proposed methods of classification, two specially constructed sets were used: learning and testing sets. Each set contained a description of one-minute of cell behavior, the vector *CMI*. These values were evaluated using a cell monitoring system[6,10] that collected microscopic images in 2–sec time increments, segmented cells, traced their contours, calculated elliptical models of their shapes, measured their areas and distances between two consecutive positions, and based on 30 cell positions, calculated *CMI* for $\Delta t = 1$ minute.

The learning sets, consisting of 25 movement sequences, came from the experimental material collected from normal subjects classified by humane expertise. The testing set consisted of the previously registered experimental material coming from healthy children and the Chediak–Higashi syndrome patient.[1] The sequences of neutrophils behavior were compared with the classification by an operator during the video film examination.

RESULTS

The reclassification error of the k-NN method had the lowest value (8%) for five nearest neighbors examined during the learning test. However, the classification error, calculated on the testing set in comparison with the classification done by an operator, had the higher value of 22%. An error of one time unit in the cells behavior examination should fall in the next step of classification, where the distribution of neutrophils' behavior of all cells in the sample was considered. The sample level of the classification was not verified because of a small number of samples collected. However, the single cell behavior classification was different from the operator classifications in 15% cases.

Two samples, one from healthy children and one from the Chediak–Higashi syndrome patient, were examined. Both show results similar to those obtained in the previous investigations.

CONCLUSION

The proposed methods of classification are independent of the type of the neutrophils ineffectiveness in movement. They can be adjusted to the particular syndrome, according to the introduction of experimental studies. The methods can be also be adjusted to the changes in the behavior time unit in the examination.

REFERENCES

1. SKOPCZYŃSKA, H., A. KORZYŃSKA, J. MICHAŁKIEWICZ, et al. 1997. Immunological abnormalities in a patient with Chediak–Higashi syndrome. Proc. 4th Int. Symp Clinical Immunology, Amsterdam. 49.
2. KORZYŃSKA, A. 1993. A method of cell movement investigation. In Selected Topics in Biomedical Image Processing. J. Kulikowski, Ed.: 71–93. PAS IBBE, Warsaw.
3. HOPPE, A., et al. 1999. A computer system for the analysis of neutrophil movement. Med. Biol. Eng. Comp. 37(Suppl. 2): 1000–1001.
4. ZICHA, D., et al. 1995. An image processing system for cell behaviour studies in subconfluent cultures. J. Microscopy 179: 11–21.
5. KORZYŃSKA, A. 1992. The mathematical background software used in investigation of the movement of leukocytes. Biocybernet. Biomed. Eng. 12: 87–99.
6. KORZYŃSKA, A., A. NECHAY, P. MAZUR, et al. 1997. Computer aided microscopy system in investigation of cell's motility. Proc. 4th Eur. Conf. Engineering and Medicine, Warsaw. 360–361.
7. KORZYŃSKA, A. 2000. The method of neutrophils activity description. medical physics. 27(6): 1389. (Also CD-publication.) Proc. World Congr. on Medical Physics and Biomed Eng., Chicago.

8. KORZYŃSKA, A. 1996. Random walk evaluation for neutrophils granulocytes in physiology and in Chediak–Higashi syndrome using video enhanced microscopy. Proc. Int. Seminar of Statistics and Clinical Practice, Warsaw. 36–39.
9. KORZYŃSKA, A. 1999. Evaluation of neutrophils activity. Abstracts, 5th Conf. Eur. Soc. Eng. and Med., Barcelona.
10. KORZYŃSKA, A. 1998. A method of segmentation of neutrophils images observed in the cell movement monitoring system. Abstracts MEDICN'98. 99. (Also CD-publication ISBN: 9963-607-13-6.) Limassol.
11. KORZYŃSKA, A. 2001. Computer Aided Neutrophil Granulocytes Movement and Shape Assessment. Ph.D. Thesis, Prace Instytutu Biocybernetyki i Inz'ynierii Biomedycznej, 57, ISBN: 0239-7455. In press. (Polish).

An Algorithm for Estimating Chaos in Mechanoemission of Blood and Magnetic Resonance Imaging in Patients with Gastric Cancer

VALERIY E. OREL, ANDREY V. ROMANOV,
NATALIE N. DZYATKOVSKAYA, AND YURI I. MEL'NIK

Institute of Oncology, Physics-Technical Laboratory, Kiev, Ukraine

ABSTRACT: The paper describes an algorithm and computational system used to estimate the spatial chaos of magnetic resonance images of patients with gastric cancer and the quantum chaos of mechanoemission in blood.

KEYWORDS: gastric cancer; MRI analysis; tribology; chaos; electromagnetic; mechanoemission; conformational changes; oncogenesis

INTRODUCTION

Several aspects of nonlinear dynamics have been explored recent years in medical engineering and physics. In general, cancer reflects dynamic and multistage processes. Complex and dynamic systems can be described mathematically by chaos theory. The concept of deterministic chaos is biohierarchical in contemporary ideas about the role of chaos in potential applications to oncology.[1] The use of nonlinear dynamics to diagnose tumors at the imaging level magnetic resonance (MR) has a major impact on our understanding of how cancer develops.[2] The known physicomathematical principles of medical diagnosis provide limited information on the interrelation between visualized information of the electromagnetic field in MR imaging (MRI) and the chaos of electromagnetic mechanoemission (ME) processes that reflect conformation and configuration changes in biological macromolecules.[3,4] This paper describes an algorithm for estimating interrelation ME chaos in blood and spatial chaos in MR images of gastric cancer patients.

MATERIALS AND METHODS

Mechanoemission Studies

We studied ME in two groups of male patients aged 45–60. Group 1 included patients with gastric cancer (22 persons), Group 2, healthy control subjects ($n = 21$). Cancer patients were evaluated at Stage II–IV with the diagnosis confirmed by morphology and histology.

Address for correspondence: V.E. Orel, Institute of Oncology, Physics-Technical Laboratory, 33/32 Lomonosov Street, Kiev, Ukraine 03022.

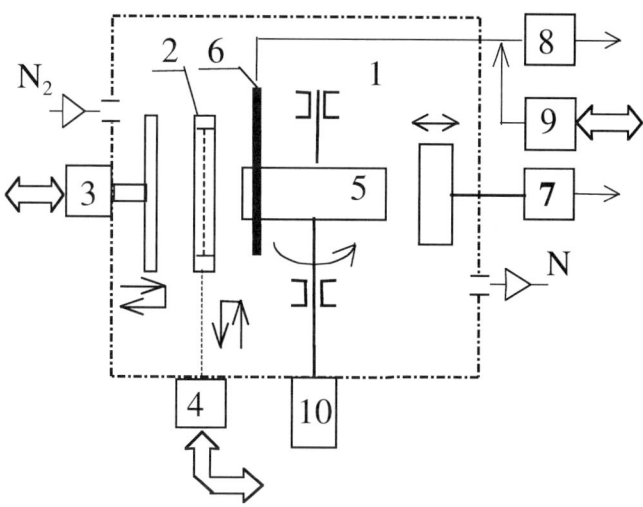

FIGURE 1. Flow diagram of the logger block with the radiosensor: 1, chamber; 2, frame with the investigated sample; 3, device of the frame pressing to the electret probe; 4, device of the frame passing to the chamber; 5, electret probe; 6, aerial; 7, device of the working surface clearance; 8, aerial amplifier; 9, aerial polarizer; 10, electric motor.

Whole blood samples, 0.02 ml, were spread on FN-2 chromatographic paper (Filtrak, Germany), which was placed in the slide frame. The blood was dried at 45% relative humidity and triboelectrified by using a 30 mm diameter rotating Teflon electret probe at 1,400±50 rpm with enhanced mode of 1.2 N in an atmosphere of highly purified gaseous nitrogen under excessive pressure of 1 kPa at 37±0.5°C. Correlation analysis was made to evaluate the quantitative pattern of periodic values in whole blood electromagnetic ME amplitude within 12.5 msec.

The radiosensor of tribogeneration system, with the electret probe, is schematically presented in FIGURE 1. The rolling unit is rotated by an electric motor. The sample frame is moved to the electret probe using a pressing device. The receiver aerial for recording electromagnetic ME generated by the electret probe and the friction of the sample is located near the sample contact zone and the rotating probe. The electrical signals enter the input of the aerial amplifier and, after amplification in the frequency range of 1 Hz to 0.1 MHz, are transmitted to the input of the analogue-digital converter.

MRI Analysis

Understanding chaotic or unstable mathematics in cancer and the interrelation of ME in blood and MR images may lead to similar strategies in the diagnosis of cancer. MR images were acquired using a 0.5 T magnet (GE Medical Systems). MR images reported by Portnoj et al.[5] of healthy and gastric cancer patients were analysed for optical density.

THE ALGORITHM FOR CHAOS ESTIMATION

A pseudophase space method was used for the analysis of the ME chaos of blood and the spatial chaos of MR images.[6] Let $I_0, I_1, \ldots, I_N$ be the ME value measured at discrete times $t_0, t_1, \ldots, t_N$, $t_i = t_0 + i\tau$, $i = 1, 2, \ldots, N$ in the case of estimating ME chaos, or let $I_0, I_1, \ldots, I_N$ be the optical density of MR image pixels $t_0, t_1, \ldots, t_N$, $t_i = i$, $i = 1, 2, \ldots, N$ ordered in a certain manner in the case of estimating spatial chaos. In order to build a continuous phase trajectory, we interpolate values $I_0, I_1, \ldots, I_N$ with a cubic spline $f(t)$, $t_0 \leq t \leq t_N$ of defect 1. The cubic spline $f(t)$ was determined by the conditions

$$\begin{cases} f(t_i) = I_i, \ i = 0, 1, \ldots, N \\ f'(t_0 + 0) = I_1 - I_0 \\ f'(t_N - 0) = I_N - I_{N-1}. \end{cases} \quad (1)$$

The analytical expression for the cubic spline is

$$f(t) = \frac{1}{6}M_{j-1}(t_j - t)^3 + \frac{1}{6}M_j(t - t_{j-1})^3 \\ + \left(I_{j-1} - \frac{1}{6}M_{j-1}\right)(t_j - t) + \left(I_j - \frac{1}{6}M_j\right)(t - t_{j-1}) \quad (2)$$

$$t \in [t_{j-1}, t_j], \ j = 1, 2, \ldots, N,$$

where the coefficients $M_j, j = 0, 1, \ldots, N+2$ are determined by the system of equations

$$2M_0 + M_1 = 0$$
$$M_{j-1} + 4M_j + M_{j+1} = 6(I_{j-1} - 2I_j + I_{j+1}), \ j = 1, 2, \ldots, N-1 \quad (3)$$
$$M_{N-1} + 2M_N = 0$$

This system has a tridiagonal matrix. The sweep method was used to solve it.

The phase trajectory is defined as the multitude

$$\{(x, y) | \ x = f(t + \tau), \ y = f(t), \ t_0 \leq t \leq t_{N+1}\}.$$

Let $\Phi = \{(x, y) | \ x = \varphi(t), \ y = \psi(t), \ t \in [\theta_1, \theta_2]\}$ be the phase diagram of a certain process, functions φ and ψ being continuously differentiable on segment $[\theta_1, \theta_2]$. Let,

$$x_{min} = \min\{\varphi(t) : t \in [\theta_1, \theta_2]\},$$
$$x_{max} = \max\{\varphi(t) : t \in [\theta_1, \theta_2]\},$$
$$y_{min}(x) = \min\{\psi(t) : t \in \varphi^{-1}(x) \cap [\theta_1, \theta_2]\},$$
$$y_{max}(x) = \max\{\psi(t) : t \in \varphi^{-1}(x) \cap [\theta_1, \theta_2]\}.$$

Define the spread parameter $S(\Phi)$ of phase diagram Φ by

$$S(\Phi) = \int_{x_{min}}^{x_{max}} (y_{max}(x) - y_{min}(x)) dx. \quad (4)$$

The physicomathematical interpretation of the parameter S is approximately the area of the figure outlined by the envelope of the phase trajectory.

Definition of the functions $y_{max}(x)$ and $y_{min}(x)$, $x \in [x_{min}, x_{max}]$ lies in finding the preimage set $\varphi^{-1}(x)$ of element x for all $x \in [x_{min}, x_{max}]$, which can involve certain

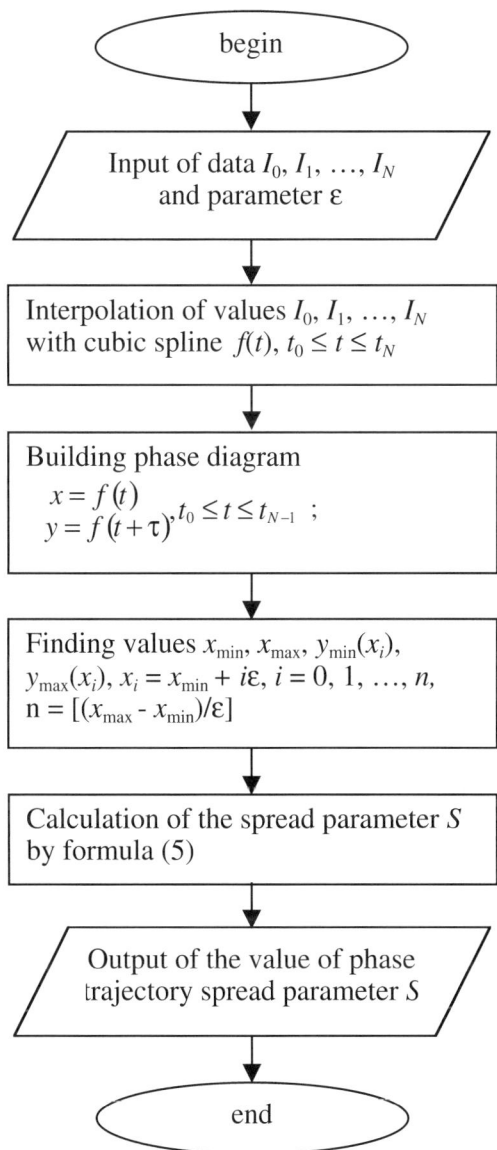

FIGURE 2. The flowchart of algorithm for calculation of phase trajectory spread parameter.

difficulties. To approximately calculate the integral on the right in Equation (4), we use the quadrature rectangle equation:

$$S(\Phi) \approx S_\varepsilon(\Phi) = \sum_{i=0}^{n-1} (y_{max}(x_i) - y_{min}(x_i)) \cdot \varepsilon, \qquad (5)$$

where $n = [(x_{max} - x_{min})/\varepsilon]$, $x_i = x_{min} + i\varepsilon$, $i = 0, 1, \ldots, n$. In this case, finding inverse values of function φ is necessary only in nodes of the partition x_i, $i = 0, 1, \ldots, n$.

The values

x_{min}, x_{max}, $y_{min}(x_i)$, $y_{max}(x_i)$, $x_i = x_{min} + i\varepsilon$, $i = 0, 1, \ldots, n$, $n = [(x_{max} - x_{min})/\varepsilon]$

were calculated using the following method. Sequentially advance along the phase trajectory with step $\Delta t_{(j)} = t_{(j+1)} - t_{(j)}$, $t_0 = t_{(0)} \leq t_{(1)} \leq \ldots \leq t_{(K)} = t_N$, the next phase diagram point $(x_{(j)}, y_{(j)}) = (f(t_{(j)}), f(t_{(j)} + \tau))$ is determined. The value $\Delta t_{(j)}$ is chosen based on the condition $\Delta x_{(j)} = |x_{(j+1)} - x_{(j)}| \leq \varepsilon$. This is the case when $\Delta t_{(j)}$ does not exceed $\varepsilon / |f'(t_{(j)})|$. Indeed,

$$x_{(j)} = |x_{(j+1)} - x_{(j)}| = |f(t_{(j+1)}) - f(t_{(j)})|$$
$$= |f(t_{(j)} + \Delta t_{(j)}) - f(t_{(j)})| = |f'(t_{(j)})| \cdot \Delta t_{(j)} + o(\Delta t_{(j)})$$

To avoid indeterminacy in the case when $f'(t_{(j)})$ is small, let

$$\Delta t_{(j)} = \frac{\varepsilon}{|f'(t_{(j)})| + 100\varepsilon}.$$

Among abscissas $x_{(j)}$, $j = 0, 1, \ldots, K$, we find the minimum x_{min} and the maximum x_{max}. In the retrieved set of phase diagram points $\{x_{(j)}, y_{(j)} | j = 0, 1, \ldots, K\}$, substitute abscissas $x_{(j)}$ by their approximate values

$$x^*_{(j)} = x_{min} + [(x_{(j)} - x_{min})/\varepsilon] \cdot \varepsilon.$$

The value of $x^*_{(j)}$ corresponds to the node x_i, $i = [(x_{(j)} - x_{min})/\varepsilon]$ in the partition $\{x_i | x_i = x_{min} + i\varepsilon, i = 0, 1, \ldots, n, n = [(x_{max} - x_{min})/\varepsilon]\}$.

As long as $\forall 0 \leq j \leq K - 1$ $\Delta x_{(j)} \leq \varepsilon$ and the phase trajectory is continuous,

$$\forall 0 \leq i \leq n \ \exists 0 \leq j \leq K : x_i = x^*_{(j)}.$$

Passing over all numbers $y_{(j)}$ for which $x^*_{(j)} = x_i$, we find numbers

$y_{min}(x_i)$, $y_{max}(x_i)$, $x_i = x_{min} + i\varepsilon$, $i = 0, 1, \ldots, n$, $n = [(x_{max} - x_{min})/\varepsilon]$.

A special algorithm was developed for the calculation of the blood ME phase trajectory spread parameter S. The flowchart of the algorithm for estimating the phase trajectory spread parameter is given in FIGURE 2.

RESULTS AND DISCUSSION

FIGURE 3 shows phase diagrams of ME of blood. It can be seen that cancer patients have a greater spread parameter S than healthy individuals. Our findings allow viewing oncogenesis as an expression of impairments of molecularly determined chaos of homeostasis.

T1-weighted MR images of the stomach are shown in FIGURE 4. Analysis of the research results of T1-weighted MR images is shown in FIGURE 5. As can be seen, the cancer patients exhibited a higher distribution of the spread parameter S range

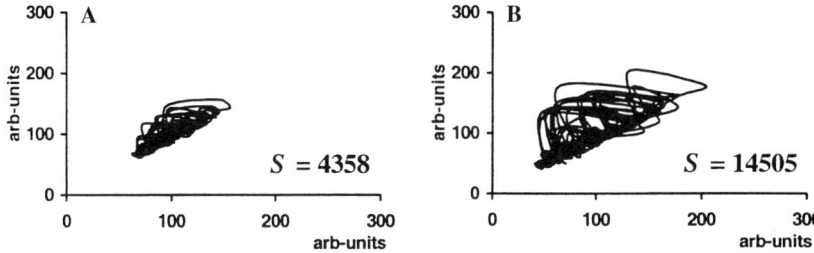

FIGURE 3. Phase diagram of ME patients: **(A)** healthy individuals; **(B)** gastric cancer patients.

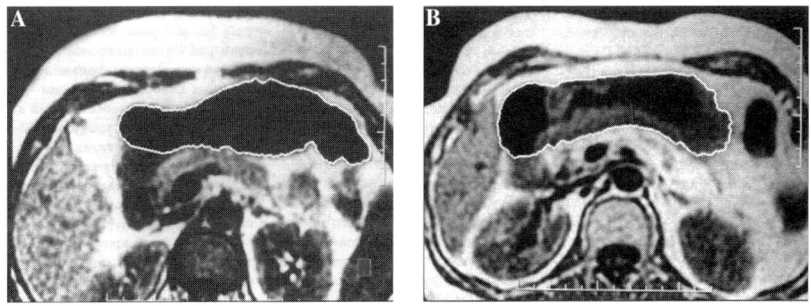

FIGURE 4. Ti-weighted MR images of the stomach: **(A)** healthy subject; **(B)** cancer patient.

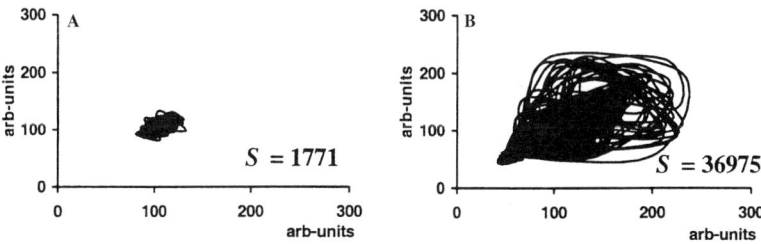

FIGURE 5. Phase diagram of MR images of the stomach: **(A)** healthy subjects; **(B)** cancer patients.

than healthy individuals. These results also permit us to consider oncogenesis as a kind of an impairment of the homeostatic deterministic chaos at the organ level.

SUMMARY

As shown here, the blood ME dynamics during a malignant tumor process is characterized by higher chaos than that of healthy individuals. These results allow to consider oncogenesis as a kind of impairment of a homeostatic deterministic chaos at the molecular level. The MR images of cancer patients also exhibited higher values within the spread parameter S range than healthy individuals. This phenomenon permits us to consider oncogenesis as a kind of increase of the homeostatic deterministic chaos at the organ level. It can, thus, be suggested that there is a contrast between normal tissue and the distinct variability of tumor tissue, which tends toward chaos. The diversity (heterogeneity) of tumor cells results from genetic instability.[7] The existence of significant unbalanced ME of blood and MR images of cancer patients, as demonstrated by the algorithm for estimating chaos in cancer patients, testifies to the particular features of oncogenesis chaos, and may find application in the development of new medical diagnostic equipment. Utilizing the algorithm to estimate the chaos of blood ME parameters and the spatial chaos of MRI shows a potential for estimation and follow-up of the prognosis of treatment efficiency of patients with gastric cancer.

ACKNOWLEDGMENT

The authors would like to thank Prof. V. Chornyi for his assistance with this paper.

REFERENCES

1. SEDIVY, R. & R.M. MADER. 1997. Fractals, chaos, and cancer: do they coincide? Cancer Invest. **15**(6): 601–607.
2. GODDARD, P., J. HOGAN, C. COLLINS & E. SHEFIELD. 1996. Fractal patterns and medical images. Develop. MR **2**(1): 22–29.
3. PRIGOGINE, I. & I. STENGERS. 1994. Time, Chaos, Quantum. Moscow (in Russian).
4. OREL, V.E. & N.N. DZYATKOVSKAYA. 2000. Mechanoemission of blood and oncogenesis. Biophotonics and coherent systems. L. Beloussov., R. Van Wijk. (Eds.), Moscow University Press. 347–363.
5. PORTNOJ, L.M., L.B. DENISOVA & G.A. STASHUK. 2000. Nefedova B.O. Magnetic resonance tomography in the diagnosis of gastric cancer. X-ray versus MRI anatomic findings, Russian J. Radiobiol. **1**: 26–40.
6. MOON, F. 1987. Chaotic Fluctuation. John Wiley & Sons, Inc.
7. COFFEY, D.S. 1998. Self-organisation complexity and chaos the new biology for medicine. Nature Med. **4**: 882–885.

Microscopic Study of Crystal Growth in Cryopreservation Agent Solutions and Water

LE-REN TAO AND TSE-CHAO HUA

Institute of Refrigeration and Cryogenics,
Shanghai University of Science and Technology, Shanghai, China

> ABSTRACT: Ice formation inside or outside cells during cryopreservation is evidently the main factor of cryoinjury to cells. In the study described here a high voltage DC electric field and a cryomicroscopic stage were used to test DMSO and NaCl solutions under electric field strengths ranging from 83 kV/m to 320 kV/m. Dendritic ice crystals became asymmetric when the electric field was activated. This change in the ice crystal shape was more pronounced in the ionic NaCl solution. In addition, ice growth of distilled water without an electric field was tested under different cooling rates.
>
> KEYWORDS: cryopreservation; crystal growth; electric field; microscopy

INTRODUCTION

During the cryopreservation procedure, the preserved biomaterials undergo cooling and thawing phases. Usually, the preserved cells/tissues are cooled together with the cryopreservation agent (CPA) solution and are kept in liquid nitrogen at a temperature of −196°C for long-term preservation. Unavoidably, ice crystals form and grow when the cells/tissues are cooled. During the thawing procedure, when the biomaterials are removed from the liquid nitrogen, the energy of the water molecules increases as the temperature goes up, the ice embryos in cells/tissues grow again and form visible ice crystals. It is generally believed that the survival of cell tissues is directly affected by the extra- and intracellular ice, as well as the so-called "solution damage" during the cryopreservation processes.[1–3] The reduction or elimination of ice formation has been the main concern of cryobiologists for some years now.

Hanyu *et al.*[4] introduced microwave irradiation at 2.45 GHz during which the test samples were cooled by so called metal-contact freezing. Compared with non-irradiated samples, the irradiated samples had a deeper ice-free layer. The mechanism of the microwave effect was assumed to be one in which electromagnetic radiation broke the H_2O molecule chain and thereby disrupted crystal nucleation. A similar technique was later used by Jackson *et al.*[5] to identify the behavior of CPA solutions after the treatment with microwave irradiation. A larger vitreous-like region was found in pictures of irradiated samples than in non-irradiated samples. The frequency of the microwaves used in these two experiments was the same as that

---

Address for correspondence: Le-Ren Tao, Ph.D., Institute of Refrigeration and Cryogenics, Shanghai University of Science and Technology, 516 Jungong Rd., Shanghai, 200093, China.
taurusq@public4.sta.net.cn

used in a domestic microwave oven, which agitates water molecules and kills the cells. A DC electric field was used in this study to avoid this danger.

MATERIALS AND METHODS

The CPA samples used in this study were DMSO and NaCl aqueous solutions. The concentration of the solutions was 5% (mass). Distilled water was also tested without an electric field. The test samples were held between two microscopic glass coverslips, forming a very thin liquid film for which nearly two-dimensional ice crystal growth could be observed. One pair of electrodes was placed in parallel at each end of the test sector. The electrodes were made from copper and their surfaces were carefully polished to avoid possible occurrence of sparks when the high voltage was applied. To measure the cooling rates during the tests, two pairs of copper–contantain thermocouples were glued to the surface of the upper coverslip. Because metal thermocouples cannot work properly under the influence of electric field, the temperature was measured only when the electric field was off. A unique temperature distribution was obtained when measuring the temperatures at various places on the coverslip surface.

A special apparatus was built for a DC field working with a cryomicroscopic system. Low temperature nitrogen gas was used to cool the test samples. The gas, which comes from a high-pressure vessel through a pressure regulator and a flowmeter, was cooled by a plate-fin heat exchanger that was immersed in a liquid nitrogen (LN_2) Dewar. The lowest temperature the gas could reach was $-160°C$ in this experiment. By using the pressure regulator together with the flowmeter, the gas flow, and hence the cooling rates, could be controlled.

RESULTS AND DISCUSSION

Freezing of DMSO Solution

The images of heterogeneous ice growth without electric field are shown in FIGURE 1. A very tiny piece of ice, left after melting of the preliminary crystallization, was used as the embryo in this case (first picture). The round seed turned into an hexagonal crystal (second picture), developing into the symmetric hexagonal dendritic ice crystal growth shown in the remaining pictures.

FIGURE 2 represents the morphologic pictures of ice crystal when the DMSO solution was frozen under the influence of the electric field. The direction of the electric field was vertically downward. All of these pictures were taken at same cooling rate and the ice crystals were grown from a very tiny embryo. Initially, the ice crystal growth was the same as in FIGURE 1. The round embryo turned to a hexagon with time. FIGURE 2(3) shows clearly that two main branches parallel to the electric field grew faster than the other branches. This asymmetric ice growth can also be seen in FIGURES 2(4) and (5).

The effects of the electric field on ice crystal growth were discussed numerically by Svishchev and Kusalik.[6–7] Their computational experiments of water freezing suggest that an electric field as strong as 105 kV/m is needed to change the ice lattice

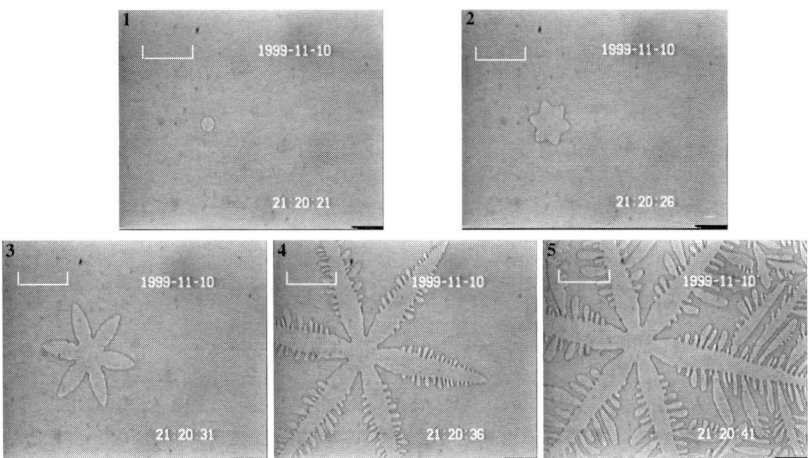

FIGURE 1. Ice growth in 5% (mass) DMSO solution without an electric field. (*Scale bars*, 0.1 mm.)

from normal ice (I_h) to cubic ice (I_c). However, the strength of electric field used in our experiments was only 1/400 of the computed value and does not seem that the ice morphology change in our study is caused by an ice lattice change. In our experiments, the electric field may cause various molecules in DMSO solution to have different characteristics. The polar water molecules/clusters may be torqued and rearranged under the influence of the electric field. The crystal growth procedure can be regarded as a process during which molecules are added on the crystal lattice. This process has equal probability for addition in all directions in the normal case.

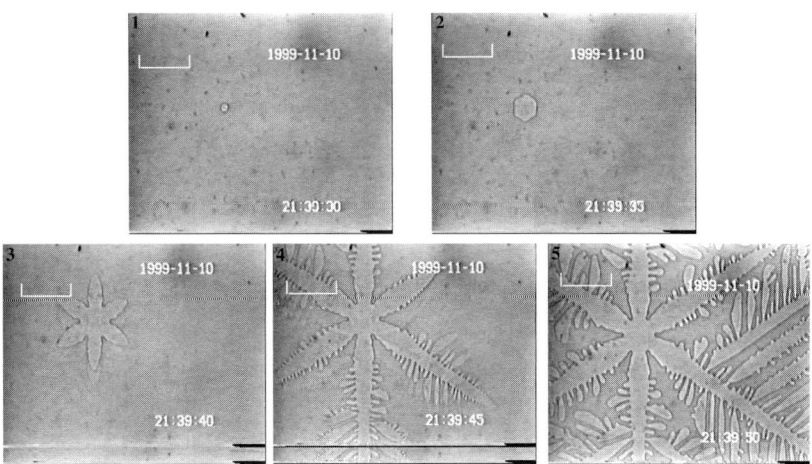

FIGURE 2. Ice growth in 5% DMSO solution under 250 kV/m. (*Scale bars*, 0.1 mm.)

When a high voltage field is applied, the water molecules/clusters are forced to join the lattice in a special orientation and position. The net result is that different growth rates may occur in various directions.

Under the action of an electric field the molecules may rearrange and line up with shared edges. From the point of view of a crystal structure, the well-ordered water molecules/clusters may seem crystal-like or quasicrystalline. In this case, the water molecules/clusters present an ideal situation for rapid crystal growth. This may be the reason why the main branches, which are parallel to the direction of the electric field, grow faster than the other branches shown in FIGURE 2.

Freezing of NaCl Solution

An ice crystal forming in the NaCl solution under the effect of an electric field strength of 320 kv/m at a relatively slow cooling rate is shown in FIGURE 3. The time interval between each picture is about 20 seconds. A more asymmetric crystal growth than that shown in FIGURE 3 can be clearly observed.

The relatively high electric field used here could be a reason for the ice morphology change. An alternative explanation is that the NaCl solution, unlike DMSO solution, is an ionic solution with numerous cations and anions. These positive and negative ions perform completely different movements when the electric field is applied. The ions are separated and gather near the corresponding electrodes. This yields a non-uniform ionic distribution in the solution, and different crystal growth rates can be expected in different regions of the solution, since the sizes of Na^+ and Cl^- are quite different: The radius of Cl^- is much bigger than that of Na^+. This may lead to a rather different effect on the ice morphology.

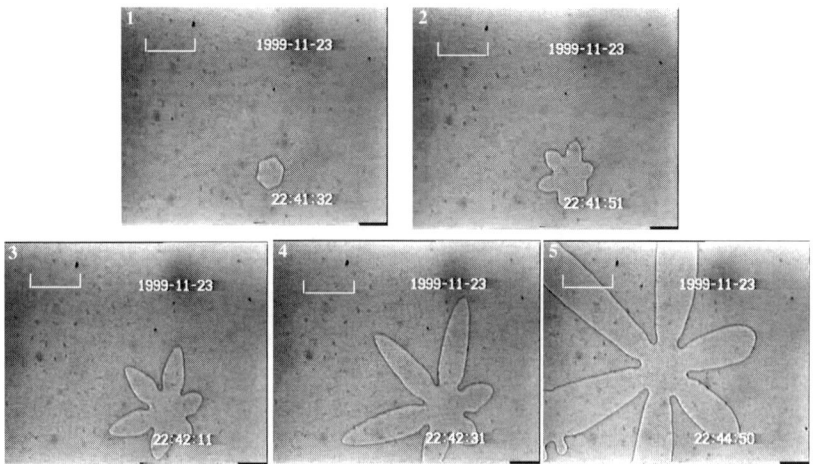

FIGURE 3. Ice crystal growth in 5% NaCl solution under 320 kV/m. (*Scale bars*, 0.1 mm.)

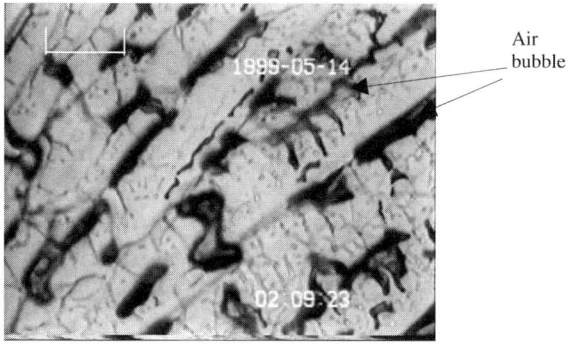

FIGURE 4. Typical freezing image for a water concentration that stops continued formation of an air bubble. (*Scale bar*, 0.1 mm.)

Freezing of Distilled Water

The air bubbles formed during the freezing of water have attracted a number of studies. Carte,[8] as well as Bari and Hallett,[9] discussed the formation air bubbles in ice, hoping that the study can help explain the formation of hailstone. Kõrber et al.[10] found that air bubbles were formed in cooling cell suspensions. This is harmful in the preservation of cells and tissues, since the existence of air bubbles may cause different stress distributions in the ice and lead to the braking of ice, which may kill the preserved cells.

Twice distilled water was used in all the experiments. The water was saturated with air in room temperature of 30°. FIGURE 4 shows a typical crystallization image for water. The black dots in the picture are air bubbles. Air bubbles appear at the ice-water boundary when the ice crystals grow and the leading edge of the ice expels the air molecules that were originally dissolved in the water. The expelled air molecules concentrate at the leading edge and eventually become supersaturated. The air bubbles are formed when the concentration of air reaches a critical value. The formation of the bubble causes a sharp decrease in the local air concentration. If ice crystal growth continues, the same process is repeated until the end of crystallization. This leads to periodic bubble formation. Consequently, the bubbles are proportionally spaced and have nearly the same size. Because of the density difference between water and air, the air bubbles are seen to be floating on the ice surface.

The main parameter that influences bubble size and distribution is the cooling rate. When the cooling rate increases, the ice growth rate also increases. FIGURE 5 demonstrates a sequence of the bubble formation process at a cooling rate of 3.4 K/sec (frames 1–5). As seen, the air bubbles are equally distributed throughout the region. In addition, some 'threads" can be seen between different pieces of ice. These threads are considered to be made of tiny bubbles that may not have had time to coalesce into big bubbles because the quick motion of the ice frontier entrapped them locally. More tiny bubbles were found when the cooling rate was increased. Indeed, no big bubbles were found when the cooling rate was about 8.2 K/sec (FIG. 5, frame 6) where the fast growth of ice quickly captures the expelled air.

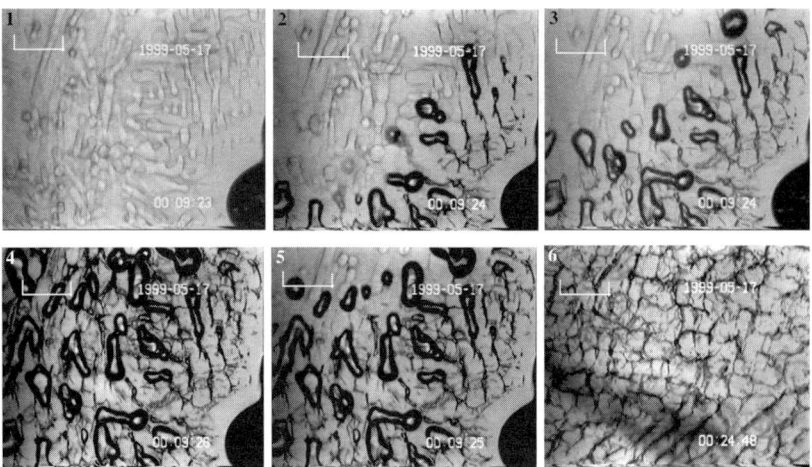

FIGURE 5. Air bubble formed in water at various cooling rates (frames 1–5 at 3.4 K/sec and frame 6 at 8.2 K/sec). (*Scale bars*, 0.1 mm.)

CONCLUSIONS

The morphology of freezing 5%DMSO and 5%NaCl solutions in DC electric field was studied. Electric fields of 250 kV/m were added to the DMSO solution. At a field strength of 250 kV/m, two main branches parallel to the field direction grew faster than others. Adding an electric field may force the water molecules/clusters to rearrange and align, causing the ice growth rate to differ in various directions. Freezing a NaCl solution in a 320 kV/m high electric field showed a more asymmetric ice crystal growth. Ion separation due to the electric field was probably the main reason for the observed phenomena. The formation of ice crystals in water expels the air out of the ice and forms visible air bubbles. The cooling rate was found to be the main factor affecting air bubble size and distribution. At higher cooling rate, the ice crystals grow fast and only small air bubbles are formed.

ACKNOWLEDGMENT

The research is supported by China Natural Science Foundation (No. 50076028).

REFERENCES

1. MAZUR, P. 1977. The role of intracellular freezing in the death of cells cooled at superoptimal rates. Cryobiology **14:** 251–264.
2. WANG, D.R., T.C. HUA & S.Y. WU. 1988. Influence of unfrozen fraction to the survival of leucocytes during slow cooling. Cryobiology **25:** 510–514.
3. ISHIGURO, H. & B. RUBINSKY. 1994. Mechanical interactions between ice crystal and red blood cells during directional solidification. Cryobiology **31:** 483–500.

4. HANYU, Y., et al. 1992. An improved cryofixation method: cryoquenching of small tissue blocks during microwave irradiation. J. Microsc. **165:** 225–235.
5. JACKSON, T.H., et al. 1997. Novel microwave technology for cryopreservation of biomaterials by suppression of apparent ice formation. Cryobiology **34:** 363–372.
6. SVISHCHEV, I.M. & P.G. KUSALIK. 1994. Crystallization of liquid water in a molecular dynamics simulation. Phy. Rev. Lett. **73**(7): 975–978.
7. SVISHCHEV, I.M. & P.G. KUSALIK. 1998. Electrofreezing of liquid water: a microscopic perspective. J. Am. Chem. Soc. **118:** 649–654.
8. CARTE, A.E. 1961. Air bubbles in ice. Proc. Phys. Soc. **77:** 757–768.
9. BARI, S.A. & J. HALLETT. 1974. Nucleation and growth of bubbles at an ice-water interface. J. Glaciol. **13:** 489–520.
10. KÖRBER, CH., S. ENGLICH, P. SCHWINDKE, et al. 1985. Low temperature light microscopy and its application to study freezing in aqueous solutions and biological cell suspensions. J. Microsc. **141:** 263–276.

Temperature and Flow Visualization in a Simulation of the Czochralski Process Using Temperature-Sensitive Liquid Crystals

JELENA ALEKSIC, PAUL ZIELKE, AND JANUSZ A. SZYMCZYK

Department of Thermofluid-dynamics and Turbo Machines, University of Applied Sciences of Stralsund, Stralsund, Germany

ABSTRACT: There is no doubt today that thermal and thermocapillary convection play a dominant role in momentum, heat, and mass transfer in the Czochralski crystal growth method. Because of the complexity of the problems, measurements in one point of the volume are not sufficient to illuminate the flow topography or to compare the experimental results with real or numerically simulated data. Therefore, it is of great interest to measure the temperatures and velocities in the whole field in order to qualitatively analyze thermally driven convection. The new experimental particle image thermometry method, based on computer-aided color analysis of the TLCs reported here, enables the simultaneous determination of the temperature and velocity fields.

KEYWORDS: Czochralski process; particle; particle image thermometry; particle image velocimetry; liquid crystals; thermometry; temperature-sensitive liquid crystals

INTRODUCTION

In the Czochralski crystal growth method,[1] a polycrystalline structure is melted in a crucible at a temperature that is a little higher than the melting point. A germ bud is then brought to the surface of the melt. The melt solidifies on the germ bud and creates a monocrystalline structure. The growth of the monocrystal into the melt is prevented by pulling the monocrystal slowly and uniformly at 0.1 to 100 mm/h, and rotating it at 1 to 100 rpm. A regular crystal rod is built up and new layers are attached to it. To increase the quality of the crystal and to optimize the process, it is important to investigate the types of convection that occur within the operating range of parameters and conditions under which transient convection appears.

The Czochralski process, as we know it today, was at first described by Dash[2] at the 1958 Conference on Crystal Growth. Most reported investigations since then have been of an experimental and theoretical nature. Numerical simulations of the most widely used semiconductors have already been made. Ferland *et al.*[3] obtained the temperature and velocity fields for ice-crystal growth with the

---

Address for correspondence: Jelena Aleksic, Dipl. Ing., Department of Thermofluid-dynamics and Turbo Machines, University of Applied Sciences of Stralsund, Zur Schwedenschanze 15, 18435 Stralsund, Germany.
  jelena.aleksic@fh-stalsubd.de

Czochralski method by using TLCs. The surface of the crystal is strongly influenced by the rotation rates, especially by the crystal rotation rate. The crystal surface is highly convex (almost like a peak) in the direction of the melt at high crucible rotation rates. The crystal surface becomes convex at the edge and concave in the middle as the rotation rate of the crystal is raised. The surface is destroyed and no growth is possible at extremely high rotation rates. The surface is mostly convex at large aspect ratios. Some numerical and experimental studies have shown that the angular velocity of the convection roll in the melt is almost the same as the rotation rate of the crucible.[4] This indicates that the flow in a silicon melt is not quite turbulent, as was previously thought. The melt becomes homogeneous and the crystals purer as the melt is stirred.[5]

EXPERIMENTAL SETUP

Our experimental equipment was designed to best match the original Czochralski process.[1] To provide the thermal and kinematic boundary conditions, the experimental equipment allowed the temperatures of the stick, the surrounding fluid bath, and the measuring cubicle bottom to be set. It is also possible to heat the upper margin of the experimental fluid, which is considered to be a modification of the real process. In addition, it is possible to rotate the measuring cubicle and the stick independently at variable speeds and rotation direction. Glycerine–water mixtures and silicone oil were used as experimental fluids. TLCs were put in the experimental fluid. These little particles change color with temperature (highlighting the temperature field), and act as tracers for velocity measurements.

Optical Flow Measurement Systems

The aim of the measurement systems associated with particle image thermometry (PIT) is a quantitative determination of the temperature map in a convective flow field. PIT is based on the property of certain TLCs, with a diameter of 10–100 μm, to reflect light of various colors depending on their temperature and the observation angle. This property was used in the Czochralski method simulation together with particle image velocimetry (PIV). In PIT, the light sheet from the white light source was recorded by a digital camera. The light source for a 1.5 mm thin light sheet was a halogen lamp. The evaluation methods and their software were successfully developed in house. The same TLCs were used in PIV as tracers, to determine the flow velocity.

RESULTS

It appears that three kinds of convection in the liquid are involved in the process at hand: (1) Marangoni convection—thermocapillary flow, due to the liquid–air interface stress; (2) thermal convection, due to density variation with temperature in the fluid layers, and (3) forced convection, due to the rotation.

The effects of changing the aspect ratio, the Prandtl (Pr), Marangoni (Ma), or the rotatory Froude (Fr) numbers on the flow topography could be identified by changing the thermal and kinetic boundary conditions and determining the temperature and velocity fields. FIGURE 1 shows the influence of Pr on the temperature and velocity fields. In the case of a very large Pr value (FIG. 1A) the temperature field is quasisymmetric. A large Ma value suggests strong boundary driven convection. It is also an indicator for entranced heat transfer with decreasing Pr and, consequently, larger warm flow spread through the volume (FIG. 1D). In the case of a very large Pr value (FIGS. 1B and C), a strong Marangoni convection built up and the velocity field was laminar and rotationally symmetric. A swiftly, cold, downward "jet" flow seemed to form in the middle of the cubicle and was reinforced by the thermal convection. The flow was steady and stable with a stable topography. A reduction in Pr resulted in the destruction of the stability and flow symmetry (FIGS. 1E and F). A secondary temperature gradient formed, which destroyed the stable density layers, and a second convection roll formed. FIGURE 2 shows a sample of the horizontal velocity profile, u, at various cubicle heights, H, and cubicle radii, r. In the left side of the experimental cubicle the horizontal velocity values (negative cubicle radius r) were positive in the upper part and negative in the lower part of the cubicle. This resulted from thermocapillary flow that was caused by a temperature gradient on the surface of the fluid and subsequent fluid flow from the edge to the middle of the cubicle. FIGURES 3A and C show the form of the boundary layer for the same configuration. The thickness of the boundary layer was determined graphically with the help of horizontal velocity profiles. The convection roll changed direction at the point where the horizontal velocity value was zero (where it cut the ordinate axis), and the value of the boundary layer thickness could be read for the particular cubicle radius. The

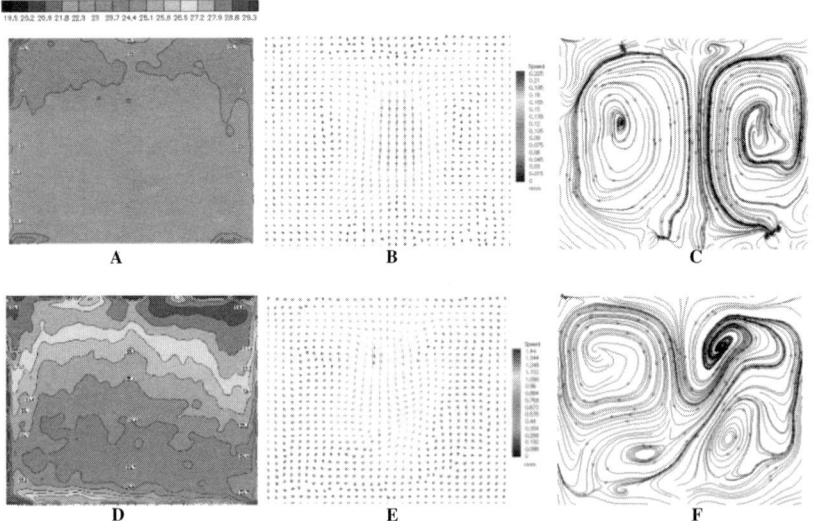

FIGURE 1. Boundary conditions (dimensionless numbers) with the appendant temperature and velocity fields and streamlines: (**A** and **D**) temperature fields; (**B** and **E**) velocity fields; (**C** and **F**) streamlines. (**A–C**) $Ma = 591$, $Pr = 8,227$, $Re = 0.07$. (**D–F**) $Ma = 43,020$, $Pr = 40$, $Re = 1,841$.

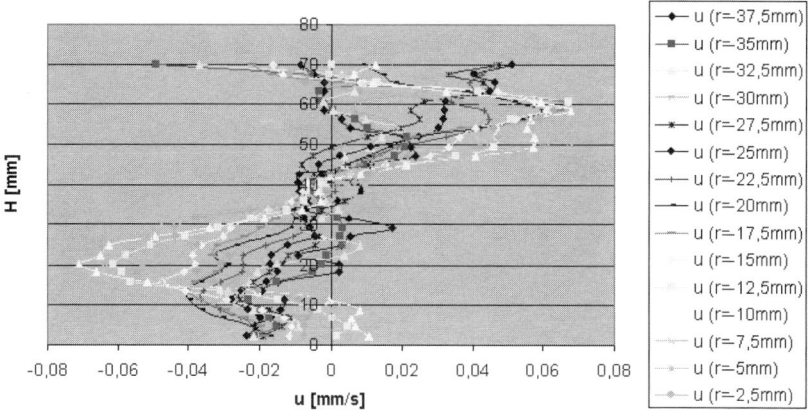

FIGURE 2. Profile of the horizontal velocities from $r = -2.5$ mm to $r = -37.5$ mm for $Ma = 591$, $Pr = 8,227$, and $Re = 0.07$.

process was repeated for every cubicle radius and a diagram of boundary layer thickness was produced. As the cold stick was dipped at the surface of the fluid, there was no free surface in this part of the cubicle and no boundary layer could be formed. The boundary layer was significantly thicker in the configuration with small Ma and larger Pr (FIG. 3C). FIGURE 4 shows the growth of the boundary layer thickness at larger Pr values. The growth of the boundary layer thickness had a positive influence

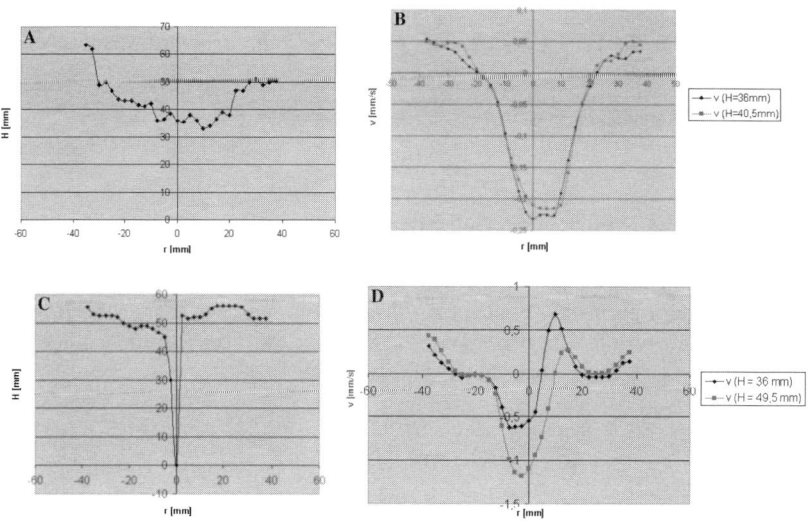

FIGURE 3. (**A** and **C**) Form of the boundary layer; (**B** and **D**) profile of the vertical velocities.

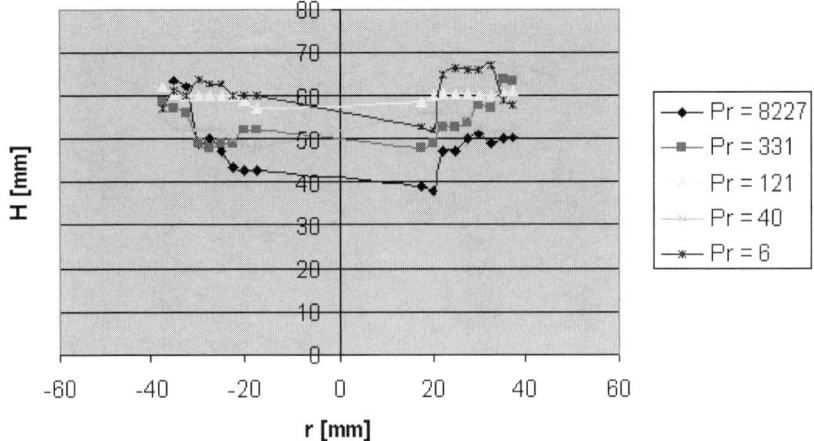

FIGURE 4. Boundary layers for various Pr values.

on the stability of the flow. The thinner boundary layer was more unstable and easier to destroy, for example, at cubicle rotations, as was seen in other experiments of ours not presented in this paper.

The vertical velocity profiles (v) are presented for half the cubicle filling height (36 mm) and for the height of the center of the convection rolls (40.5 and 49.5 mm, respectively) (FIG. 3B and D). The profile is symmetric at large Pr. The values of velocity were very small in the middle of the cubicle marking the cold jet, which had a thickness of approximately 40 mm at both heights. The velocity profile was asymmetric at smaller Pr. The cold jet had a much larger velocity, but it was thinner and unstable.

SUMMARY

The flow was laminar for all investigated values of Pr, Ma, and Re. and the temperature and velocity fields were coupled. The convection flows determined the heat transfer. Instability of the flow can be expected at small Pr with an unstable devolution and a permanently changing flow topography. An increasing Re value also yielded this result. It has yet to be proven if this flow turns into some other mode and if it leads to an oscillatory buoyancy driven flow.

ACKNOWLEDGMENT

The authors would like to thank the Federal Ministry of Education and Research, for promoting the project, under number: 1704598. The third author would like to dedicate this work to Professor Siekmann, who nurtured his scientific development for years.

REFERENCES

1. CZOCHRALSKI, J. 1918. Ein neues Verfahren zur Messung der Kristallisationsgeschwindigkeit der Metalle. Z. Physik. Chem. **92:** 219–221.
2. DASH, W.C. 1958. The growth of silicon crystals free from dislocations. Proc. Intl. Conf. on Crystal Growth. Cooperstown, New York. 361–385.
3. FERLAND, M., D. MISHRA & V. PRASAD. 2000. Study of solidification in a simulated Czochralski system via liquid crystal tracers. Proc. NHTC2000 12050, 34th National Heat Transfer Conf, Pittsburgh, Pennsylvania, USA.
4. KAKIMOTO, K. 1995. Heat and mass transfer in semiconductor melts during single-crystal growth processes. J. Appl. Phys. Rev. **77**(5): 1827–1842.
5. KOKH, A. 1998. Crystal growth through forced stirring of melt or solution in Czochralski configuration. J. Crystal Growth **191**(4): 774–778.
6. ALEKSIC, J., J.A. SZYMCZYK, A. LEDER & T.A. KOWALEWSKI. 2001. Experimental investigations on thermal, thermocapillary and forced convection in Czochralski crystal growth configuration. CMEM 2001, Alicante, Spain, Computational Methods and Experimental Measurements X: 627–636. WIT Press.
7. ALEKSIC, J., P. ZIELKE, J.A. SZYMCZYK & A. DELGADO. 2000. Diagnose des Geschwindigkeits- und Temperaturfeldes in einer Czochralski Simulation, 8. Fachtagung "Lasermethoden in der Strömungsmeßtechnik—neuere Entwicklungen und Anwendungen" GALA 2000 Tagungsband, 37. Shaker Verlag.

Visualization of Droplet Boiling on Heated Transparent Solid Surfaces with Various Thermal Properties

SHIGEAKI INADA

Department of Mechanical System Engineering, Gunma University, Kiryu, Japan

ABSTRACT: The purpose of the research described in this paper is to elucidate the phenomenon of miniaturization boiling, which involves the scatter of a large number of minute liquid particles from a droplet surface to the atmosphere when the droplet impinges on a solid heating surface. This behavior was photographed from the underside of the heating surface, with special attention to the liquid–solid contact. Transparent solid surfaces with various thermal properties included sapphire, glass with indium tin oxide coating, quartz, and quartz with chromium coating.

KEYWORDS: miniaturization boiling; impinging droplet; transparent heater; liquid–solid contact frequency; heterogeneous nucleation; microbubble

INTRODUCTION

A technique for observing the boiling behavior of a droplet that impinges on a heated solid quartz surface from the backside of the quartz, is one of the reported methods[1,2] for observing the state of the liquid–solid contact surface. Nishio and Hirata[2] reported that a liquid–solid contact surface (wetting surface) formed on a quartz surface, even if it was in the high temperature range at which the spheroidal state of the drop was maintained on the solid surface. They also reported that, similar to nucleate boiling, a dry spot, regarded as an attached spot of a vapor bubble, appeared on the quartz surface and that liquid–solid contact situations on quartz and metal surfaces were very different. Inada and Yang[3,4] reported that there are two distinguished types of splashing behavior of liquid microglobules depending on the heated surface temperature. One is nucleate-boiling type splashing, caused by the generation of a *small-size vapor bubble*, as observed at the maximum evaporation rate point. The other type of splashing involves the vapor–liquid interfacial pressure difference following the formation of a *microbubble* inside the drop. A violent dispersion of minute liquid clusters that are much finer than those seen in the generation of the small size vapor bubble, is observed. The former takes place when the contact duration is equal to or greater than the drop residence time, whereas the latter occurs within the residence time, but at a higher heated surface temperature range than the former. The dispersion phenomenon of the droplet is not recognized at heated surface temperatures between these two ranges. The drop is in a state of *latent heat*

Address for correspondence: Shigeaki Inada, Ph.D., Dept. of Mechanical System Engineering, Gunma University, 1-5-1, Tenjincho, Kiryu 376-8515, Japan.
inada@me.gunma-u.ac.jp

transport and is characterized by the appearance of tiny bubbles throughout the entire drop. Therefore, it can be assumed that the phenomenon in the second type is induced by instant vaporization (explosive boiling) resulting from a direct contact between part of the drop and the heated surface.

The purpose of our study using the droplet impinging boiling system is to observe the difference between the generation behavior of the *small-size vapor bubble*, which arises in the initial liquid–solid contact surface (wetting surface), and the generation behavior of the *microbubble*. However, as the aforementioned researchers reported, it is difficult to identify the difference between *small-size vapor bubbles* and *microbubbles* when a heating quartz surface and a transparent solid surface with a thermal property other than quartz is used.

Inada *et al.*[5] reported that miniaturization boiling, which originates from the generation of the *microbubble*, shows a higher heat transfer characteristic than the type with the generation of *small-size vapor bubble*, and is capable of cooling the blanket or divertor plates in a nuclear fusion reactor. Therefore, it is of great interest to learn how this phenomenon can be stably and steadily maintained. Zhang and Yang[6] reported various aspects of drop evaporation on the heated surface and concluded that the mechanisms for triggering drop explosion include the spontaneous nucleation bubble growth phenomena and the destabilization of film boiling.

To clarify why the generation of this microbubble phenomenon brings high heat removal ability, this research focuses on the heterogeneous nucleation and the collapse of the thin vapor film, which seems to trigger miniaturization boiling.

EXPERIMENTAL SETUP

The various heating surfaces used in the boiling experiments were: (1) an artificial white sapphire of 30 mm diameter and 15 mm thickness; (2) an indium tin oxide (ITO) conductive film plane of 0.15 μm thickness sputtered on the surfaces of glass plate 1.8 mm thick; and (3) a chromium coated plane of 0.15 μm thickness sputtered on a quartz cylinder of length 40 mm and diameter 50 mm. These heating blocks were inserted into the center of a heating cylindrical copper block in which cartridge-type heaters were built, at equal intervals in the circumferential direction, so that the block could be uniformly heated from its periphery by heat conduction. The ITO-heating surface was heated by putting it on the quartz heating cylinder. The upper surface of the heating block, onto which droplets were vertically dropped in the center, was maintained horizontal. Degassed and distilled water droplets at a temperature of 16°C and 3.5 mm in diameter were dropped from a height of 65 mm. The Weber impingement number was $We = 61$, and the free vibration period τ_r was about 19 msec. The temperature of heating surface was measured by attaching the thermocouple with ceramic adhesive at two points on the surface. The behavior of droplets impinging on the heating surface was photographed in real time with a high-speed video camera from the underside of the heating block.

The static solid–liquid contact angle was measured at room temperature, by placing a water droplet on the surface of the quartz, sapphire, ITO film plane, and chromium-coated plane. The measured contact angles were 21°, 30°, 30°, and 71°.

RESULTS AND DISCUSSION

FIGURE 1 indicates the lifetime of vaporizing droplets in relation to the initial surface temperature of the heating block (T_{w0}). A stopwatch and a high-speed camera were used to measure the lifetime. A quartz plane with a small thermophysical property shows that the period with the short lifetime expands to the high temperature region, and the maximum evaporation rate point is not clearly evident. The dashed line in FIGURE 1 indicates the liquid–solid contact time (τ_t) at the nucleate boiling range, which was calculated by Makino and Michiyoshi,[7] accounting for the thermal property of metals, diameter of droplet, and heating surface temperature. We adopted this contact time as the lifetime (τ_e) related to the diameter of the droplet and the surface temperature of quartz. The evaporation lifetime property of nuclear boiling (boiling in condition whereby the heated surface and liquid film sufficiently contact each other) in the ITO film plane and the chromium coating plane is similar to the data obtained in the quartz plane. As is evident from this figure, a sapphire plane with 4.9 times the thermal property value of quartz, approaches the lifetime curve of the metal surface, and the heating surface temperature range shifts to the low temperature

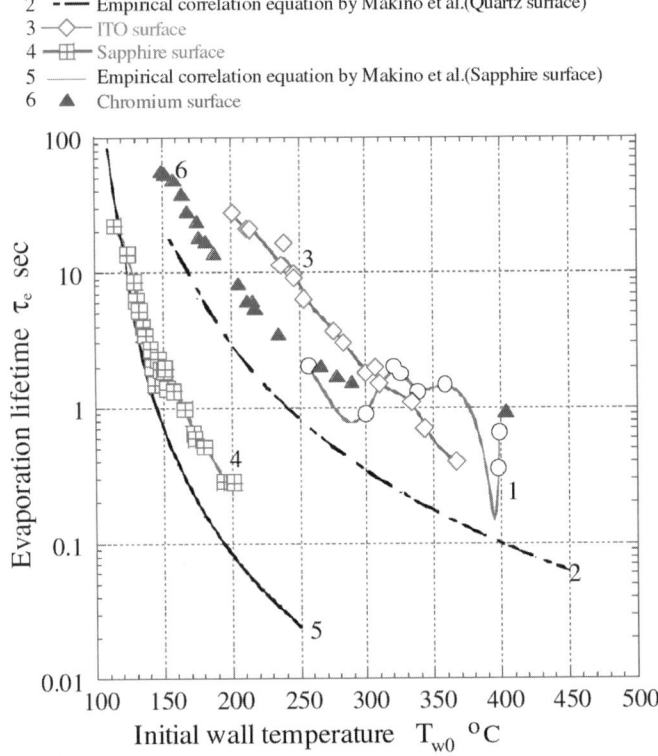

FIGURE 1. Evaporation lifetime curves.

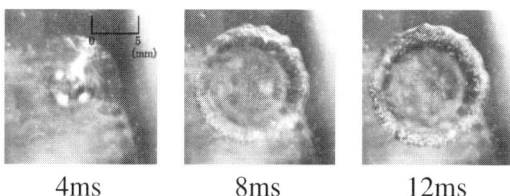

FIGURE 2. Boiling of a droplet impinging on an ITO surface ($T_{w0} = 202°C$).

region. The solid line in FIGURE 1 was applied to the sapphire plane by the same method as for the dashed line.

FIGURE 2 shows the boiling behavior of the droplet on the ITO plane, with a surface temperature of 202°C. The number shown under each photograph is the elapsed time after the drop impingement. The drop expands keeping a round shape after impingement, and reaches its largest diameter at 12 msec. A liquid film shows the crater shape as the circumference rises and the drop expands in this interval. Subsequently, a large number of white spots, which are considered to be the generation of the minute vapor bubble in a liquid film, are observed.

FIGURE 3 shows the boiling behavior of the droplet on the sapphire plane when the surface temperature was 380.9°C. The pictures on the left side at each elapsed time in FIGURE 3 were photographed from the horizontal direction, and the right side was taken from the underside. Part of the white ring is clearly projected in the right picture at 1 msec after the drop impingement because light from the upper part has converged and emitted to the visual field of the camera. This seems to be a lens effect, with the result that a droplet displays the meniscus-like shape at the bottom. The pictures from the underside show that the droplet expands in a round shape, and the generation of a comparatively large vapor bubble, which is observed in nucleate boiling, is not recognized. (However, a large vapor bubble was recognized in the case of heating surface temperatures of 320°C or less.) The boiling seems to evolve to a situation in which a thin vapor film is formed and the droplet exists on the film. Judging from the photograph taken from the horizontal direction at 4 msec, the minute liquid particles are rapidly scattering from the surface of a liquid film. However, the expected changes that bring about the dispersion of these minute liquid particles and should be evident on the liquid–solid contact surface, are not clearly reflected in the photograph from the underside. FIGURE 4 shows the boiling behavior after 4 msec,

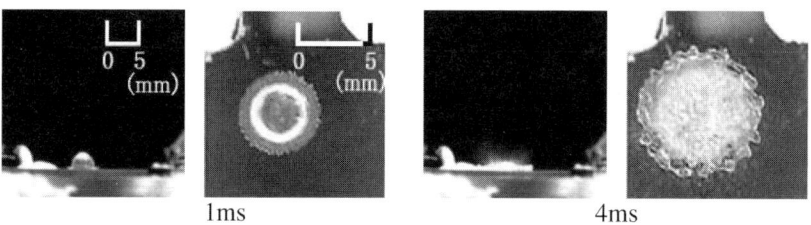

FIGURE 3. Boiling of a droplet impinging on a sapphire surface ($T_{w0} = 380.9°C$).

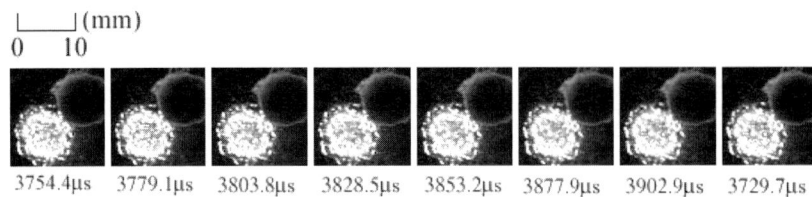

FIGURE 4. Boiling of a droplet impinging on a sapphire surface ($T_{w0} = 370°C$).

photographed from the back side of the sapphire at $T_{w0} = 370°C$. We consider that FIGURES 3 and 4 both represent a similar phenomenon, although the temperature in FIGURE 4 is lower than in FIGURE 3. Both shutter speed and photography speed were increased more in FIGURE 4 than in FIGURE 3, so that the generation of many minute vapor bubbles can be recognized. However, the resolution is bad, and a vapor bubble of 0.2 mm or less is not projected since a pixel corresponds to about 0.2 mm in this picture. Focusing our attention on the circular and white central part, it is evident that the minute vapor bubbles are locally and randomly generated in the short period of about 100 microseconds. It is, therefore, considered that these minute vapor bubbles grow from the microsize bubble and then leave from the liquid–solid contact surface.

FIGURE 5 shows the droplet when photographed from the back side of the sapphire, in case of $T_{w0} = 350°C$. A Fig. 5 concentric circular wave, which seems to arise in superheated liquid, is observed. The duration of the wave is about 100 microseconds. The oscillatory wave lapses after about 100 microseconds, and this phenomenon is repeated at intervals of about 100 microseconds. This wave seems to be a pressure wave, generated by heterogeneous nucleation. Although the generation of the minute bubble observed in FIGURE 4 is not observed in this photograph, we believe that microbubbles should have been generated and ejected.

According to the classical molecular kinetic theory,[8,9] the incipient boiling temperature in heterogeneous nucleation is about 315°C for water at atmospheric pressure. At the instant of droplet collision with the heated surface, the heated surface temperature should drop a little. This temperature drop could hardly be noted in the case of the quartz, but it has been confirmed that the temperature drop increases as the thermal property of the heated surface increases. Therefore, the temperature drop should occur in the case of the sapphire. Considering FIGURE 5 at $T_{w0} = 350°C$, it can be assumed that the actual heated surface temperature can at some moments become almost equal to the above mentioned theoretical incipient temperature, if the temperature drop is considered. It was reported[10] that a microbubble is generated by

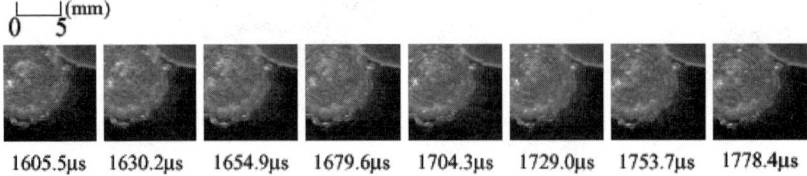

FIGURE 5. Boiling of a droplet impinging on a sapphire surface ($T_{w0} = 350°C$).

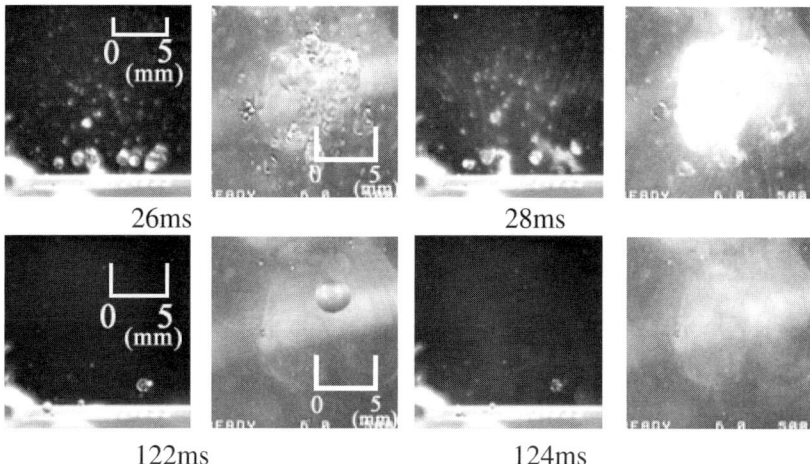

FIGURE 6. Boiling of a droplet impinging on a chromium surface ($T_{w0} = 531.7°C$).

instantaneously heating a very thin heating surface in the water. The generation of a microbubble is based on the idea that it is induced by fluctuation nucleation. Using droplet boiling system that collides with a heated surface having a comparatively large heat capacity, should clarify the induction of a microbubble by fluctuation nucleation. FIGURE 6 shows the boiling behavior of the droplet on the chromium coating plane, where the surface temperature is 531.7°C. At 450°C or less, nucleate boiling accompanied by a comparatively large vapor bubble is observed, the droplet separates and the divided droplets scatter. However, the boiling characteristics change once the heating surface temperature exceeds 450°C. Part of the divided droplets keeps contact with the heating surface through the thin vapor film. The explosive dispersion of the minute liquid particle is generated at the moment of collapse of the vapor film. FIGURE 6 shows the phenomenon at that time, with one arising between 26 msec and 28 msec, and the other generated between 122 msec and 124 msec. The radial explosive dispersion is projected in the photograph from the horizontal direction. On the other hand, in the photograph from the underside, the picture is projected white because of the incidence of the irregular reflection light strengthened by the dispersion of the minute liquid particle. The reason for the dispersion of these minute liquid particles may originate from the generation of a microbubble, which arises in the direct liquid–solid contact accompanied by the collapse of the thin vapor film.

SUMMARY

The study of a water droplet boiling following collision with a heated surface having a comparatively large heat capacity proves the possibility that the boiling can be generated by the occurrence of heterogeneous nucleation and the direct liquid–solid contact accompanied by the collapse of the thin vapor film. The generation of number of minute vapor bubbles with rapid frequency, which seem to grow from the microbubble, was also confirmed.

ACKNOWLEDGMENTS

This study was funded by the Ministry of Education, Science, Sports and Culture in Japan, Grant-in-Aid for University and Society Collaboration, No. 12792005. The author thanks Mr. M. Furuse, on the staff of PHOTRON Co., Ltd. for his help in high-speed photography.

REFERENCES

1. TAMURA, Z. & Y. TANAZAWA. 1958. Evaporation and combustion of a drop contacting with a hot surface. Proc. 7th Symposium on Combustion. 509–522. Butterworths Scientific Publications, London.
2. NISHIO, S. & M. HIRATA. 1978. Study on the leidenfrost temperature—2nd report, behavior of liquid-solid contact surface and leidenfrost temperature. Trans. Japan Soc. Mech. Engrs. **44:** 1335–1346.
3. INADA, S. & W.J. YANG. 1993. Effects of heating surface materials on a liquid–solid contact state in a sessile drop-boiling system. Trans. ASME J. Heat Transfer **115:** 222–230.
4. INADA, S. & W.J. YANG. 1993. Mechanisms of miniaturization of sessile drops on a heated surface. Int. J. Heat Mass Transfer **36**(6): 1505–1515.
5. INADA, S., Y. MIYASAKA & K. NISHIDA. 1985. Transient heat transfer for water drops impinging on a heated surface (1st report, effects of drop subcooling on the liquid–solid contact state). Bull. JSME **28:** 2675–2681.
6. ZHANG, N. & W.J. YANG. 1983. Evaporation and explosion of liquid drops on a heated surface. Exper. Fluids **1:** 101–111.
7. MAKINO, K. & I. MICHIYOSHI. 1984. The behavior of a water droplet on heated surfaces. Int. J. Heat Mass Transfer **27:** 781–791.
8. VOLMER, M. 1939. Kinetics der Phasenbildung. Steinkopff.
9. SCRIPOV, V.P. 1974. Metastable Liquids. John Wiley and Sons.
10. IIDA, Y., K. OKUYAMA & K. SAKURAI. 1994. Boiling nucleation on a very small film heater subjected to extremely rapid heating. Int. J. Heat Mass Transfer **37:** 2770–2780.

Ice/Water Slurry Blocking Phenomenon at a Tube Orifice

TAKERO HIROCHI,[a] SHUICHI YAMADA,[b] TUYOSHI SHINTATE,[c] AND MASATAKA SHIRAKASHI[b]

[a]*Electronic Mechanical Engineering Department, Toba National College of Maritime Technology, Toba, Japan*

[b]*Department of Mechanical Engineering, Nagaoka University of Technology, Nagaoka, Japan*

[c]*Tohoku Epson Corporation, Sakata, Japan*

ABSTRACT: The phenomenon of ice-particle/water mixture blocking flow through a pipeline is a problem that needs to be solved before mixture flow can be applied for practical use in cold energy transportation in a district cooling system. In this work, the blocking mechanism of ice-particle slurry at a tube orifice is investigated and a criterion for blocking is presented. The cohesive nature of ice particles is shown to cause *compressed plug type blocking* and the compressive yield stress of a particle cluster is presented as a measure for the cohesion strength of ice particles.

KEYWORDS: blocking; ice/water slurry; district cooling; cohesion

INTRODUCTION

A new cold energy transportation system using ice/water slurry flow is under development for the purpose of raising cold energy density in district cooling systems. Many experimental studies[1-4] have been carried out to clarify the flow and heat transfer characteristics of ice/water slurries and to develop devices and instruments to be applied to the system. These studies report that blocking of pipeline elements such as reducers, valves and distribution manifolds is likely to occur due to the formation of compressed ice particle clusters.[4-7] This type of blocking is quite different from blocking due to the arching or particle deposition observed in more usual solid-particle/water slurry transportation pipelines. It is commonly assumed that the cohesive nature of ice-particles is the main cause of ice/water slurry blocking. However, very few studies have been carried out to date on ice particle blocking and no data has correlated the cohesion force of the ice particles in water with this phenomenon.

The specific aim of this work was to clarify the blocking mechanism in an ice/water slurry flow at an orifice in a pipeline and to find a criterion for the occurrence of this phenomenon.

Address for correspondence: Takero Hirochi, Electronic Mechanical Engineering Department, Toba National College of Maritime Technology, 1-1, Ikegami, Toba, 517-8501, Japan.
hiro@twing.toba-cmt.ac.jp

EXPERIMENTAL

Test Particles

Three kinds of ice particles with various mean diameters d_p and density ρ were used as samples, since the particle cohesion strength depends strongly on particle size. Non-cohesive polypropylene beads of comparable size and density were used as a reference for comparison.

The samples used were: (1) chip ice ($d_p = 10\,\text{mm}$, $\rho = 917\,\text{kg/m}^3$), made in a common refrigerator; (2) granulated snow ($d_p = 3\,\text{mm}$, $\rho = 917\,\text{kg/m}^3$), gathered from local spots in winter or made by cracking chip ice, sieved out in either case (the sieve mesh size was 5 mm); (3) fresh snow ($d_p = 1\,\text{mm}$, $\rho = 917\,\text{kg/m}^3$), gathered within 24 hours after it fell on the ground; and (4) polypropylene beads ($d_p = 4\,\text{mm}$, $\rho = 908\,\text{kg/m}^3$). This is used for reference because of its non-cohesiveness.

Cohesion Strength Test

A compression test was carried out to estimate the cohesion strength of particles that cluster in ice/water slurry flow through pipe. A cylindrical test piece was formed by softly packing the particles in a circular pipe with an inner diameter $d_0 = 65\,\text{mm}$ and a height $h_0 = 65\,\text{mm}$. The sample was then submerged in water at 0°C and an axial load w was abruptly imposed on it.

In the case of chip ice, the test piece scattered into particles due to the buoyancy when it was submerged in water. Hence, the yield stress σ_y of chip ice can be estimated from the following equations:

$$\sigma_y \leq \frac{4V_i(\rho_w - \rho_i)g}{\pi d_0^2}, \tag{1}$$

$$V_i = C\left(\frac{\pi d_0^2}{4}\right)h_0, \tag{2}$$

where V_i is the substantial volume of ice in the test piece, C is the ice packing volume fraction (0.5), ρ_w is the density of water, and ρ_i is the density of ice.

In the case of the granulated and fresh snow, the test piece was compressed to a height h, keeping its diameter unchanged at d_0 when the load w was smaller than the respective critical value. When w was larger than this critical value, the test piece was completely crushed. The compression ratio r_h and the compression stress σ were defined by $r_h = (h_0 - h)/h_0$ and $\sigma = 4wg/(\pi d_0^2)$, respectively, where h_0 is initial test piece height, h the test piece height after adding a load w, and d_0 is the initial test piece diameter. FIGURE 1A shows the result of the compression test; $r_h = 1$ means that the test piece was completely crushed. The compression yield stress σ_y was determined as the minimum value of σ at $r_h = 1$; that is, the lowest value of σ causing the test piece to be crushed. The compression yield stress of test pieces thus determined were: (1) chip ice, $\sigma_y \leq 0.03\,\text{kPa}$; (2) granulated snow, $\sigma_y \leq 0.7\,\text{kPa}$; and (3) fresh snow, $\sigma_y \leq 1.9\,\text{kPa}$.

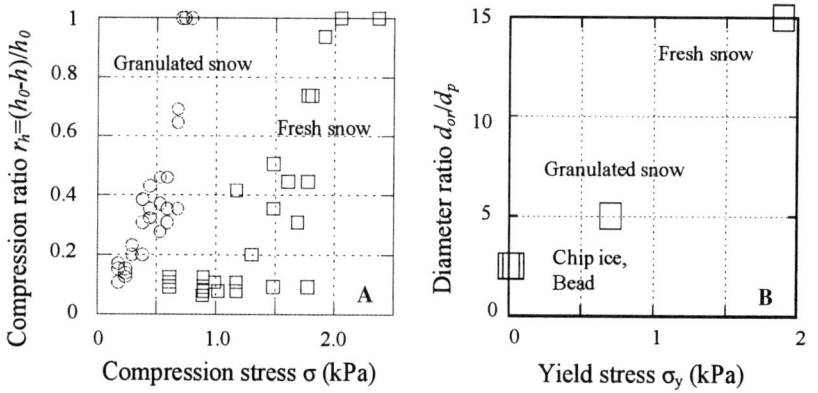

FIGURE 1. Compression characteristics of ice particles: (**A**) compression ratio and compression stress; (**B**) blocking diameter ratio versus compression yield stress.

Apparatus and Methods

To clarify the blocking mechanism in a piping element isolated from a pump–pipeline system, a test apparatus was devised and used to study the blocking phenomenon at the tube orifice. The test section was a transparent acrylic pipe with inner diameter $d = 52$ mm and length $l = 985$ mm, attached to the test orifice at one end. Before a test run, test particles were weighed and put into the test section uniformly along the pipe axis. Ice-particle/water mixture filling the test section set in this way was driven out through the test orifice by a piston-cylinder unit. The blocking process was observed and recorded by using a videocamera. The pressure difference $p_1 - p_2$ between two points upstream, one by the orifice and the other at an axial distance of 254 mm, and the volumetric ice fraction C at 90 mm upstream of the orifice, were recorded[8] simultaneously. Three orifices with different diameters were tested, and the velocity u in the test section was varied in the range 0–1 m/sec. The initial particle fraction C_0 was 0–32%. The pressure was generated by a hydraulic cylinder and a piston located at the end of the transparent test sections, opposite the orifice.

RESULTS AND DISCUSSIONS

Since the cohesive nature of ice particles in water is considered to be a governing factor for *compressed plug type blocking* (defined below), the ratio of the maximum orifice diameter for the blocking to the particle size is plotted against the compression yield stress in FIGURE 1 B. This shows that the higher the yield stress, the larger is the maximum orifice diameter relative to the particle size for a required occurrence of blocking. This result infers that the cohesive nature of the particles estimated by σ_y can be used as a measure for the occurrence of blocking.

FIGURE 2 A and D show the blocking of the ice chips and the bead slurries, respectively. No blocking of non-cohesive particles occurred irrespective of the initial fraction Co and the flow velocity u when the orifice diameter d_{or} was larger than the

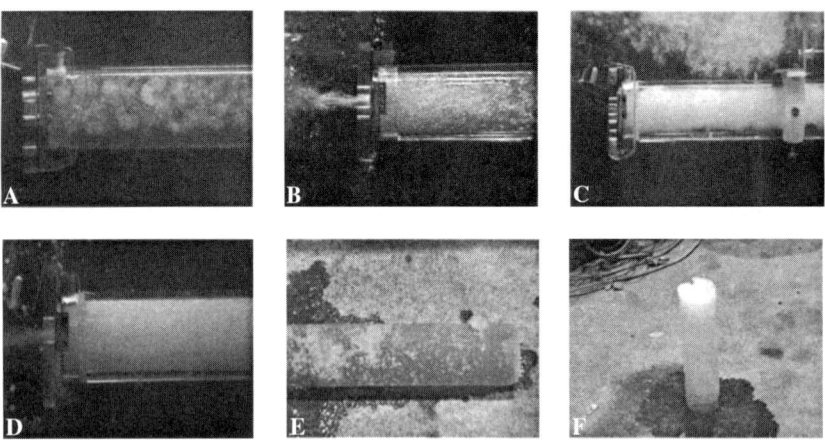

FIGURE 2. Blocking at an orifice, with flow direction from right to left. (**A**) Chip ice, $u = 0.13$ m/sec, $C_0 = 25\%$. (**B**) Granulated snow, $u = 0.18$ m/sec, $C_0 = 35\%$. (**C**) Fresh snow, $u = 0.13$ m/sec, $C_0 = 31\%$. (**D**) Beads, $u = 0.42$ m/sec, $C_0 = 35\%$. (**E**) Plug-like cluster of granulated snow. (**F**) Plug-like cluster of fresh snow.

respective critical values, $d_{or} > 25$ mm for the ice chips and $d_{or} > 10$ mm for the beads. When d_{or} was smaller than this value and blocking occurred, the particle cluster could not keep its shape and the particles separated from each other when taken out of the test pipe. Hence, the blocking of these particles is considered to be caused by arching.

In the case of fresh and granulated snow, blocking occurred when $d_{or} \leq 15$ mm, as can be seen in FIGURE 2B and C. Blocking occurred after a small particle-cluster stagnated at the orifice and grew until a plug-like particle cluster filling the pipe was formed. The particle-plug had considerable strength, like a solid column, as can be seen in FIGURE 2E and F. Hence, we name this type of blocking, *compressed plug type blocking*.

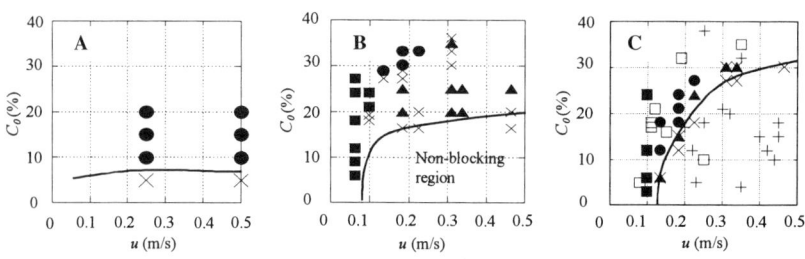

FIGURE 3. Blocking condition. (**A**) Chip ice, $d_p = 10$ mm, $d_{or} = 25$ mm. (**B**) Granulated snow, $d_p = 3$ mm, $d_{or} = 15$ mm. (**C**) Fresh snow, $d_p = 1$ mm, $d_{or} = 15$ mm. Symbols: ×, particles pass through; ■, particles stagnate; ▲, blocking observed; ●, blocking occurs every time; □, stagnate in pipeline; +, particles pass through in pipeline; *curved line*, criterion curve.

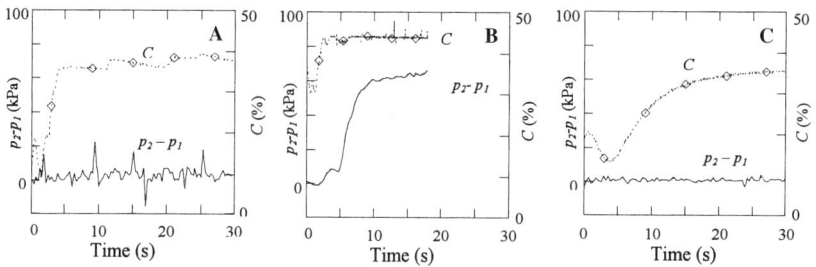

FIGURE 4. Particle fraction and pressure difference ($u = 0.11$ m/sec): **(A)** chip ice, **(B)** granulated snow, and **(C)** fresh snow.

The criterion for the blocking obtained through the element test is presented in FIGURE 3 as a curve in the u–C_0 plane. In FIGURE 3A, blocking occurs when the initial fraction C_0 exceeds a certain value and is independent of the velocity u. For granulated snow (FIG. 3B), stagnation of the particle cluster occurs at the lowest value of C_0 when the velocity u is lower than a certain value. Since stagnation may lead to blocking if the orifice is installed in a pipeline for cold energy transportation, this velocity is taken as the threshold for the non-blocking condition. The threshold value of u for fresh snow is higher as can be seen by comparing FIGURE 3B with FIGURE 3C. FIGURE 3C includes results for the blocking pipeline tests (see below) of fresh snow.

FIGURE 4 shows the typical behavior of ice fraction and pressure difference at the orifice for the three types of slurry when blocking occurs. For chip ice, FIGURE 4A, the particle fraction C increases steeply and soon attains a constant value nearly equal to that of the natural packing. Since the gaps among the coarse particles permit water flow through them with small resistance, the pressure difference through the cluster is small. For granulated snow, FIGURE 4B, the profile of the fraction is similar to that for chip ice. Because of the small size of particles, the pressure difference increases as a compressed plug grows at the orifice. For fresh snow, FIGURE 4C, because of the softness of the particles, the fraction C increases gradually after a small snow particle cluster stagnates at the orifice until a plug-like particle cluster filling the pipe is formed. Since a water path or channel is formed through the cluster of fresh snow, the pressure difference in FIGURE 4C is small even when the plug has formed. However, this would not be the case in a real pump–pipeline system, because the particles are supplied continuously and will plug the paths.

BLOCKING IN PIPELINE

The orifices used in the test element were attached in a pipeline of a slurry transportation system, and blocking was examined by lowering the velocity u while keeping C constant. The test section was made of transparent acrylic or PVC pipe with an inner diameter 50 mm and had a length upstream of more than 10 m. Test particles and water were mixed in a mixing tank and driven into the pipeline by a vortex pump. Particle fraction was regulated at an arbitrary desired value by a fraction regulator.[6]

The results for blocking in the pipe of the fresh snow are compared in FIGURE 3C with those for the test element. The criterion curve for the latter is seen to present the condition for occurrence of blocking of ice/water slurry in a pump–pipeline system.

SUMMARY

The blocking characteristics of ice-particle/water slurry at an orifice isolated from a pipeline were investigated using an apparatus specifically devised for this purpose. The same orifices were also attached in a pump–pipeline system and the threshold values of the velocity and the ice fraction for blocking were examined. The major conclusions are:

1. Blocking of chip-ice slurry and the polypropylene beads slurry occur due to arching. In this case, the governing factor of blocking is the ratio of particle size to the orifice diameter. The blocking of the granulated snow slurry and the fresh snow slurry are caused by a compressed ice-particle cluster, like a plug, that forms upstream from the orifice. The main cause of the compressed plug type blocking is the cohesive nature of ice-particles. Flow velocity and the ice fraction are the dominant parameters for its occurrence.

2. For a fixed ice fraction, compressed plug type blocking occurs when the velocity u is lower than a certain critical value u_c. The yield stress of an ice particle cylinder obtained by the compression test gives a measure for this type of blocking.

3. The critical condition for blocking in fresh snow slurry at the orifice in the pump–pipeline system correlates well with the criterion curve obtained through the pipeless test element.

REFERENCES

1. LARKIN, B. & J.C. YOUNG. 1989. Influence of ice slurry characteristics on hydraulic behavior. Proc. 80th Annual Conf. IDHCA. 340–351.
2. ONOJIMA, H., et al. 1991. A study on the hydraulic transportation of ice–water mixture for district cooling system. Proc. 1st ASME/JSME Fluids Eng. Conf. **118:** 241–246.
3. KAWADA, K., S. YAMADA, K. YOSHIDA, et al. 2001. Concept design of a district cooling system utilizing ice/water mixture flow and development of the storage tank. J. Jpn. Soc. Snow Ice **63**(1): 35–48.
4. WINTERS, P.J. & R.J. KOOY. 1991. Direct freeze ice slurry district cooling system evaluation. Official Proc. Annunciation Conf. IDHCA **18:** 381–398.
5. UMEMURA, T., M. NAKAYAMA, A. UCHIYAMA, et al. 1986. Blocking of snow/water mixture flow and criterion of stagnation of snow at pipe orifice. J. Jpn. Soc. Snow Ice **48**(4): 207–214.
6. SHIRAKASHI, M., M. TAKAHASHI & A. UCHIKURA. 1990. Hydraulic conveying of snow. Jpn. J. Multiphase Flow **4**(1): 61–71.
7. KAWADA, K., M. SHIRAKASHI & S. YAMADA. 2001. Flow characteristics of ice/water slurry in a horizontal circular pipe. J. Jap. Soc. Snow Ice **63**(1): 11–19.
8. KITAHARA, T., M. SHIRAKASHI & Y. KAJIO. 1993 Development of a snow-fraction meter based on the conductometric method. Ann. Glaciology **18:** 60–64.

Convective Heat Transfer and Infrared Thermography

GIOVANNI M. CARLOMAGNO, TOMMASO ASTARITA,
AND GENNARO CARDONE

University of Naples – DETEC, P. le Tecchio, Naples, Italy

ABSTRACT: Infrared (IR) thermography, because of its two-dimensional and non-intrusive nature, can be exploited in industrial applications as well as in research. This paper deals with measurement of convective heat transfer coefficients (h) in three complex fluid flow configurations that concern the main aspects of both internal and external cooling of turbine engine components: (1) flow in ribbed, or smooth, channels connected by a 180° sharp turn, (2) a jet in cross-flow, and (3) a jet impinging on a wall. The aim of this study was to acquire detailed measurements of h distribution in complex flow configurations related to both internal and external cooling of turbine components. The *heated thin foil* technique, which involves the detection of surface temperature by means of an IR scanning radiometer, was exploited to measure h. Particle image velocimetry was also used in one of the configurations to precisely determine the velocity field.

KEYWORDS: convective heat transfer; infrared thermography; gas turbine cooling

INTRODUCTION

Since the introduction of gas turbine engines in the 1940s the entry temperature has increased from about 900 K to above 2,000 K. This increase has improved the thermal efficiency of the thermodynamic cycle, resulting in improved overall performance of the engine. Although some of the increase in the turbine entry temperature was made possible through the use of more sophisticated materials, a great part of it occurred because of developments in cooling technology. To reduce the temperature of turbine components, a certain amount of air, derived from a compressor, is expended to cool the nozzle guide vanes, rotating blades, and disks the blades are attached to. For efficiency reasons, the maximum temperature increase corresponds an increase of the compression ratio with a consequent increase in the cooling air temperature.

It can be easily understood that cooling a modern gas turbine engine is both indispensable and technologically intricate. From a strictly engineering point of view, the parameters that interest the designer are essentially the stress level, due to both mechanical and thermal loads, as well as the fatigue life of the engine components. However, these quantities can be only computed on the basis of a detailed evaluation

---

Address for correspondence: G.M. Carlomagno, University of Naples, DETEC, P. le Tecchio, 80–80125 Naples, Italy.
carlomagno@unina.it

of the local temperature distribution in the involved component, which, in turn, requires knowledge of the local convective heat transfer coefficient under various operating conditions.

EXPERIMENTAL TECHNIQUES

Steady State Heated Thin Foil Technique

Measurement of the local convective heat transfer coefficient is performed by means of the steady state *heated thin foil* technique associated with the detection of wall temperature by means of infrared thermography.[1–3]

The heated thin foil method consists of uniformly heating a thin metallic foil (or a printed circuit board) by Joule effect and measuring under steady-state conditions, the convective heat transfer coefficient h from the foil to the stream flowing on it by means of the relationship:

$$h = \frac{q - q_l}{T_w - T_r}, \qquad (1)$$

where q is the known Joule heating flux, q_l represents the thermal losses that are mainly due to tangential conduction along the foil and radiation, T_w is the wall temperature measured when the foil is heated (*hot image*), and T_r is a reference temperature which, for channel flows, is equal to the local bulk temperature and, in case of jets, coincides with the adiabatic wall temperature. The latter may be obtained by measuring, with the IR scanning radiometer, the wall temperature of the foil when this is not heated (*cold image*).

The foil is generally thermally insulated on its back face (i.e., the face opposite to that the stream is flowing on). When insulation cannot be accomplished (e.g., for optical access reasons) additional thermal losses, such as natural convection and radiation, must also be taken into account. In this case, the measurement can be performed on both sides of the foil; indeed, for small Biot numbers, the foil can be considered isothermal across its thickness.

Tangential conduction, which modulates the thermal signal, can be evaluated by means of the second derivative of the wall temperature. In principle, this seems relatively easy to perform by taking into account the fact that the camera provides a very large number of temperatures. However, it needs to be stressed that spurious effects, linked to noise, must be avoided by carefully filtering the temperature signal. In the steady state, such as in the case of the heated thin foil technique, noise can be consistently reduced by averaging a large number of thermal images.

When the heated thin foil is made of a printed circuit board, the bulk tangential thermal conductance along the foil is non isotropic. In fact, if the circuit is obtained with several electrical conducting tracks arranged in a Greek fret mode, the thermal conductance along the tracks is generally higher than in the perpendicular direction. While reducing data, this may be considered by allowing for two different bulk thermal conductances, one along and the other perpendicular to the tracks.[4] Further details on the application of the heated thin foil technique can be found elsewhere.[3]

The thermographic system used was the AGEMA Thermovision 900LW. The field of view was scanned by the Hg–Cd–Te detector in the 8–12μm infrared window. Nominal sensitivity, expressed in terms of noise equivalent temperature difference, was declared to be 0.07°C when the scanned object was at ambient temperature. The scanner spatial resolution was 235 instantaneous fields of view per line at 50% slit response function. Each thermal image was digitized at 12 bits per frame of 136×272 pixels. Application software was employed for each image, which generally involves noise reduction by numerical filtering, computation of temperature maps, evaluation of radiation and tangential conduction losses, and heat transfer correlations.

Particle Image Velocimetry

The particle image velocimetry (PIV) technique permits the evaluation of instantaneous fluid velocities by recording the positions of the image of very small tracers suspended in the fluid and illuminated by a laser light sheet at two successive time instants. The mean flow, turbulence intensity, and Reynolds (Re) shear stresses could also be obtained by performing a statistical analysis over several instantaneous sample fields. The primary elements of the apparatus were the fluid seeding device and the laser system. A conventional olive oil atomizer was used to seed the fluid; oil particles stayed in air at rest for hours and did not significantly change in size. Indeed, their mean diameter was about 1 µm. The light source used to generate the light sheet consisted of two pulsed Nd:YAG lasers. Each light beam had the following characteristics: wavelength 532 nm, pulse duration 6 nsec, energy per pulse about 200 mJ, and light sheet thickness in the test region about 1 mm.

To display, acquire, and record digital images, the following components were used: a video camera with a CCD sensor (Kodak Megaplus model ES 1.0), a PC equipped with an acquisition board and dedicated PIV software, and a laser-acquisition system synchroniser. The camera was able to catch pairs of images at a rate of 15 double frames/sec, each composed of 1,008×1018 pixels at 256 grey levels. For each test, 200 images were acquired, sufficient to give significant information about the effective mean flow by operating an ensemble average.

The interrogation method of the frame pairs used in this study consisted of an upgraded version of the windows displacement iterative multigrid (WIDIM) algorithm.[5] Briefly, WIDIM calculates the cross-correlation of two homologous windows in which each full frame is subdivided. Once the cross-correlation map is obtained, a peak detection operation is performed over the map to determine the precise location of the peak through a subpixel interpolation routine. In this way, it is possible to compute an initial displacement field of the particles in the entire image. Subsequently, the method makes use of a translation, a rotation, and a deformation of each interrogation window of the second frame. The predicted displacement is corrected by means of an iterative procedure. In addition, while iterating, the algorithm allows for a reduction of the size of the window areas to obtain better spatial resolution. In these experiments, the starting square area had a linear dimension of 64 pixels and a final one of 16 pixels.

EXPERIMENTAL RESULTS

Three flow configurations were examined here and detailed heat transfer measurements were performed. PIV was used to determine the velocity field in the case of the jet impinging on the wall.

180° Sharp Turn

Cooling turbine blades by means of internal forced convection generally involves supplying cooling air from the compressor into the blade interior and, after flowing through a serpentine passage, discharging the air at the blade trailing edge. The serpentine passages are mostly made of several adjacent straight ducts, spanwise aligned along the blade and connected by 180° turns. The presence of these turns causes flow separation and reattachment and induces secondary flows. These events produce large variations of h with consequent increased thermal stresses in the blade wall. The use of rib turbulators in the ducts enhances the heat transfer efficiency and may also reduce the thermal stresses.

The effects of rib turbulators have been investigated by many researchers; in particular the work of Ekkad and Han,[6] Ekkad et al.,[7] Park et al.,[8] Akella and Han,[9] and Arts et al.[10] are acknowledged. The present tests dealt with a square channel, 80 mm side length, 2,000 mm long, which was tested both with and without ribs. The ribs were symmetrical placed on the two side planks with respect to the channel midplane. The ribs, made of aluminium, had an inclination of 30° with respect to the channel axis and a square cross section with side length $e = 8$ mm. Two different rib pitches P are herein reported, 80 and 160 mm. Experimental results are presented in the form of local values of the normalized Nusselt number (Nu), which is the ratio between the measured Nu and that evaluated according to the Dittus and Boelter correlation.[11] The normalized Nu represents, in practice, the ratio between the measured h and the mean value corresponding to a fully developed flow. Here, only results relative to $Re = 16,000$ and to a symmetric heating boundary condition are reported.

The efficiency of thermography in performing this type of measurement is well proven by the normalized Nu maps of FIGURE 1, which are related to three different geometries. The top map refers to a smooth channel without ribs. By moving streamwise along the channel, the quasiregular trend of the Nu distribution across the inlet duct, proves the presence of a fully developed flow there. Three high heat transfer regions follow in sequence: the first is located near the end wall, the second occurs downstream of the second outer corner and extends for about two diameters, the third is located two diameters after the second inner corner and is attached to the partitioning wall. These high heat transfer zones are due to the *jet* effect of the flow through the bend, which has also been observed previously. In FIGURE 1A it is also possible to notice relatively low heat transfer zones, namely in the first outer corner and near the tip of the partition wall.

The other two maps refer to ribbed channels where the ribs are clearly visible due to the higher heat transfer rates occurring on them. The presence of the ribs completely alters the flow field both in the straight channels and in the turning region. For both cases, the flow is completely developed at the channel entrance and, in particular for $P/e = 10$, the contours after the first and second ribs are practically

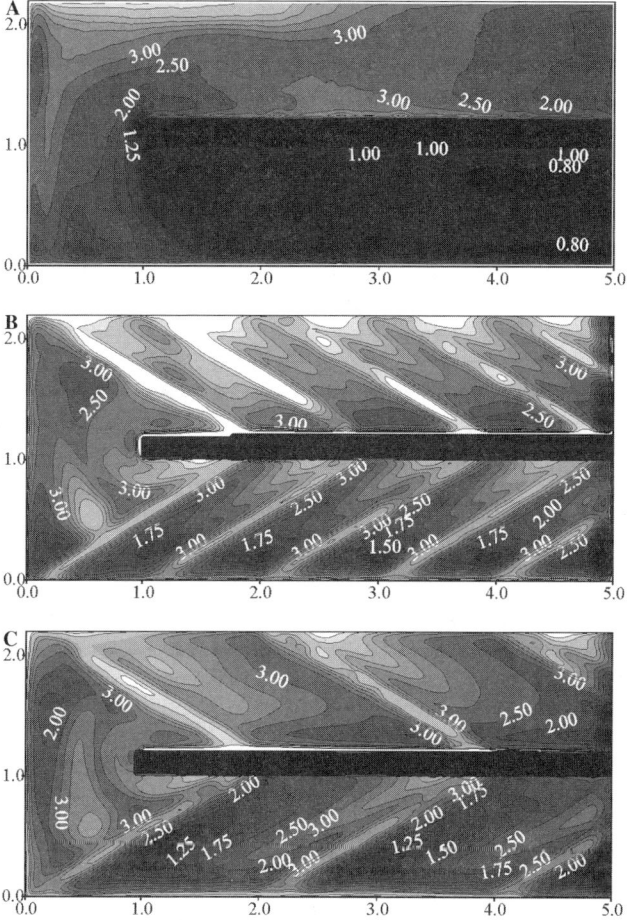

FIGURE 1. Normalized Nusselt number maps in channels connected by an 180° turn. (**A**) smooth channel, (**B**) ribbed channel ($P/e = 10$), (**C**) ribbed channel ($P/e = 20$).

identical, the only differences being due to edge effects at the duct inlet. The rib angle causes the formation of secondary flows in the form of two counter rotating vortices. In the inlet channel, the main flow, in the proximity of both bottom and top wall planks, entrapped by the ribs tends to accelerate towards the external wall. The secondary flow, after licking the external wall, go back to the partition wall and create a nearby jet effect. This explains the asymmetry of the maps; in fact, the jet effect tends to increase h near the partition wall, compared to that at the external wall. In the outlet channel, the secondary flow is reversed and, therefore, h is much higher at the external wall compared to that near the partition wall. The reattachment line, downstream of the ribs, is clearly identified by the locus of the normalized Nu local maxima and its distance from the previous rib increases for the higher rib pitch.

The Nu maps for the ribbed channels in the turn zone show a completely different behavior compared to that for the smooth channel. For $P/e = 10$, in proximity of the first external corner, it is again possible to notice a low heat transfer zone, due to a recirculation bubble, as already observed for the smooth channel. The interaction between the secondary flow and the sharp turn produces, just after the last rib, two high heat transfer zones and, between them, a local minimum. A similar configuration is also found for $P/e = 20$, but the minor relative importance of the secondary flow reduces the difference. A relatively low heat transfer zone is also observed in the second part of the bend and just before the outlet channel first rib. For $P/e = 20$ this zone is placed near the second external corner, whereas for the smaller rib pitch it tends to move towards the partition wall. Of course, the overall heat transfer rate increases for increasing rib presence.

For both rib pitches, the overall increase of the turbulence due to the bend globally induces higher values of the normalized Nusselt number in the outlet duct but the percentage increase is lower than that for the smooth channel because of the presence of turbulence already induced by the ribs. This is because of the lower additional turbulence induced by the bend and, in practice, should cause a decrease the thermal stresses in a real turbine blade.

Jet in Cross Flow

A second important cooling mechanism in turbine engines is transpiration cooling. Air taken from the internal channels is ejected from holes placed nearby the blade leading edge and cools its external part. Each singular jet may be treated as a jet in cross flow and this explain the second tested configuration. Obviously, in the actual situation the external flow field is not constant and the inclination and interaction of adjacent jets cannot be neglected. However, the intrinsic features of such a flow configuration justify the need for a complete understanding of its fundamental aspects, including heat transfer at the wall.

A jet discharging normal to a cross flow gives rise to a complex interaction between the two streams, mainly resulting in the deflection of the jet in the direction of the cross flow. The jet cross-sectional area increases as it entrains fluid from the external stream and it tends to assume a horseshoe shape with a pair of counter rotating trailing vortices.[12–15] Furthermore, ring-like vortices are present at the wall shear layer, which become distorted with streamwise distance; the wake region beneath the downstream side of the jet contains streamwise wall vortices that lie over the flat wall, as well as vertically oriented shedding vortices.

Attention is focused on the measurement of h over the flat wall the jet is issuing from. In particular, the combined effects of the free stream and the jet perpendicularly injected into it for a given stream velocity, U, by varying the injection ratio, $R = U_j/U$ (where U_j is the mean velocity at the jet nozzle exit) are analyzed. Tests were performed in a low turbulence wind tunnel, $300\times400\,\text{mm}^2$ cross section. The heated wall had a surface area of $300\times600\,\text{mm}^2$ and a nozzle, 24 mm in diameter at its center, to allow for the jet injection. Particular attention was given to maintaining, during tests, the plenum jet temperature equal to the ambient temperature that, in turn, coincided with the free stream temperature. Experimental data is again presented in terms of Nu normalized to the case without jet injection and with the hole closed to simulate the case of a flat plate.

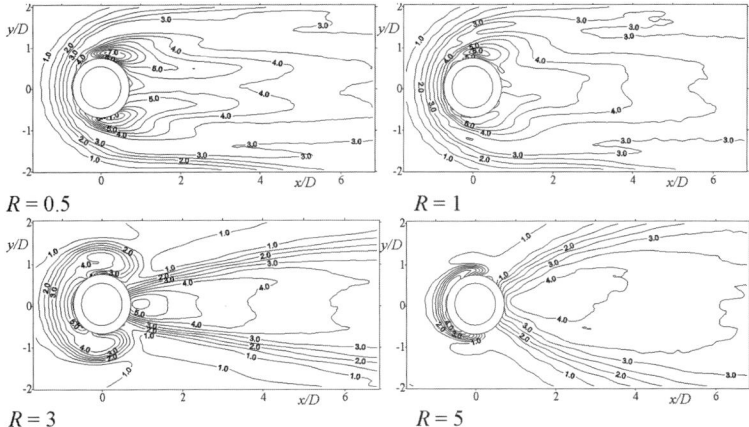

FIGURE 2. Normalized Nusselt number on the wall the jet is issuing from for various injection ratios.

The maps of the normalized Nu ratio for the lowest tested velocity $U = 5$ m/sec and $R = 0.5$, 1, 3, and 5 are shown in FIGURE 2; the main flow is from left to right. The inner circle denotes the position of the hole the jet is issuing from and data in the ring external to it are not herein presented because of edge effects. As expected, the normalized Nu takes a value equal to unity outside of the neighborhood and the jet wake. For injection rate increasing, the extension of the jet influence around the hole decreases and the wake width tends to enlarge further away. Immediately downstream of the injection hole, due to flow separation, a low heat transfer zone is present. Further downstream Nu increases and attains its maximum value. Two high heat transfer zone are also symmetrically placed on the nozzle sides, slightly displaced downstream. They seem to disappear at higher R values.

Jet Impinging on a Wall

Because of their high thermal efficiency, impinging jets, with several adaptations, are used to cool turbine engine components, including disk and blade walls. In this work an axisymmetric jet perpendicularly impinging on a plate was tested. The jet issued from a nozzle having $D = 18$ mm exit diameter. In particular, the effects on h of the vortex rings that develop for small values of the ratio between the nozzle-to-plate distance z and D were analyzed.

Experimental heat transfer data, azimuthally averaged, were reduced to dimensionless form in terms of Nu_D based on nozzle exit diameter D. Nu_D profiles, for three Re values, based on jet initial velocity and D, and $z/D = 2$ are shown in FIGURE 3 where r/D is the radial dimensionless coordinate. As can be seen, the Nu_D profile exhibits two maxima and two minima (one of which is at the jet axis). For $Re = 15{,}000$ maxima and minima tend to flatten out. For $Re = 60{,}000$, besides the minimum at the stagnation point ($r/D = 0$), the Nu_D distribution presents two maxima at about $r/D = 0.8$ and 2.4 and a minimum between them at about $r/D = 1.4$. As the

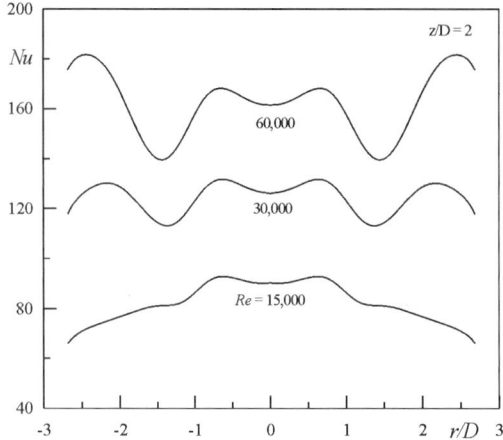

FIGURE 3. Nusselt number radial profiles of the impinging jet for various Re values.

impingement distance increases (not shown in the figure), maxima and minima become much milder and disappear for $z/D = 6$, when mixing enters the jet and only a maximum at the stagnation point is left. In fact, for increasing z/D, the inner maxima move inwards and finally coalesce, taking the place of the stagnation point minimum. At the same time, minima and outer maxima move outward and eventually disappear.

The instantaneous streamline field and the vorticity map for $z/D = 2$ and $Re = 60,000$ are shown in FIGURE 4. The flow dynamics of the wall jet is governed by the

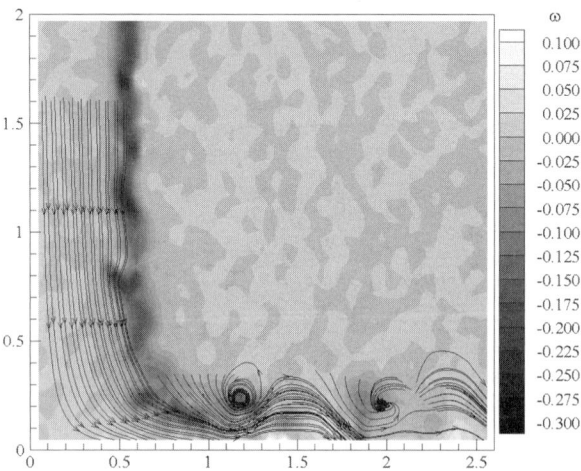

FIGURE 4. Visualization of vortex rings and vorticity map of the impinging jet; $z/D = 2$, $Re = 60,000$.

vortex rings that develop in the shear layer and depends strongly on the impingement distance. When the jet impinges on the wall with its potential core (i.e., at short distances), since the boundary layer of the wall jet is thin, separation can arise. For the values of Re tested, separation seems to occur at about $r/D = 1.3$ and rolls-up into a ring vortex. Further downstream the flow reattaches and this causes the second heat transfer maximum. The vortices in the shear layer give rise to the formation of primary vortices, which are clearly distinguishable in the map of FIGURE 4 and explain the previous heat transfer data. These results are in general agreement with those of Laperches and Buchlin.[16]

CONCLUSIONS

The heated thin foil technique coupled with measurement of surface temperature by IR thermography were used to measure the convective heat transfer coefficient in three types of flows.

Ribbed channels show spanwise asymmetry of the heat transfer maps because of the presence of secondary flows. In the turn zone, the Nu maps for the ribbed channel show a behavior completely different from the smooth one and the interaction between the secondary flow and the sharp turn produces, just after the last rib, two high heat transfer zones with a local minimum between them. For both tested rib pitches the overall increase of turbulence due to the bend induces higher values of the normalized Nu in the outlet duct, but the percentage increase is lower than the corresponding smooth channel because of the rib induced turbulence. This should decrease the thermal stresses in the turbine blade. As expected, the normalized Nu for the jet in cross flow takes values equal to unity outside the jet wake. Immediately downstream of the injection, a low heat transfer zone is present. Further downstream Nu increases and reaches a maximum.

The radial Nu_D profile, for the jet impinging on a wall at $Re = 60,000$ and $z/D = 2$, shows, besides the minimum at the stagnation point, two maxima at about $r/D = 0.8$ and 2.4 and a minimum between them at about $r/D = 1.4$. As the impingement distance increases, maxima and minima become much milder and disappear for $z/D = 6$, when mixing enters the jet and only a maximum at the stagnation point remains. The results agree with PIV measurements showing the formation of ring vortices for such a short impingement distance.

REFERENCES

1. CARLOMAGNO, G.M. & L. DE LUCA. 1989. Infrared thermography in heat transfer. *In* Handbook of Flow Visualization. W.J. Yang, Ed.: 531–553. Hemisphere.
2. CARLOMAGNO, G.M. 1993. Heat transfer measurements by means of infrared thermography. *In* Measurement Techniques. von Karman Inst. Fluid Dynamics Lect. Ser. 1993-05, Rhode-Saint-Genese. 1–114.
3. CARLOMAGNO, G.M. 1997. Thermo-fluid-dynamics applications of quantitative infrared thermography. J. Flow Visualization Image Proc. **4:** 261–280.
4. ASTARITA, T. & G. CARDONE. 2000. Thermofluidynamic analysis of the flow in a sharp 180° turn channel. Exp. Thermal Fluid Sci. **20:** 188–200.
5. SCARANO, F. & M.L. RIETHMULLER. 1999. Iterative multigrid approach in PIV image processing. Exp. Fluids **26:** 513–523.

6. EKKAD, S.V. & J.C. HAN. 1997. Detailed heat transfer distributions in two-pass square channels with rib turbulators. Int. J. Heat Mass Transfer **40**(11): 2525–2537.
7. EKKAD, S.V., Y. HUANG & J.C. HAN. 1998. Detailed heat transfer distributions in two-pass square channels with rib turbulators and bleed holes. Int. J. Heat Mass Transfer **41**: 3781–3791.
8. PARK, C.W., S.C. LAU & R.T. KUKREJA. 1998. Heat/mass transfer in a rotating two-pass channel with transverse ribs. J. Thermophys. Heat Transfer **12**: 80–86.
9. AKELLA, K.V. & J.C. HAN. 1999. Impingement cooling in rotating two-pass rectangular channels with ribbed walls. J. Thermophys. Heat Transfer **13**: 364–371.
10. ARTS, T., G. RAU, M. CAKAN, et al. 1997. Experimental and numerical investigation on flow and heat transfer in large-scale, turbine cooling, representative, rib-roughened channels. J. Power Energy **211**: 263–271.
11. DITTUS, P.W. & L.M.K. BOELTER. 1985. Heat transfer in automobile radiators of the tubular type. Int. J. Comm. Heat Mass Transfer **12**: 3–22.
12. CARLOMAGNO, G.M., A. CENEDESE & G. DE ANGELIS. 1998. PTV analysis of jet in cross-flow. Proc. 9th Int. Symp. Appl. Laser Techn. Fluid Mechanics **2**: 38.3.1–7.
13. ANDREOPOULOS, J. & W. RODI. 1984. Experimental investigation of jets in cross-flow. J. Fluid Mech. **138**: 93–127.
14. COELHO, S.L.V. & J.C.R. HUNT. 1989. The dynamics of the near field of strong jets in cross flow. J. Fluid Mech. **200**: 95–120.
15. KELSO, R.M., T.T. LIM & A.E. PERRY. 1996. An experimental study of round jets in cross-flow. J. Fluid Mech. **306**: 114–144.
16. LAPERCHES, M. & J.M. BUCHLIN. 1997. Jet Impact Analysis Using PIV and Heat Transfer. D.C. Thesis, von Karman Institute of Fluid Dynamics.

Visualization of Flow and Heat Transfer Augmentation by Oblique Impingement Jets

HIDEO KIMOTO,[a] CHAYUT NUNTADUSIT,[a] AND KENJI HAMABE[b]

[a]*Graduate School of Engineering Science, Osaka University, Osaka, Japan*
[b]*Akashi R & D Center, Kawasaki Heavy Industry Ltd, Akashi, Japan*

> ABSTRACT: Various nozzle geometries for impingement cooling jets have recently been devised and favorable designs for cooling effectiveness have been reported. However, impinging flow and the characteristics of impingement cooling are not sufficiently clear. This paper reports on an investigation of impingement jet cooling techniques. The impingement cooling characteristics by oblique jets through a rectangular nozzle have been clarified. Preliminary numerical simulations have not necessarily presented the details of heat transfer characteristics of the oblique jets.
>
> KEYWORDS: visualization; rectangular jet; liquid crystal; heat transfer

INTRODUCTION

Jet impingement is widely used in industrial applications as a method to enhance heat and mass transfer involving heating, cooling, and drying processes. These include annealing metals and glasses, cooling gas turbine components and electrical equipment, and drying textiles and films. A number of experimental and numerical studies[1–9] were carried out on impinging jets under various conditions, such as nozzle geometry, flow velocity at the nozzle outlet, distance of the nozzle to plate, thermal boundary condition of target surfaces, and turbulence in jet flow and cross flow. However, relatively few studies paid attention to a rectangular nozzle geometry. In this work, the flow and heat transfer characteristics of obliquely impinging jets through a rectangular nozzle were examined by experiments and numerical analysis.

Normal and oblique jets flowing through rectangular nozzles were caused to impinge onto a flat target plate and their cooling effects as functions of the distance between nozzle and target surface were investigated. Temperature distributions on the target surface were measured by using a temperature-sensitive liquid crystal sheet and the flow fields near the target surface were visualized by the oil film technique on a transparent plate.

The geometric center on the target plate through nozzle is the origin of the coordinate system. The upstream direction of the wall jet and the normal direction of the target plate are referred to as the *X*- and *Z*-axis, respectively. The typical results reported here were obtained at a constant jet flow of 40L/min.

Address for correspondence: Hideo Kimoto, Graduate School of Engineering Science, Osaka University, Osaka 560-8371, Japan.
  kimoto@me.es.osaka-u.ac.jp

EXPERIMENTAL APPARATUS AND METHOD

In the experiments we describe, air for the jet, supplied by a compressor, passed through a flow meter and an injection tube. The jet left a rectangular nozzle with an aspect ratio 10 (14 mm × 1.4 mm, 140 mm long) and impinged normally or obliquely on the target surface. The experiments were carried out for various distances between the nozzle and the target plate, angles of attack of the impinging jet on the target plate, and Reynolds number (Re) values for the jet and the heat flux of the target plate.

The target area was made of stainless steel foil (SUS304), 160 mm × 100 mm and 0.01 mm thick. A liquid crystal sheet was attached with a binder film onto the rear side of the impingement surface of the stainless foil. The liquid crystal sheet used in this study changed its color systematically (from red to blue) twice between 20 and 50°C. Direct current was supplied to the stainless steel foil from a power supply unit via copper bus bars that were attached to both ends of stainless steel foil and acrylic plate. A boundary condition of a uniform heat flux could be achieved on the target surface. Input power to the stainless steel foil was measured by the output current and voltage of the power supply.

An acrylic target plate with an embedded Pitot tube was used to measure the pressure distribution.

EXPERIMENTAL RESULTS

Characteristics of Impinging Jet

The velocity distribution and turbulence of the impinging jet from the rectangular nozzle were measured by using a Pitot tube and a hot wire anemometer. The pressure distribution on the target plate was measured through a small hole on the target plate by a Pitot tube.

The central velocity distributions of the jet without the target plate condition were measured along the major and minor axes of the nozzle by a Pitot tube. Measured velocity distributions along the X-axis; that is, in the direction of the nozzle width; are shown in FIGURE 1 A for various values of L, the distance between nozzle outlet and target surface normalized by the nozzle width (B). When the target plate is in place, $X = 0$ corresponds to a geometric impingement point on the target plate through the center of jet nozzle.

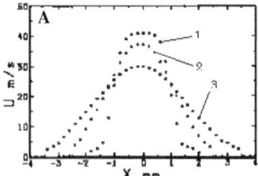 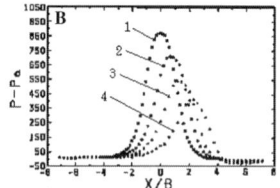

FIGURE 1. (A) Central velocity distributions along the X-axis of a rectangular free jet at various L values (1, $L = 2$; 2, $L = 6$; 3, $L = 10$). (B) Pressure distributions on the target plate along the X-axis ($L = 6$) (1, $\theta = 90°$; 2, $\theta = 60°$; 3, $\theta = 45°$; 4, $\theta = 30°$).

Pressure Distributions on the Target Plate

The pressure distribution on the target surface was measured by a Pitot tube embedded in the acrylic impinging plate and the pressure profile on the target surface was measured by traversing the impinging plate with a computer-controlled stepping motor. The wall pressure profiles on the target surface are shown in FIGURE 1B for various attack angles (θ) of the rectangular jet. As can be seen in FIGURE 1, the stagnation point of the jet impinging on the target plate moves upstream with the attack angle of the impinging jet.

Flow Pattern of Jet on the Target Plate

To visualize the flow pattern near the target plate, the oil film visualization technique was used. In this experiment a 3-mm thick glass plate was used as the target plate. A mixture of liquid paraffin and titanium dioxide was painted on the target surface. At zero-heat flux to the target plate, the change of oil film pattern was recorded every 30 seconds by a camera running for two minutes at a time.

Typical results are shown in FIGURE 2, for various values of L after one- and two-minute runs. The experiments show that the stagnation point of the jet impingement on the target plate move upstream with a decrease in the attack angle.

Heat Transfer Patterns on the Target Plate

Experiments were carried out by changing the heat flux to the target plate, the attack angle of the impinging jet, and the impinging jet velocity. Direct current was supplied to the stainless steel foil from an electric power supply unit via copper bus bars. While varying the electric current to the foil (i.e., the heat flux of the target plate) color patterns of the liquid crystal sheet were recorded by camera. Because the stainless steel foil used in this study was sufficiently thin, the temperatures on the target plate were well represented by colors of the liquid crystal sheet.

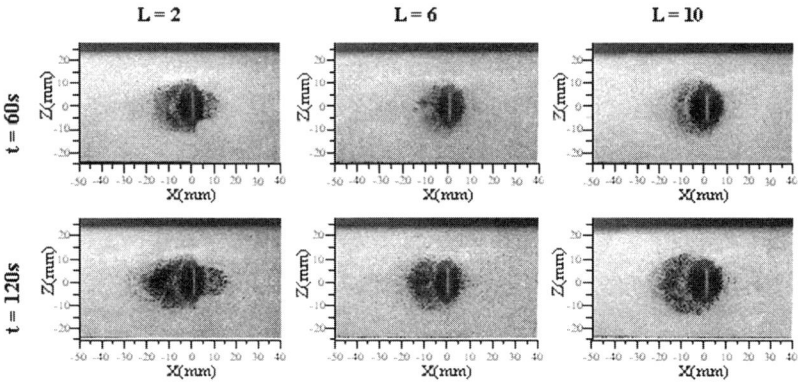

FIGURE 2. Typical wall-flow patterns by the oblique impinging jet (θ = 60°).

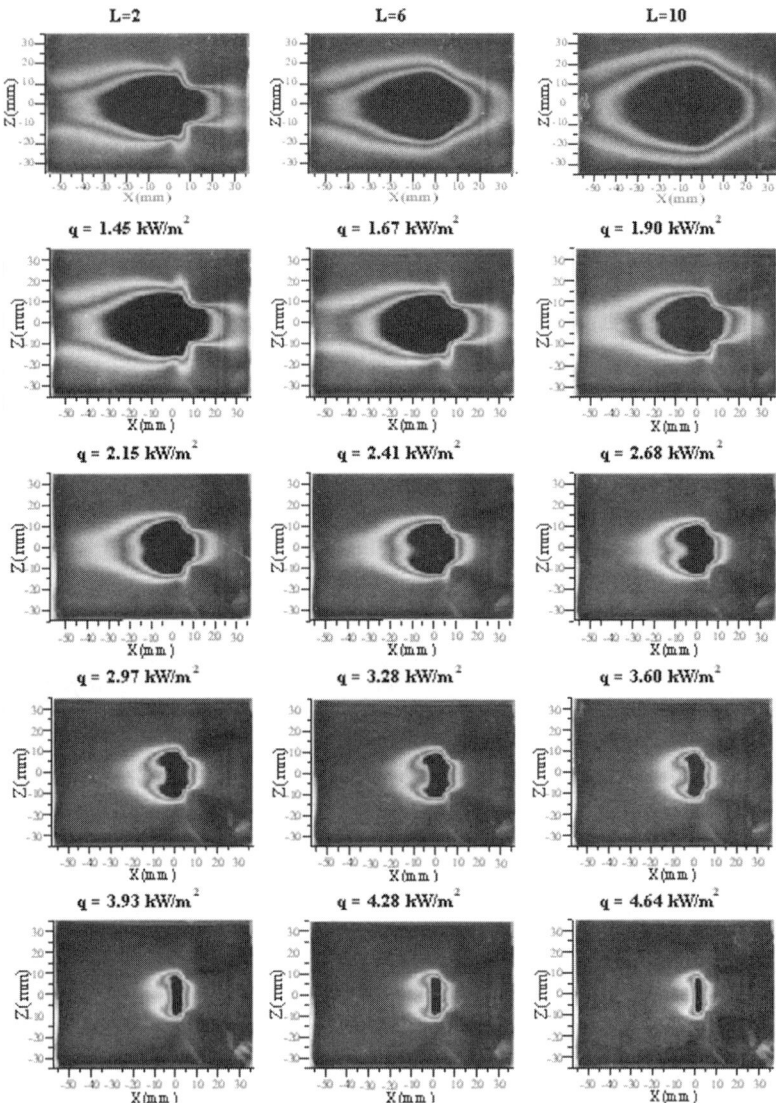

FIGURE 3. Top row presents typical color patterns of liquid crystal sheet obtained by varying L of an oblique impingement jet ($\theta = 60°$, $q = 1.45\,\mathrm{kW/m^2}$). **Bottom four rows** represent typical color patterns of the liquid crystal sheet obtained by varying the heat flux of a constant flow oblique impinging jet ($\theta = 60°$, $L = 2$). $X = 0$ corresponds to a geometric impingement point on the target plate through the center of jet nozzle.

Since the color contour lines on a liquid crystal sheet correspond to isotherms, an isotherm shows a line of constant heat transfer coefficient—that is, constant Nusselt number under constant heat flux boundary conditions. By changing the input current (i.e., the heat flux on the target surface) the position of the isotherm moved and the distribution of heat transfer coefficient could be measured by recording the position of a color temperature on the liquid crystal sheet for each change of the heat flux. Similar experiments were carried out by changing the normalized distance. Some results for θ = 60° are shown in FIGURE 3 (top). In this figure, the temperature distribution for large normalized distance L shows a typical pattern of the oblique impinging jet.

Typical results for the heat transfer patterns are shown in FIGURE 3 (bottom four rows), at various heat flux (q) values. For large heat flux, the jet cooling area was localized and the shape showed a particularly characteristic pattern. The color pattern indicated on the liquid crystal sheet was analyzed by the LED sensor method, which was developed in our laboratory. Moreover, color patterns on the liquid crystal sheet were separated into RGB components and each color component was calibrated into temperature.

NUMERICAL SIMULATION

Preliminary numerical calculations were performed in an attempt to better understand the flow field and the thermal field on the target surface. The model involved a steady state, incompressible, three-dimensional turbulent jet, injected through the rectangular nozzle impinging on a uniform heat flux surface. In the calculations, the k–ε model and the finite volume procedure were adopted to discretize the governing differential equations with a staggered non-uniform rectangular mesh. In the numerical simulation of the present experiment, an insulated target plate (100 mm × 300 mm) was partitioned into 36 × 72 meshes, which included the heated plate (42 mm × 100 mm) partitioned into 26 × 52 meshes. The constant heat flux of the heated plate was set at 1.5 kW/m^2·°C. The cross-section of the rectangular nozzle (1.4 mm × 7.0 mm) was separated into 8 × 14 meshes and the numerically analyzed height (100 mm) into 40 meshes.

A hybrid scheme was used to discretize the convection and conduction terms, and the equations were solved by the SIMPLEST algorithm.[10] Typical numerical results are shown in FIGURE 4.

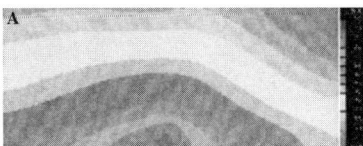 

FIGURE 4. (**A**) Numerical distribution of temperature on the target surface (θ = 60°, L = 10). (**B**) Flow velocity along the X-direction in the first mesh plane from target surface.

SUMMARY

In general, the wall jet for impingement cooling through a rectangular nozzle does not spread over the target plate along the long axis of the nozzle. The wall jets generated by oblique jets for impingement cooling through a rectangular nozzle, accelerated downstream, and decelerated in the upstream direction. Consequently, the wall jets are distorted on the target plate and the stagnation point of the impinging jet moves upstream along the target plate. Our numerical simulations did not successfully reflect the details of the temperature pattern on the target plate for small L. Obviously, better models are needed and until then, visualization techniques are remain effective and important.

REFERENCES

1. MARTIN, M. 1977. Heat and mass transfer between impinging gas jet and solid surfaces. Adv. Heat Transfer **13**: 1–60.
2. GOLDSTEIN, R.J. & J.F. TIMERS. 1982. Visualization of heat transfer from arrays of impinging jets. Int. J. Heat Mass Transfer **25**(12): 1857–1868.
3. HO, C. & E. GUTMARK. 1987. Vortex induction and mass entrainment in a small-aspect-ratio elliptic jet. J. Fluid Mech. **179**: 383–405.
4. GOLDSTEIN, R.J. & M.E. FRANCHETT. 1988. Heat transfer from a flat surface to an oblique impinging jet. Trans. ASME J. Heat Transfer **110**: 84–90.
5. KIMOTO, H., K. IIZUKA & K. HAMABE. 1992. Precise and convenient measurement of the temperature profile appearing on the temperature-sensitive liquid crystal. Heat Transfer Jpn. Res. **21**(1): 77–90.
6. HIWADA, M., et al. 1999. Heat transfer characteristics of a non-circular impinging jet. 36th Natl. Heat Transfer Symp. of Japan. **2**: 313–314.
7. NUNTADUSIT, C., H. ISHIDA & H. KIMOTO. 1999. Characteristics of some rectangular impinging jets. J. Flow Measure. **15**(21): 85–92.
8. NUNTADUSIT, C., H. KIMOTO, H. ISHIDA & K. HAMABE. 2000. Flow and heat transfer characteristics of some impinging jets. Proc. 9th Int. Symp. on Flow Visualization, Edinburgh. **165**: 1–15.
9. NUNTADUSIT, C., H. KIMOTO & K. HAMABE. 2001. On impingement cooling against heated concave surface. Proc. AFI-2001, Sendai, Japan. 348–353.
10. PATANKAR, S.V. 1980. Numerical Heat Transfer and Fluid Flow. Hemisphere Pub. Co.

Unsteady Flow Patterns in the Vicinity of Heated Wall-Mounted Transverse Ribs

GUILLAUME POLIDORI AND JACQUES PADET

Laboratoire de Thermomécanique, UTAP EA 2061, Université de Reims, Reims, France

ABSTRACT: This paper deals with experimental modeling of the unsteady junction flow features in the vicinity of an isoflux heated wall with mounted insulated rectangular ribs representing three distinctive ribbed test geometries. Both flow visualizations and surface temperature distributions show that the blockage effect upstream of the ribs, as well as the presence of complex eddy structures inside the open cavities, significantly affect the heat transfer process. All the configurations indicate degraded heat transfer performance in the area close to the ribs and an enhancement just downstream from the last rib.

KEYWORDS: unsteady free convection; flow visualization; heat transfer

INTRODUCTION

Knowledge of transient free convection from vertical surfaces with mounted large-scale obstacles is of great interest in many industrial applications (e.g., surfaces of buildings and electronic equipment). Indeed, roughness elements can induce transition to turbulence in the flow behavior and modify the local heat transfer process. However, how this modification is generated has not been fully examined because the analysis of flow patterns and thermal fields is very complex in the unsteady flow regime. Consequently, relatively few studies report on natural convection in ribbed surfaces. Burak *et al.*[1] studied the case of a vertical heated plate in the presence of one or several rectangular steps and noted a rotational flow between the ribs that intensifies the process of heat transfer. *A contrario*, Tanda[2] showed that the presence of ribs alters the heat transfer considerably. Shakerin *et al.*[3] observed that the influence of the roughness is mainly localized to within about two roughness heights above and below the roughness location. Nevertheless, their dye visualization did not show the presence of a flow-separation bubble inside the cavities between the ribs. Considering an array of heated protrusions on a vertical surface in water, Joshi *et al.*[4] presented flow visualizations with no evidence of vortex formation around the protrusions. Achyria *et al.*[5] indicated that a smooth surface yields a greater heat transfer rate when compared to ribbed geometries, due to dead regions in the near-rib space. Abu-Mulaweh *et al.*[6] focused attention on natural convection flow over a backward-facing step where the step height is significantly affects the flow and heat transfer rate. Bhavnani *et al.*[7] specify that enhancement of heat transfer

Address for correspondence: Guillaume Polidori, Ph.D., Laboratoire de Thermomécanique, UTAP EA 2061, Université de Reims, 51687 Reims, France.
guillaume.polidori@univ-reims.fr

can be accomplished by using transverse roughness elements with proper sizing and shape selection.

The purpose of the present work is to extend previous studies to the unsteady phenomena occurring in separated boundary layers from both dynamic and thermal viewpoints, in the case of isoflux heated vertical wall mounted with different insulated roughness geometries.

EXPERIMENTAL CONDITIONS

Experiments were performed in a plexiglas towing tank with water ($Pr = 7$) as the working fluid. The wall was a vertical plane thermofoil heater ($4,000 \text{W/m}^2$) on which were placed three wooden transverse rectangular ribs as insulated roughness elements with poor conduction paths. To obtain more details about the unsteady phenomena occurring in perturbed boundary layers, three complementary configurations, differing in the size of the ribs, were tested. Thus, the array of ribs was, streamwise, either regularly increasing in size (I), constant (II), or regularly decreasing in size (III). The first rib was located 54 mm from the leading edge of the wall and the others were regularly spaced at intervals of 18 mm. All the ribs were 18 mm thick and had a thickness to height ratio varying from 0,5 to 1,5. The modified Grashof number (Gr) at the entrance of the roughness region was $Gr^* = 10^8$ so that $Gr^*Pr < 10^{11}$, corresponding to a laminar domain. Thermal data acquisition was obtained with thermocouples (chromel–alumel type) calibrated to ±0.1 K. To examine the flow features, two kinds of flow visualization techniques were employed to get either streaklines or streamlines in the meridian section of the flow. For this purpose, data was obtained by an electrolytic precipitation method as well as by a suspended fine rilsan particles[8] illuminated with a laser sheet from which integrated streamlines could be drawn.

UNSTEADY FLOW PATTERNS

The objective was to characterize the flow field around the roughness elements qualitatively through streakline visualizations. An example, at time $t = 300$ sec, is given in FIGURE 1. Whatever the configuration, the presence of ribs spaced 18 mm apart induces separation of the viscous layer that penetrates inside the open cavities and follows the solid external contours of the protrusions, except downstream of the last one, where the wall reattachment phenomenon occurs. Consequently, one observes the appearance of circulation flow that is considerably altered. To obtain more details about the development in time of these vortex motions, we visualized the junction flow with track particles (rilsan) and have drawn the integrated corresponding streamlines at various times. Corresponding patterns are given in FIGURE 2 for the case $t = 112$ sec.

Early in the start-up transient, up to $t = 57$ sec, one observes the birth of vortices just behind the ribs, due to the separation of the dynamic boundary layer, whatever the configuration. At increasing times, all the open cavities present two oppositely rotating vortices and only a single cell is evidenced in the first cavity of configuration

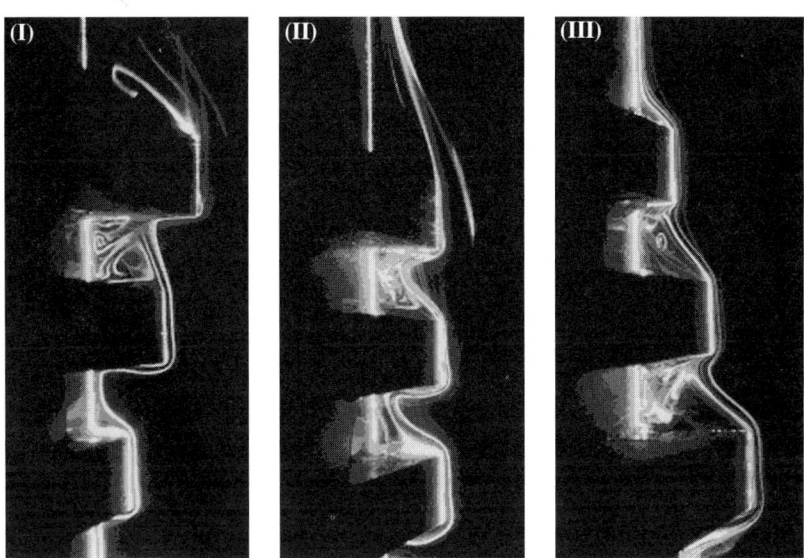

FIGURE 1. Streaklines at time $t = 300$ sec visualized with an electrolytic precipitation method.

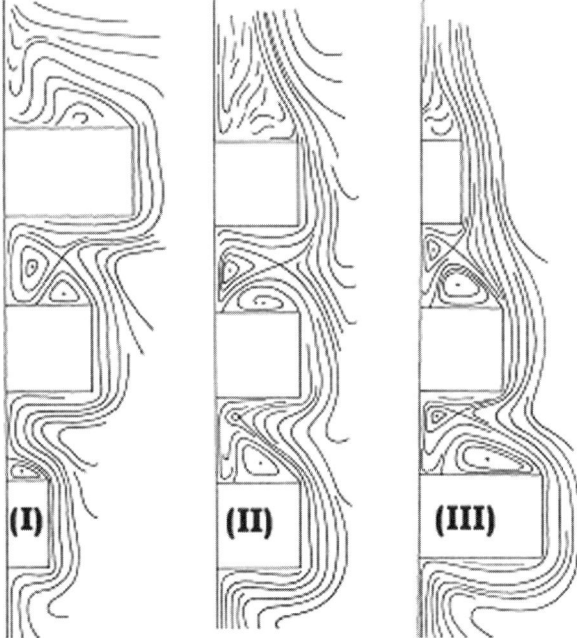

FIGURE 2. Streamline patterns at time $t = 112$ sec deduced from track visualizations.

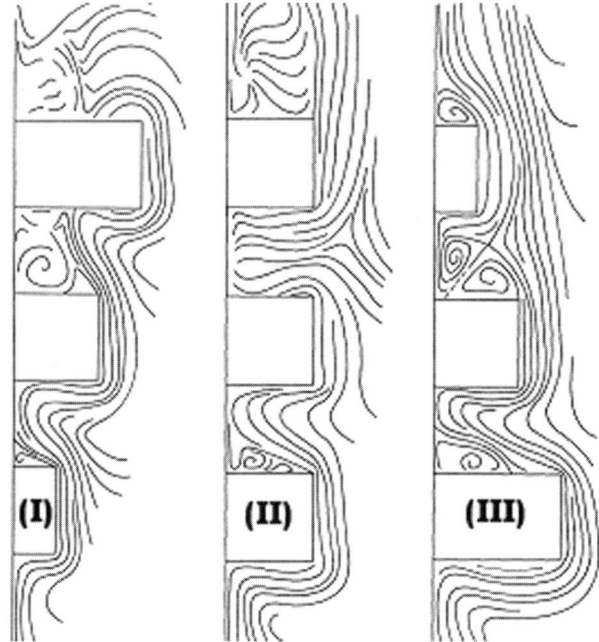

FIGURE 3. Three-dimensional behavior of the flow from streamlines patterns at time $t = 610$ sec.

(I) because of the small size of the first element. Moreover, whatever the time, the vortex motion does not occupy the entire first cavity for configuration (I). The transient vortex shedding process is evidenced in FIGURE 2, configuration (II). Indeed, one can see that the upper vortex of the first cavity is progressively shed downstream into the adjacent flow. As time passes, the flow in the vicinity of the roughness becomes very complex and tends to a three-dimensional behavior, except in configuration (III) where the structures stay almost coherent, as can be seen in FIGURE 3.

TRANSIENT WALL TEMPERATURE VARIATIONS

The tested ribbed surfaces were instrumented with ten thermocouples regularly located so that data could be obtained on the wall at the lower and upper surfaces of each element. To ensure the utility of the thermal device, a smooth surface without ribs was studied first and this was considered a reference. It appears that, with increasing time, the temperature profiles reach, with good agreement, the steady state solution of natural convection on vertical surfaces.[9] Before comparing the various configurations, in FIGURE 4 we present the time-evolution of surface temperatures for the smooth case and for configuration (II). Looking at FIGURE 4, where both graphs are presented with the same scales, one can see that spanwise temperature measurements indicate the complex effects of the protrusions on the heat transport, especially close to the ribbed region.

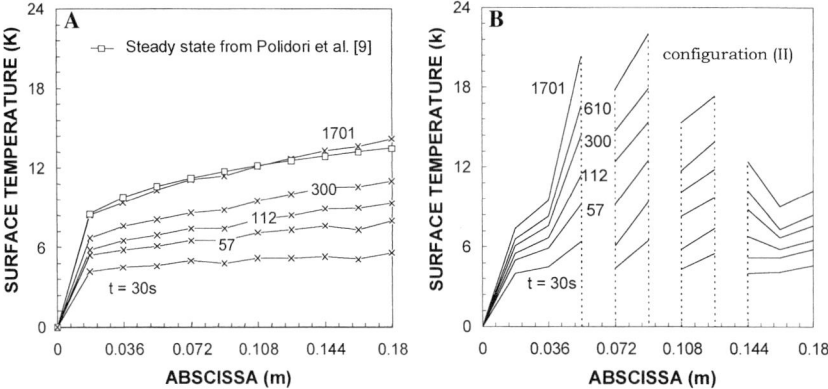

FIGURE 4. Streamwise temperature distributions: **(A)** smooth walls, **(B)** ribbed wall (configuration II).

To obtain a precise idea about how the heat transfer is modified during the transient, in FIGURE 5 we present the temperature distributions at two extreme times, namely, $t = 57$ sec and $t = 1{,}701$ sec, which correspond to the observed steady state value. To answer the fundamental questions on heat transfer enhancement in the presence of ribs, the smooth flow case is also included in FIGURE 5.

Three distinct regions of temperature profiles are evident in FIGURE 5, regardless of the configurations. The first region is from the bottom of the plate up to the lower surface of the first rib element. It appears that up to the distance of a width rib (18 mm) below the first rib, the ribbed surface has no influence on the heat transfer compared to the smooth case. Immediately below the first element the surface temperature increases significantly due to the slowing of the flow. Because the convective heat transfer is defined by $h(x,t) = \Phi/\Theta_w(x,t)$, where Φ is the heat flux density

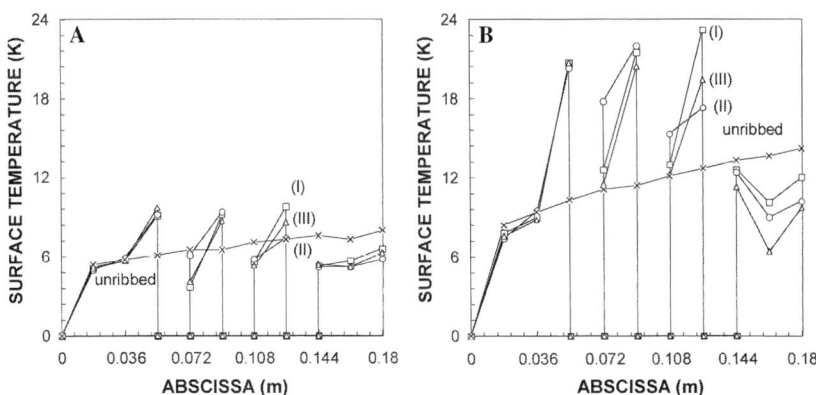

FIGURE 5. Comparisons of streamwise temperatures between the three configurations at two distinct times: **(A)** $t = 57$ sec, **(B)** $t = 1{,}701$ sec.

and $\Theta_w(x,t)$ is the surface temperature, one notes a decrease in the heat transfer phenomenon in this first region. The temperature has identical representative curves for the three configurations, indicating that the size of the first insulated element is not a dominant parameter.

The second region is defined near the protrusions up to the trailing edge of the last one. It can be seen that the ribs significantly influence the heat transfer rate, especially at the lower surface of each element, by increasing the temperature whereas a significant decrease of the temperature is noted behind each element. The reason is that the fluid flows around the thermally insulated elements and the fluid becomes colder. The observed peak in the temperature inside the cavities is a consequence of the vortex motion and the probable transition to turbulence with increasing time. Once again, the smooth surface yields a greater heat transfer rate when compared with ribbed geometries.

In the last region below the ribs one observes interesting differences from the two previous regions. Indeed, the temperature profiles here are lower than in the smooth flow case, indicating a significant enhancement heat transfer phenomenon.

Comparison of the three configurations indicates that the inside of the first cavity configuration (II) is inferior in terms of the heat transfer rate, as is configuration (I) in the second cavity. Configuration (III) is the best for heat transfer rate in downward flow on ribbed geometries.

SUMMARY

An experimental model was used to qualify the flow features and heat transfer processes in unsteady vertical free convection flow through insulated large-scale ribbed geometries, while the plate was isoflux heated. Both flow visualizations and temperature measurements indicate that the flow and thermal characteristics in the neighborhood of the roughness elements are significantly affected by the roughness. The presence of vortex structures seems to decrease the heat transfer process for the ribbed region, whereas an enhancement of heat transfer rate is observed below this region.

ACKNOWLEDGMENT

The French Agency Amélioration des Echanges Thermiques (AMETH) supported this study with a research grant.

REFERENCES

1. BURAK, V.S., S.V. VOLKOV, O.G. MARTYNENKO, *et al.* 1994. Free convective flow on a vertical plate with a constant heat flux in the presence of one or more steps. J. Eng. Phys. Thermophys. **67:** 848–853.
2. TANDA, G. 1997. Natural convection heat transfer in vertical channels with and without transverse square ribs. Int. J. Heat Mass Transfer **40:** 2173–2185.

3. SHAKERIN, S., M. BOHN & R.I. LOEHRKE. 1988. Natural convection in an enclosure with discrete roughness elements on a vertical heated wall. Int. J. Heat Mass Transfer **31:** 1423–1430.
4. JOSHI, Y., T. WILLSON & S.J. HAZARD. 1989. An experimental study of natural convection from an array of heated protusions on a vertical surface in water. J. Electronic Packaging **111:** 121–128.
5. ACHARYA, S. & A. MEHROTRA. 1993. Natural convection heat transfer in smooth and ribbed vertical channels. Int. J. Heat Mass Transfer **36:** 236–241.
6. ABU-MULAWEH, H.I., B.F. ARMALY & T.S. CHEN. 1995. Laminar natural convection flow over a vertical backward-facing step. J. Heat Transfer **117:** 895–901.
7. BHAVNANI, S.H. & A.E. BERGLES. 1990. Effect of surface geometry and orientation on laminar natural convection heat transfer from a vertical flat plate with transverse roughness elements. Int. J. Heat Mass Transfer **33:** 965–981.
8. MERZKIRCH, W. 1987. Flow Visualization, 2nd edit. Academic Press Inc., London.
9. POLIDORI, G., C.E. MLADIN & T. DE LORENZO. 2000. Extension de la méthode de Karman–Pohlhausen aux régimes transitoires de convection libre, pour $Pr > 0.6$. C.R. Acad. Sci. Paris, t. 328, **IIb:** 763–766.

Investigation of Transport Phenomena Inside a Microcapsule

LEONID GOUBERGRITS, KLAUS AFFELD, PERRINE DEBAENE, AND ULRICH KERTZSCHER

Biofluidmechanics Laboratory, Charité, Spandauer Damm, Berlin, Germany

ABSTRACT: Mass transfer within a microcapsule is enhanced by convection, achieved by introducing a movable body inside the microcapsule. This body has a different density from the fluid within the microcapsule and can be moved relative to the microcapsule by application of an external force, such as a magnetic field or acceleration/deceleration. The effect of the transport improvement was investigated as a function of the Peclet number (Pe) by using computational fluid dynamics and particle image velocimetry. The results show that the period to achieve 80% saturation in the microcapsule was reduced by 60% with a movable body with half of the diameter of the microcapsule and $Pe = 600$, compared to the mass transport in the microcapsule with a non-movable body.

KEYWORDS: particle image velocimetry; microcapsules; enhanced mass transfer; computational fluid dynamics; Peclet number

INTRODUCTION

Microcapsules are used in many chemical or biochemical processes. They consist of semipermeable membranes that enclose absorbing solutions, catalysts, or vital biomaterials, such as enzymes. In general the capsules have a spherical shape and a diameter of up to 3 mm. The microcapsules are usually dispersed in a fluid, which contains nutrient elements and other components required for the process. Diffusion and convective flow of the surrounding fluid transports the nutrients and metabolic products to and from the microcapsule membrane. These substances pass the membrane of the microcapsule by diffusion. Transport within the microcapsule normally depends on diffusion. However, slow diffusion in the microcapsules can render the whole process uneconomic.[1,2] A new approach was used to accelerate the transport process within the microcapsule by introducing convection inside the capsule. The convection conveys the substances from the membrane into the center of the microcapsule and vice versa. This was achieved by placing a movable spherical body inside the microcapsule. This body had a different density or different magnetic properties, than the fluid within the microcapsule. It could be moved relatively to the microcapsule by applying an external force using a magnetic field or by acceleration/deceleration. The flow field generated by the motion of the body inside the microcapsule, and mass transport improvement by this motion, were investigated.

Address for correspondence: Leonid Goubergrits, Dr. Ing., Biofluidmechanics Laboratory, Charité, Spandauer Damm 130, D-14050 Berlin, Germany.
  leonid.goubergrits@charite.de

METHODS

Investigation of Mass Transport Inside a Microcapsule by Computational Fluid Dynamics

Two methods were used to investigate the flow field generated by the motion of the body inside a microcapsule and its effect on mass transport improvement by the generated convective flow field: (1) computational fluid dynamics (CFD) and (2) particle image velocimetry (PIV). The latter was used to investigate the flow field generated by the movable body inside the microcapsule. The average velocity of the fluid in the microcapsule was measured in order to calculate the Peclet number (Pe) of the convective–diffusive mass transport inside the microcapsule. Transport improvement was measured by the decrease in the period to achieve an 80% saturation of the transported material inside the microcapsule.

CFD was used to investigate the effect of the mass transport improvement inside the microcapsule by the convection generated by a moving body within the microcapsule. The CFD flow program packet, FLUENT5, which creates a structured or unstructured computational mesh, is well suited for incompressible flow and transport problems. A simplified model was applied: a sphere simulating a movable body, was placed in a spherical shaped microcapsule having diameter $D_{m.c.} = 3\,mm$. Only transverse motion along the microcapsule axes, without rotation of the internal movable body, was considered. This simplification requires the solution of a two-dimensional axially-symmetric problem. Three cases using different diameters for the movable body ($d_{m.b.} = 0.6$, 1.0, and $1.5\,mm$) were generated with a geometric preprocessor, Pre-BFC. A steady convective flow field was generated inside the microcapsule. A simple steady flow model approximated the motion of the movable body in the microcapsule: the surface of the movable body was classified as inflow. This allowed the definition of the velocity magnitude and direction at all surface cells. The velocity was set according to the velocity of the movable body. This formulation of boundary conditions limits the CFD model to fixed positions of the movable body. The calculation provided the pressure scalars and the velocity vectors for all cells in the microcapsule. Next, a non-stationary transport process inside the microcapsule with a steady flow field was calculated. A constant concentration of the diffusing substance outside the microcapsule was defined as the boundary condition. During this non-stationary simulation of the mass transport process the position of the movable body was kept constant. The Peclet number ($Pe = LV/D_f$) is a dimensionless parameter that defines a relation between convection and diffusion. L is the diffusion path or characteristic length, V is the averaged velocity magnitude of the flow field inside the microcapsule, and D_f is the diffusion constant of the transported material. Pe was varied by changing the diffusion path, using various body diameters and various values of the averaged velocity produced by the movable body inside the microcapsule. The transport improvement effect was evaluated by comparing the periods needed to achieve 80% saturation in the microcapsule. Preliminary calculations were performed to study the mean velocity inside the microcapsule as a function of position, diameter, and velocity of the movable body. Two positions of the movable body were considered: the body at the centre of the microcapsule and at the lateral position. Two simulations were performed for each set of parameters for the movable body in the lateral position where the body moved to, and away, from the microcapsule wall.

Investigation of the Flow Field Inside a Microcapsule by Particle Image Velocimetry

The relation between the velocity of a movable spherical body and the average velocity in the microcapsule was investigated experimentally by using the PIV method. The experimental set-up for the flow field investigation employed a 20-times enlarged microcapsule model with a diameter of 60 mm. In this set-up the rotational movement of a motor was converted to a linear up and down motion of the movable sphere in the fixed capsule. This resulted in an axial motion of the body $V_{m.b.} = V_{max}\sin(\omega t)$, where ω is the rotational velocity of the motor. The capsule was made of transparent polysterene and filled with various solutions. The kinematic viscosities of the glycerin/water solutions, measured after the experiments with a capillary viscometer (Schott Geräte GmbH) were: 742×10^{-6} m^2/sec (I), 430×10^{-6} m^2/sec (II), and 1×10^{-6} m^2/sec (III) (water) in three different experiments. Rotational velocities 12 and 24 rpm were investigated in each experiment. In all these experiments the diameter of the movable sphere was $d_{m.b.} = 0.5 D_{m.c.}$. The particles (Vestosint 7182, Degussa AG) with mean diameter of 40 μm were added to the fluid as a tracer for flow visualization. The part of the experimental set up with the microcapsule model was placed in a transparent water bath for better optical conditions. The flow field was measured using the digital particle image velocimetry (DPIV) method. The flow was assessed with a digital camera (MX12) with $1,024 \times 1,024$ pixel resolution. A central cross section of the microcapsule was illuminated from left by a light sheet. The light sheet was generated with an Argon laser (Exel 3000, 2 W). A cross-correlation method implemented in the Visiflow™ software was used to analyze the velocity field.

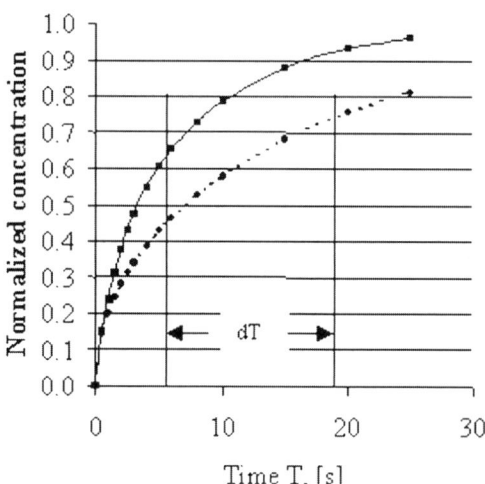

FIGURE 1. Time-dependent curves of the normalized concentration of transported material inside a microcapsule with (*black line*) and without (*dotted line*) convective flow field generated by a movable body. The body diameter was half of the microcapsule diameter. $Pe = 670$.

RESULTS

Investigation of Mass Transport Inside a Microcapsule by CFD

The results of the CFD simulations show that a transport process inside a microcapsule can be improved by the introduction of movable body even without generating of a convective flow field. For example, the period to achieve 80% saturation in the microcapsule was reduced by 40% with a stagnant body with a diameter of half the microcapsule diameter. This effect with the stagnant body is a result of reducing the diffusion path length L, which defines the characteristic time T of the mass transport process, since the characteristic time T for the diffusion is proportional to $L^2/(4D_f)$. The generation of a flow field in the microcapsule results in further acceleration of the mass transport process. In the case of a convective flow field, the material is transported to the center of the microcapsule by convection. This results in acceleration of the mass transport and reduces the period needed for the transport process. The period can be further reduced by 60% ($Pe = 670$) in comparison with the transport process with a stagnant body inside the microcapsule. FIGURE 1 shows

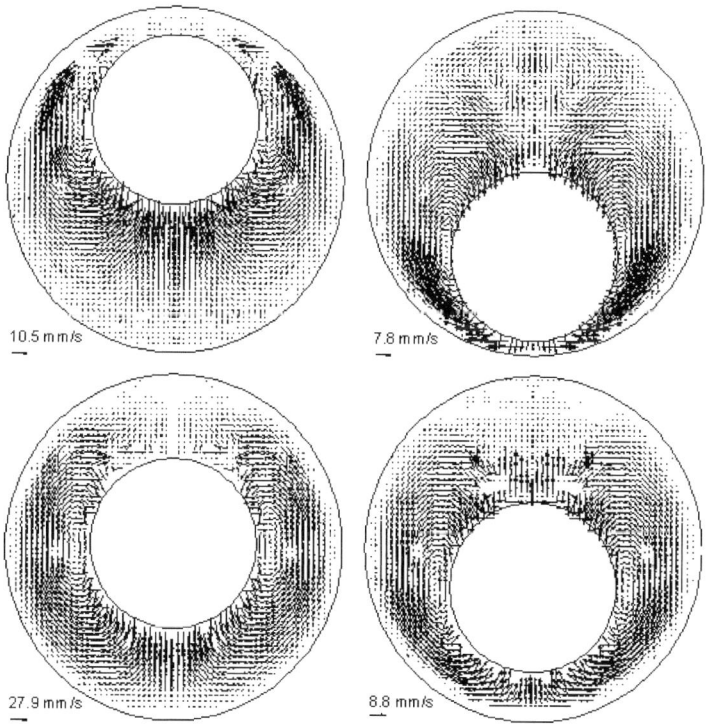

FIGURE 2. Sequence of four velocity fields generated by the movable body in the microcapsule. The PIV flow field was measured only in the left half of the model. The right half was mirrored. The velocity field shows an expected generation of the ring vortex[3] in the microcapsule.

results of two numerical simulation with the CFD program FLUENT. Two time courses of the averaged transported material concentration inside the microcapsule with (black line) and without (dotted line) a convective flow field are depicted.

As can be seen, the transport process can be accelerated 4–5 times by the introduction of a movable body inside a microcapsule and by the generation of convection inside them. The curves in FIGURE 1 show that the effect of the mass transport period reduction increases with time and is more effective for transport processes with a saturation of more than 30–40%. The effect increases with higher Pe values achievable by larger velocities of the movable body or larger body diameters. The data shows that at very low Pe ($Pe \approx 8$), the effect of the period reduction is only 10%. The effect increases to 30% at $Pe \approx 80$ and 60% at $Pe \approx 600$. The results show that the generated flow field depends on the diameter of the movable body and its position in the microcapsule.

Investigation of the Flow Field Inside a Microcapsule by PIV

The main aim of the PIV investigation was to determine the mean velocity of the flow field in the microcapsule generated by the movable spherical body. This parameter could not be determined by the CFD method, because a very simple steady flow model was used to simulate a flow field in the microcapsule. FIGURE 2 shows some examples of flow fields generated by a movable sphere inside a microcapsule with glycerin solution (II). The rotation speed of the motor was 24 rpm. The maximal Reynolds number (Re) was 2.4. The experiments showed that the relation between the average velocity in the microcapsule V_{mean} and movable body velocity $V_{m.b.}$ is about $V_{mean}/V_{m.b.} = 0.07$ with a diameter of half the microcapsule diameter, independent of the movable body velocity for $0 < Re < 500$. FIGURE 3 shows the calculated V_{mean} for one period of the movable body motion in the microcapsule. The curve reflects the sine function characteristics of the movable body velocity.

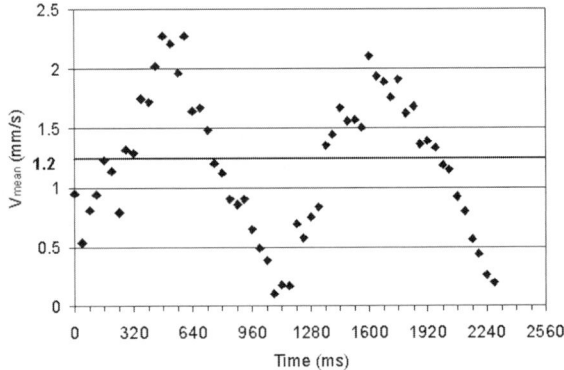

FIGURE 3. Mean velocity magnitude in the microcapsule by motion of the movable body with maximum $Re = 2.4$ (maximum velocity $V_{m.b.} = 35$ mm/sec). Movable body diameter was half of the microcapsule diameter. *Thicker black line* marks velocity averaged over the entire cycle.

SUMMARY

The results of the CFD investigation of the mass transport inside a microcapsule show that a significant improvement in mass transport can be achieved by the generation of a convective flow field inside the microcapsule. The resulting effect depends on the generated flow field inside a microcapsule, investigated experimentally by the PIV method, and depends on size and velocity of the internally movable body. The results show that the generation of the flow field. in the microcapsule by a movable body was independent of Re for a constant body diameter. An experimental study is planned to validate the results of the CFD estimation of mass transport improvement by the generation of the convective flow field inside the microcapsule. Work is also planned to investigate potential methods for generating a convective flow field inside the microcapsule.

REFERENCES

1. WIESMANN, R., W. ZIMELKA, H. BAUMGARTL, *et al.*1994. Investigation of oxygen transfer through the membrane of polymer hollow spheres by oxygen micro-electrodes. J. Biotechnol. **32**(3): 221–229.
2. PARK, J.K., G.S. JEONG & H.N. CHANG. 1997. The effect of oxygen transfer on the activity of encapsulated whole cell β-galoctosidase. Bioprocess Eng. **17**: 197–202.
3. TIETJENS, O. 1970. Strömungslehre. Bewegung der Flüssigkeiten und Gase. 36–89. Springer-Verlag, Berlin.

Measurement of the Density of CO_2 Solution by Mach–Zehnder Interferometry

YONGCHEN SONG,[a] MASAHIRO NISHIO,[b] BAIXIN CHEN,[a] SATOSHI SOMEYA,[b] TSUTOMU UCHIDA,[b] AND MAKOTO AKAI[b]

[a]*Research Institute of Innovative Technology for the Earth, AIST-Division of RITE, Research Institute of Energy Utilization, Tsukuba-shi, Japan*

[b]*National Institute of Advanced Industrial Science and Technology, AIST-Division of RITE, Research Institute of Energy Utilization, Tsukuba-shi, Japan*

ABSTRACT: The density of CO_2 solution was measured by using Mach–Zehnder interferometry in the pressure range from 5.0 to 12.5 MPa, at temperatures from 273.25 to 284.15 K, and CO_2 mass fraction in solution up to 0.061. It was found that the density difference between the CO_2 solution and pure water at the same pressure and temperature is monotonically linear with the CO_2 mass fraction. The slope of this linear function, calculated by experimental data fitting, is 0.275.

KEYWORDS: CO_2; density; Mach–Zehnder; interferometry; greenhouse gases; ocean; sequestration

INTRODUCTION

Global warming induced by growing concentration of greenhouse gases has attracted increasing importance in the past two decades. Environmental impact and climatic changes play a key role in sustainable economic development. Mitigating CO_2 gas emitted from fossil fuel burning and other major greenhouse gases (CH_4, N_2O, etc.) in the atmosphere is undoubtedly an urgent task aimed at stabilizing the atmosphere concentration of greenhouse gases at a certain level and to prevent dangerous disruption of the climate system due to human activities. Several options have been proposed, including biological sequestration, geological sequestration, and ocean storage.[1,2] The latter is considered to have advantages in cost, efficiency, capability, and environmental impact.

There are two ideas for CO_2 ocean storage. One is *dented ocean floor storage*, by which a large amount of liquid CO_2 could be stored in an ocean basin deeper than 3,000 m in order to guarantee that the density of CO_2 is larger than that of seawater. Another idea is the *middle depth dilution*. At a depth of about 1,000 m, LCO_2 might be directly injected into the ocean by either of a set of fixed nozzles or a method called *moving-ship*[3,4] to form two plumes of LCO_2 droplets and CO_2 enriched seawater. The CO_2 droplets should dissolve into the seawater as they rise due to positive

Address for correspondence: Yongchen Song, Ph.D., Research Institute of Innovative Technology for the Earth (RITE), AIST, 1-2-1 Namiki, Tsukuba-shi, Ibaraki 305-8564, Japan. y.song@aist.go.jp

buoyancy. In this way, it is expected that the local lowest pH could be controlled to a level with acceptable biological impact.[5]

For both ideas, knowledge of physical and chemical mechanism and properties of this CO_2–seawater system (e.g., the mechanism and characteristics of hydrate formation, the density of CO_2 solution, the solubility, and surface tension) are indispensable for engineering design and for developing a reasonable numerical model to predict the ocean environmental impacts on CO_2 sequestration. The density of CO_2 solution will not only influence the structure of the plumes near the release nozzle, but also the further evolution of CO_2 enriched plume. For gaseous CO_2 saturated solution, Parkinson[6] measured the density data in pressure range from 1.0 to 3.4 MPa at temperatures from 273.15 to 313.15 K. Nighswander[7] added some new data for pressures from 2.0 to 10.0 MPa at temperatures from 353.15 to 473.15 K. Focusing on CO_2 ocean storage investigation, Ohsumi[8] gave density data at lower concentrations by using a vibration density meter. Recently Aya[9] reported CO_2 solution density varied with respect to CO_2 mass fraction, by using a weighting technology.

In this paper, we report the last experimental results of CO_2 solution density at pressures and temperatures ranging from 5.0 to 12.5 MPa and 273.25 to 284.15 K, and CO_2 concentration (in mass fraction) up to 0.061. These experiments were carried out by using a high-pressure cell, with standard safely pressure of 15.0 MPa, and Mach–Zehnder Interferometry.

METHODOLOGY

Fundamental Equations

The density of CO_2 solution in fresh water was measured based on the method of Mach–Zehnder interferometry. The fundamental principle of Mach–Zehnder interferometry is to make use of the physical phenomenon of the difference of the refractive index indicated by interference from two identical laser beams that are divided by a half-silvered mirror. One beam passes through the CO_2 solution test cell and the other is a reference beam. From optical physics, we can relate the fringe shift between the CO_2 solution and the fresh water to the difference between these refractive indexes[10]

$$\Delta n \delta = (n - n_{h2o})\delta = \lambda \Delta s, \tag{1}$$

where Δn is the difference of refractive indexes, δ is the distance inside of test cell, λ is the laser wave length, ΔS is the fringe shift between CO_2 solution and fresh water, and n_{h2o} and n are the refractive indexes of water and CO_2 solution, respectively.

For a pure substance, we have the Lorentz–Lorenz formulation[11]

$$\frac{n_i^2 - 1}{n_i^2 + 2} = \frac{R_i}{M_i}\rho_i, \tag{2}$$

where ρ is the density, M is the molar mass, and R is the molar refractivity. Index i indicates the ith kind of substance. Applying Equation (2) to pure water and CO_2, the following relations can be derived straightforwardly:

$$\frac{n_{h2o}^2 - 1}{n_{h2o}^2 + 2} = \frac{R_{h2o}}{M_{h2o}} \rho_{h2o}, \qquad (3)$$

$$\frac{n_{co2}^2 - 1}{n_{co2}^2 + 2} = \frac{R_{co2}}{M_{co2}} \rho_{co2}. \qquad (4)$$

According to Maxwell electric-magnetic theory[12]

$$\frac{\varepsilon - 1}{\varepsilon + 2} = \frac{4}{3}\pi \sum_i N_i \alpha_i, \qquad (5)$$

where ε is the electric constant, α is the electronic polarizability of atom, and N is the atomic number in unit volume. Subscript i indicates the ith kind of atom of the molecule. The equation to relate the density to n for a CO_2 solution can be obtained from

$$\frac{\varepsilon - 1}{\varepsilon + 2} = \frac{4}{3}\pi \left(\sum_i N_i \alpha_i\right)^{co2} + \frac{4}{3}\pi \left(\sum_i N_i \alpha_i\right)^{h2o} = \frac{\varepsilon_{co2} - 1}{\varepsilon_{co2} + 2} + \frac{\varepsilon_{h2o} - 1}{\varepsilon_{h2o} + 2} \qquad (6)$$

Note that (6) coincides with Feynman's[13] arguments on the refractive indexes of a mixture. Substituting $\varepsilon_i = n_i^2$ into (6),[12] we finally obtain the equation applicable to the CO_2–water system

$$\frac{n^2 - 1}{n^2 + 2} = \frac{n_{co2}^2 - 1}{n_{co2}^2 + 2} + \frac{n_{h2o}^2 - 1}{n_{h2o}^2 + 2} = \frac{6.68}{44}\rho_{co2} + \frac{3.71}{18}\rho_{h2o}, \qquad (7)$$

where ρ_{h2o} and ρ_{co2} are the densities of water and CO_2, respectively. The density and the refractive index of CO_2 can be calculated by solving Equations (1) and (7), since the fringe shift between CO_2 solution and fresh water can be directly measured and the other parameters appearing in these two equations (the density of water, the distance inside of test cell, laser wave-length, and refractive index of water) are known. Consequently, the density of CO_2 solution and the mass fraction of CO_2 are obtained from

$$\rho = \rho_{co2} + \rho_{h2o}, \qquad (8)$$

$$C = \frac{\rho_{co2}}{\rho} = \frac{\rho_{co2}}{\rho_{co2} + \rho_{h2o}}. \qquad (9)$$

The Experimental Setup

The pressure vessel in this experiment was designed to safely withstand 15.0 MPa of pressure. Three circular windows with diameter of 20.0 mm were placed on the vessel. Two of them were located on opposite sides for optical measurement, with a distance of 35.0 mm between them. The third, is a window for monitoring, placed perpendicular to the measurement windows at same horizontal level. The temperature of water or CO_2 solution inside of vessel was adjusted by a heat exchange system with temperature fluctuations to within ±0.2 K.

The experiment was carried out as follows. Having been evacuated, the pressure vessel was fed by fresh water to a considered initial state of pressure and temperature. This state was recorded by pressure and temperature sensors and monitored continually to guarantee that steady state was reached. Those initial parameters were used to estimate initial density of pure water, which was assumed constant as the experiment progressed, if the leakage and the expansion of the vessel could be neglected. LCO_2 was then injected into the water by an up-down nozzle with a diameter of 1.3 mm to form a droplet. Because of the density difference between LCO_2 and water, and the density of LCO_2 under these experimental conditions is less than that of water, the injected LCO_2 droplet could be fixed at the nozzle exit during the period of being dissolved into the surrounding water. Except for recording of pressure and temperature, the fringe shifts induced by the LCO_2 injection and LCO_2 dissolution were acquired by a CCD Camera and stored for further analysis. Once the

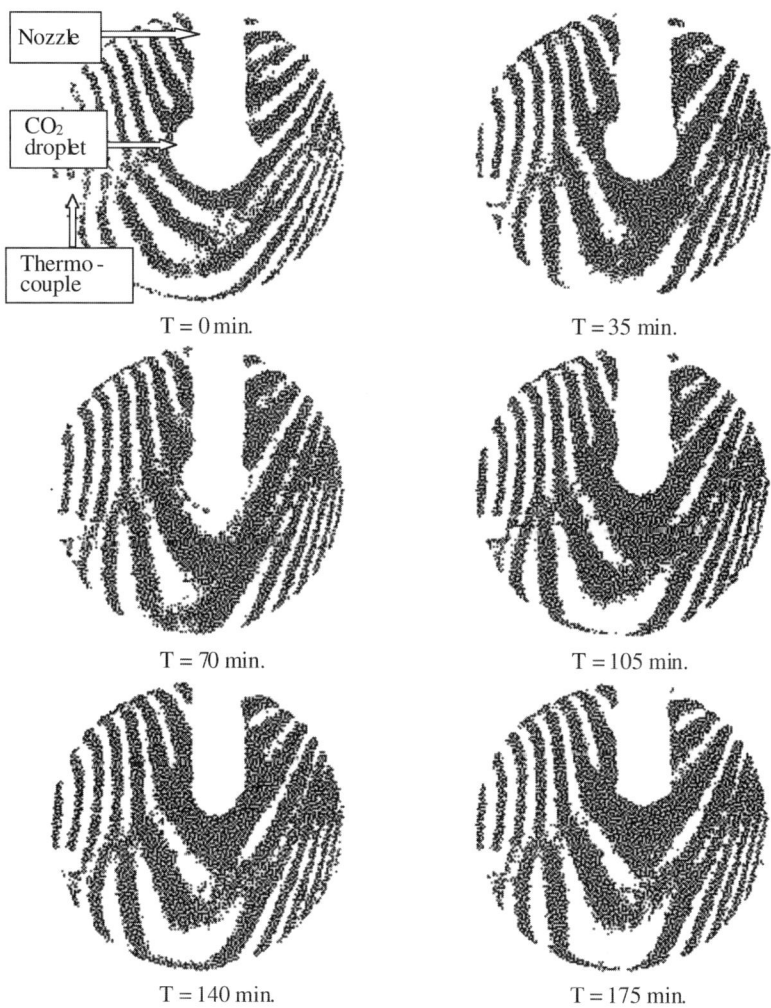

FIGURE 1. Time evolution of a CO_2 droplet dissolution in water.

LCO$_2$ droplet was completely dissolved, another was injected. This cyclic process continued until the solubility approximately approached its limit. The density of CO$_2$ solution and the CO$_2$ mass fraction inside of the vessel for each cycle were obtained by solving the coupled Equations (**1**), (**7**)–(**9**) using the measured data for fringe shift.

The time evolution of the CO$_2$ drop dissolution into water was recorded and is shown in FIGURE 1 as an example. The CCD video recorded the number of fringe shifts due to density changes and also the CO$_2$ injection.

RESULTS AND DISCUSSION

Two sets of data were obtained for pressures ranging from 5.0 to 12.5 MPa, temperatures from 273.25 to 284.15 K, and CO$_2$ mass fraction in water up to 0.061. These data indicate that the density of CO$_2$ solution is nonlinearly proportional to the CO$_2$ mass fraction (see FIGURE 2 left). However, the ratio of CO$_2$ solution density to that of pure water at the same pressure and temperate, or the difference between these two densities, appear to be a monotonically linear relation with CO$_2$ mass fraction and seem to be independent of pressure and temperature under the experimental conditions listed here. The slope of this linear function, 0.275, was calculated by data fitting. FIGURE 2 (right) compares three sets of data obtained by Ohsumi,[8] Aya,[9] and this study. The agreement is evident.

The mechanism of increasing the CO$_2$ solution density while CO$_2$ is gradually dissolved in water may be expressed in terms of the interaction between water and CO$_2$ molecules. In fact, the size of a CO$_2$ molecular is less than the distance between two water molecules, which allows the former to penetrate into the gaps between the water molecules as they dissolve. Furthermore, the molecular density of CO$_2$ solution becomes higher than that of pure water and causes the CO$_2$ solution density to increase. It is interesting to look at the pressures at the time that liquid CO$_2$ begins

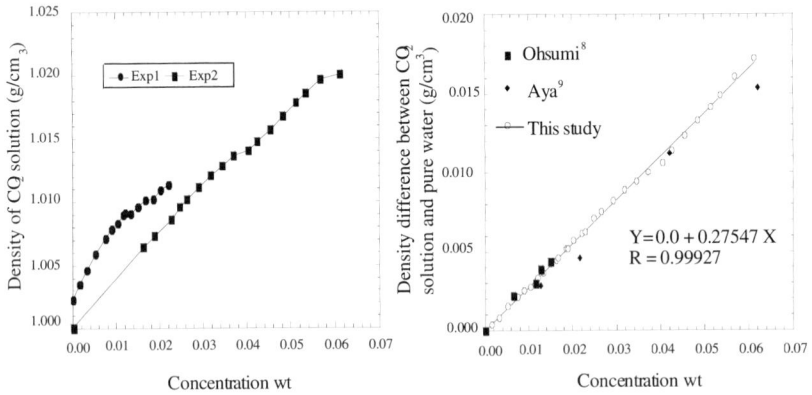

FIGURE 2. CO$_2$ solution density as function of CO$_2$ mass fraction (wt). (**A**) CO$_2$ solution density for two sets of experiment of this study. (**B**) Density difference between CO$_2$ solution and pure water obtained by this study along with data of Ohsumi[8] and Aya.[9]

to be injected P_b, at the end of injection P_a, and at the time of complete dissolution P_s. They exhibit the relation $P_b < P_s < P_a$, which tells us that the volume of CO_2 solution is less than the individual volumes of pure water and pure CO_2 at same state. It also means that the distance between the water molecules increases when CO_2 is dissolved into water. On the other hand it can, in general, be estimated that the density of CO_2 solution is a function of pressure, temperature (as that of pure water or pure CO_2), and the mass fraction of CO_2 dissolved. This is true when CO_2 solution density was not normalized by pure water density (FIG. 2 left). For CO_2 ocean sequestration, as mentioned above, this linear density difference relation between CO_2 solution and water with CO_2 concentration, (FIG. 2 right), indicates that the additional negative buoyancy induced by CO_2 dissolution is a constant at the same CO_2 mass fraction and independent of the depth (i.e., pressure and temperature). This also confirms the suggestion that CO_2 enriched seawater will break down the original ocean stratification state and produce a gravity wave.

SUMMARY

CO_2 solution density was successfully measured by using a high-pressure cell and Mach–Zehnder interferometry in a pressure range of 5.0 to 12.5 MPa, at temperatures from 273.25 K to 284.15 K, and CO_2 mass fraction in solution of up to 0.061. Residual air and leaking from the cell must be carefully handled in order to guarantee the measurement accuracy. The experiment results indicate that the linear relationship between the density ratio of CO_2 solution to that of pure water and CO_2 mass fraction has a slope of 0.275.

ACKNOWLEDGMENT

This study is a part of the investigation of the CO_2 Ocean Sequestration Project managed by Research Institute of Innovative Technology for the Earth (RITE) and founded by New Energy and Industrial Technology Development Organization (NEDO), Japan.

REFERENCES

1. MARCHETTI, C. 1997. On geoengineering and the CO_2 problem. Climate Change **1:** 59–68.
2. STEINBERG, M., H.C. CHEN & F. HORN. 1980. Brookhaven Net. Labo. Rep. OE/CH/00016, Upton, New York.
3. HAUGAN, P.M., F. THORRILDSEN & G. ALENDAL. 1995. Dissolution of CO_2 in the ocean. Energy Conservation Management **36:** 461–466.
4. LIRO, C., E.E. ADAMS & H.J. HERZOG. 1992. Modeling the release of CO_2 in the deep ocean. Energy Conservation Management **33:** 667–674.
5. NAKASHIKI, N., T. OHSUMI & N. KATANO. 1995. Technical view on CO_2 transportation on the deep ocean floor and dispersion at intermediate depth. *In* Direct Ocean Disposal of Carbon Dioxide. N. Handa & T. Ohsumi, Eds.: 183. Terre Pub., Tokyo.
6. PARKINSON, W.J., *et al.* 1969. Partial molal volume of carbon dioxide in water solution. Ind. Eng. Chem. Fundam. **8:** 709–712.

7. NIGHSWANDER, J.A., et al. 1989. Solubilities of carbon dioxide in water and 1% wt NaCl solution at pressure up to 10 MPa and temperature from 80 to 200°C. J. Chem. Eng. Data **34:** 355–360.
8. OHSUMI, T., et al. 1992. Density change of water due to dissolution of carbon dioxide and near field behavior of CO_2 from a source on deep-sea floor. Energy Conservation Management **33:** 685–690.
9. AYA, I. 2000. Direct measurement on CO_2 solution density in the hydrate region. Proc. Japanese Chemical Engineering Symp. Miyazaki, (Japanese).
10. UCHIDA, T., et al. 1999. Interferometric observations of CO_2 hydrate formation and growth. Second Int. Symp. On Ocean Sequestration of Carbon Dioxide. 28–63.
11. THE CHEMICAL SOCIETY OF JAPAN. 1984. Chemical Handbook, 3rd edit. Maruzen Pub. Company, Tokyo.
12. KOUBE, K. 1977. Light and Molecule. Kyoritsu Pub. Comp., Tokyo.
13. FEYNMAN, R.P., R.B. LEIGHTEN & M.L. SANDS. 1965. The Feynman Lectures on Physics. Adison-Wesley Pub. Comp.

Particle Image Velocimetry and Thermometry for Two-Phase Flow Problems

TOMASZ A. KOWALEWSKI

Department of Mechanics and Physics and Fluids,
Institute of Fundamental Technical Research, Polish Academy of Sciences,
Warsaw, Poland

ABSTRACT: Image processing in fluid mechanics has become an important quantitative tool for flow analysis. The feasibility of simultaneous measurements of instantaneous velocity and temperature fields, as well as for tracking interfaces, creates a functional tool to describe thermally driven flows accompanied by phase change. The information thus gathered is essential for the verification and validation of computation models. The paper reviews the image processing methods developed by the author and his coworkers to analyze typical problems of two-phase flow. Simultaneous measurement of temperature and velocity fields, obtained by using liquid crystal tracers, are applied to elucidate solidification and growing vapor bubble problems.

KEYWORDS: particle image velocimetry; thermometry; liquid crystals; thermography

INTRODUCTION

Understanding fluid flow accompanied by thermal effects is of great interest in a number of manufacturing and environmental problems. Progress in numerical techniques allows the simulation and modeling of complex flow configurations involving thermal effects, including phase change. Nevertheless, due to the problem complexity, direct application of numerical methods to many engineering problems is not a trivial task. Errors appear due to limited accuracy of various numerical methodologies and due to inevitable simplifications introduced in the models. Hence, the experimental verification of the models has gained special importance. However, there are only few techniques that allow undisturbed and simultaneous monitoring of the relevant characteristics of the temperature and velocity fields. One such procedure consists of seeding the flow with thermochromic liquid crystals (TLCs). Small particles of the liquid crystal material suspended in the fluid ideally play a role of tracers following the flow pattern. Using the standard particle image velocimetry (PIV) technique, the local velocity of the flow can be measured by cross-correlating two sequential images. In addition, these particles change color with temperature.[1] Hence, after proper calibration, they behave as small thermometers, simultaneously monitoring local fluid temperature.

Address for correspondence: Tomasz A. Kowalewski, Ph.D., Department of Mechanics and Physics and Fluids, Institute of Fundamental Technical Research, Polish Academy of Sciences, IPPT PAN, Swietokrzyska 21, PL 00-049 Warszawa, Poland.
tkowale@ippt.gov.pl

Several studies have been made using TLC to measure flow velocity and temperature. However, due to the experimental difficulties, most of them are limited to qualitative flow visualization experiments. The application of 3-CCD RGB camera and digital image analysis gave impulse to developing the feasible concept of digital particle image thermometry.[2] The possibility to combine PIV and particle image thermometry (PIT) was demonstrated by Hiller *et al.*,[3] who first used the same suspension of TLC particles for digital evaluation of both temperature and velocity fields. Recently Park *et al.*[4] documented in detail the application of the joint PIV and PIT method to turbulent flow. Their uncertainty analysis clarifies the major limitations of the temperature calibration methods used by several authors. It follows that careful analysis of the acquired data is necessary when this method is used for quantitative measurements of complex thermal and velocity fields.

Here, we describe the application of the PIV and PIT method to experimental investigations of thermally driven flows, such as natural convection in small cavities, solidification, and boiling. The main aim of these experiments is to generate full field temperature and velocity data that can be directly compared with their numerical counterparts. We believe that despite of the limited accuracy of the measurements, the non-intrusive character and the possibility of instantaneous and full field measurements of velocity and temperature create a valuable extension to the traditional PIV technique.

EXPERIMENTAL METHODS

Temperature visualization is based on the property of some cholesteric and chiral-nematic liquid crystal materials to refract light of selected wavelengths as a function of the temperature and the viewing angle.[5] Hence, in specific temperature ranges they appear as small color spots following the flow. Their color change ranges from clear at low temperature, through red as temperature increases and then to yellow, green, blue, and finally clear again at the highest temperature. These color changes are reproducible and reversible as long as the liquid crystals are not physically or chemically damaged.

A light sheet illuminates the investigated flow. The arrangement is similar to that used for classical PIV experiments. However, white light is necessary to obtain the selected color refraction from the TLC particles. Since the color of light refracted by TLCs depends on the temperature and the observation angle,[5] it is important that the investigated flow is illuminated by a well-defined light plane and observed by a camera from a fixed direction. To minimize color variation within the illumination plane, the camera-viewing angle should be kept small.

The density of the TLC material is very close to that of water and, in most cases, the TLC tracers can be treated as neutrally buoyant. The size of particles is another important issue. Large particles (0.1 mm and more) produce strong, clear colors. As the size decreases, their color quickly fades due to the increasing effects of light scattering. Also, the camera resolution starts to play an important role in color degradation when the particle images approach the pixel dimensions of the sensor. Hence, some compromise is necessary for optimal selection of the particle size. The mean diameter of the unencapsulated TLC tracers used in our experiments was usually about 50μm. Particles of such size guarantee bright, easily visible colors of the refracted light and are still small enough to follow the flow. Smaller particles (10μm and less) were used for microscopic observations of the temperature field in the vicinity of a vapor bubble.

The TLC material can be commercially obtained as raw greasy mixtures or in a microencapsulated form. The encapsulated particles are chemically resistant and easy to use as tracers. However, the polymer shell used for encapsulation distorts the light and produces additional light scattering. Therefore, we preferred to use TLC particles produced by dispersing the raw material in liquids. The major drawback of this approach is that in practice such particles were found to be chemically stable only in water and glycerol.

The temperature measurements are based on the digital color analysis of red, green, and blue (RGB) images of the liquid crystal seeded flow field. To evaluate the temperature, the incoming RGB video signals are transformed pixel by pixel into a hue, saturation and intensity (HSI) color map. The temperature is determined by relating the hue to a temperature calibration function. A simple formulation introduced by Hiller *et al.*[3] is used to evaluate hue with 8-bit resolution. However, the color–temperature relationship is strongly non-linear. Hence, the accuracy of the measured temperature depends on the color (hue) value, and varies from 3% to 10% of the full color play range. For the liquid crystals typically used, it results in the absolute accuracy of 0.15°C for lower temperatures (red–green color range) and 0.5°C for higher temperatures (blue color range). The most sensitive region is the color transition from red to green and takes place for a temperature variation less then 1°C.

Compared to surface thermography,[6] the use of TLC as dilute suspension in a fluid bears additional problems. First, the color images of the flow are discrete—that is, they represent a non-continuous cloud of points. Second, the overall color response may be distorted due to the camera properties, secondary light scattering, and reflections from the sidewalls and internal cavity elements. Hence, the use of specifically developed averaging, smoothing, and interpolating techniques are indispensable to remove ambiguity in the resulting isotherms. Furthermore, every experimental setup needs its own calibration curve, obtained from the images using the same fluid, the same illumination, acquisition, and evaluation conditions. This is a serious drawback of the PIT method. Whereas the qualitative analysis of the temperature field can be relatively easily done, additional support of a few point measurements (e.g., thermocouples) is necessary to obtain quantitative measurements.

The full field velocity measurements are performed by PIV using the same color images. For this purpose, the color images of TLC tracers are transformed to greyscale intensity images. After applying special filtering techniques, bright images of the tracers, well suited for PIV, are obtained. To improve the resolution of the velocity field, the recently developed optical flow based method of image analysis has also been used.[7] For a typical displacement vector of 10 pixels, the relative accuracy of the single point velocity measurement is better than 5%.

In a typical experiment, the flow is illuminated by a xenon flash or a halogen lamp and observed at 90° by a high-resolution 3-CCD color camera. The 24-bit RGB images are acquired using a three-channel frame grabber and stored on a computer disk in digital form for further analysis. In the experiments described here we use a

PCI-based 32-bit AM-STD module (ITI) and a Pentium computer with 128 MB memory. This setup permits us to acquire in real time more than 50 RGB images with 768 × 564 pixel resolution, before they are saved on disk. A system of step-motors combined with a mirror was used to acquire quickly images of several cross-sections of the convective flow in small cavities. Due to the relatively slow variation of the flow structures, transient recording of the main three-dimensional flow features was possible.

The flow images were used to evaluate the shape and location of the phase front. These measurements were performed using image analysis software. As for the solidification front, the edge detection, supported by manual intervention, appeared to be the most efficient way to find the interfacial profiles. By integrating this information, the volumetric growth rate of the solid phase was evaluated. Concerning bubbles and droplets, a precise description of the interface is necessary for further analysis of their dynamics. Hence, to improve the accuracy, an additional high-speed backlight illumination was applied. The investigated objects were then recorded by a camera and appear as dark shadows, which are easy to detect and separate from the background. By selecting an appropriate edge detection routine, a sequence of pixels is extracted along the interface. In a second step, these pixels are used to find a functional representation of the bubble (or droplet) shape. This procedure allowed us to find a smooth description of the interface, necessary to evaluate its local velocity and curvature, and to evaluate the inner pressure.

The application of the liquid crystal tracers in a few cases investigated in our laboratory is illustrated here. The employment of TLCs seemed to be very useful in understanding the flow structure and helps us to discover effects that are difficult to find using point measurements. Application of the digital image analysis allowed us to quantify measured temperature and velocity fields. Furthermore, comparison with the numerical counterparts let us identify discrepancies that partly originated from the simplifications made in the numerical models.

INVESTIGATED CONFIGURATIONS

Solidification

We consider convective flow in a differentially heated cubic box. This configuration is often used as a benchmark for testing reliability of numerical solutions. In our case, two opposite vertical walls of the box are made of metal and assumed isothermal, the remaining four walls being made of plexiglas. One of the isothermal walls is held at temperature 10°C, the opposite wall is held at −10°C. The cavity inner size is 38 mm and it is filled with distilled water. The natural convection of water in the vicinity of the freezing point differs significantly from the well-known patterns of the benchmark solutions. The competing effects of positive and negative buoyancy forces result in a flow with two distinct circulations (see FIGURE 1). There is a *normal* clockwise circulation, where the water density decreases with temperature (upper-left cavity region) and an *abnormal* convection with the opposite density variation and counter-clockwise rotation (lower-right region). At the upper part of the cold wall, the two circulations collide, intensifying the heat transfer and effectively decreasing the interface growth. Below, the abnormal circulation limits the convective heat transfer

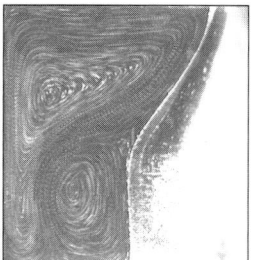 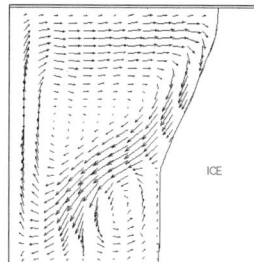 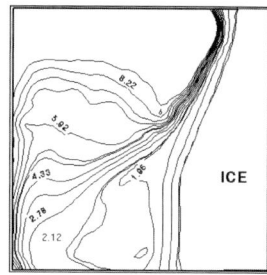

FIGURE 1. Modeling directional solidification; water freezing in a differentially heated cube shaped cavity with a hot isothermal wall (left) at $T_h = 10°C$ and a cold isothermal wall (right) at $T_c = -10°C$. **Left**, particle tracks; **center**, velocity field obtained by PIV evaluation of liquid crystal tracer displacements; **right**, variation of temperature evaluated from liquid crystals color.

from the hot wall, separating it from the freezing front. Hence, the phase front is only initially flat. As time passes it deforms strongly, getting a characteristic "belly" at its lower part.

It is noteworthy that the region separating the *normal* and *abnormal* circulations overlaps with the isotherms of the density maximum. Our numerical simulations performed for the freezing problem[8] show severe discrepancies when compared with the experimental data. It turns out that this flow structure, with the two competing circulations, is very sensitive to the thermal boundary conditions at the sidewalls. Full field flow measurements led us to discover important discrepancies and indicated directions to improve the model. Despite of improvements in the numerical model we used, the computational results still differ in detail from their experimental counterparts.

In the second configuration, the top wall of the cavity is isothermal and kept at low temperature T_c. The other five walls are non-adiabatic, allowing a heat flux from the fluid surrounding the box. There is no well-defined "hot wall" in this configuration. The temperature at the internal surfaces of the cavity adjusts itself depending on both the flow and the heat flux through and along the walls. In order to define the non-dimensional parameters describing the flow, the external temperature T_h is used to calculate the temperature difference. The lid-cooled cavity was selected to investigate the convective flow with and without a phase change (freezing of water at the top wall). When the phase change occurs, it resembles to some extent a directional solidification in a Bridgman furnace used for crystal growth. Physically this configuration bears some similarity to the Rayleigh–Bénard problem. However, due to the altered thermal boundary conditions at the sidewalls, the flow structure is different. The cubic symmetry of the enclosure imposes a strong downward flow along the vertical axis of symmetry. Several oscillatory changes in its pattern are observed before a stable final flow structure is achieved. Numerical simulations confirmed this instability.[9] It appears that the initial cold thermal boundary layer at the lid is unstable and breaks down to several plumes falling down along the sidewalls. Depending on the experimental disturbances or the numerical noise present, the flow pattern exhibits several strongly asymmetric transitions before a final configuration with a single cold *jet* along the cavity axis and a reverse flow along sidewalls establishes.

The formation of ice was studied by decreasing the lid temperature to −10°C. The complicated flow pattern, which is established, also becomes visible in the structure of the ice surface. It was found that the creation of the ice layer at the lid has a stabilizing effect on the flow. This follows from the symmetry of the ice solid surface, which imposes the direction and character of the flow, eliminating the instabilities observed in the pure convection case. There is also a density inversion under the lid that decelerates the main *jet* and limits strong vortex generation in that region. Due to the stochastic development of the flow pattern, the direct comparison of transient experimental and numerical results is difficult in the early time steps. Hence, another arrangement was used to minimize the uncertainty of the initial conditions. We call it the *warm start*, because the freezing starts after a steady convection pattern is established in the cavity. This initial flow state corresponds to natural convection without phase change, with the lid temperature set to 0°C. A regular flow pattern is then seen, with a central, stable cold jet at the cavity axis. FIGURE 2 shows the temperature and velocity field evaluated at time step 3,600 sec for this case.

To simulate the main flow characteristics accompanying casting processes, a simple experimental model was investigated using the PIV and PIT technique.[10] A hot fluid was injected under high pressure into an inclined rectangular box (38 × 38 × 110 mm) through a bottom inlet (see FIGURE 3 left). The fluid propagated inside the box between two cold isothermal walls, maintained at 10°C, passing two divisions simulating the internal complexity of a mould. Due to the sudden cooling of the fluid, the TLC tracers changed color from dark blue (hot) to red (cold regions).

The main features of the experiment (such as flow acceleration and deceleration at the obstacle, free surface flow, and sudden increase of the viscosity as the fluid cools down) are typical for the solidification of a melt in a mould. Moreover, contrary to a real casting, this experimental configuration allows for full control of the experimental conditions and for full field measurements of the temperature and velocity fields. The collection of the quantitative transient data of the flow should permit verification and validation of numerical models used for typical casting problems. The main goal of these investigations is to create an experimental benchmark

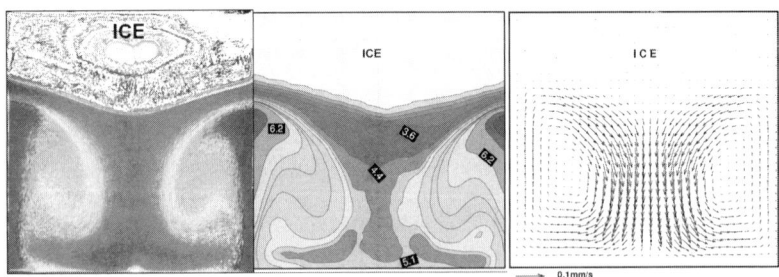

FIGURE 2. Modeling crystal growth: freezing of water from the top in the lid cooled cavity observed at 3,600 sec after cooling starts. Isothermal lid temperature $T_c = -10°C$, external bath temperature $T_{ext} = 20°C$. **Left**, recorded image of TLC tracers with central cold jet (dark color); **center**, PIT evaluated isotherms; **right**, PIV evaluated velocity field. (COLOR PLATE 6.)

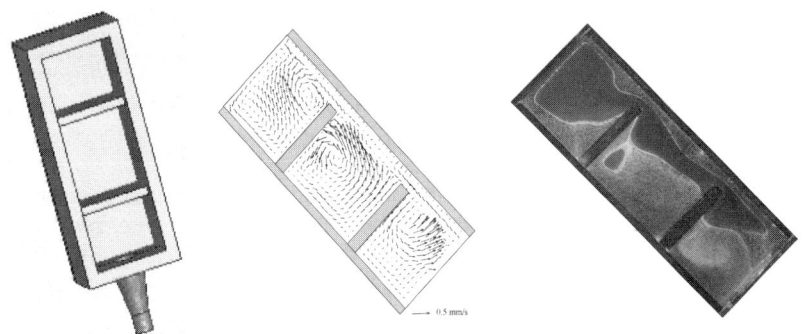

FIGURE 3. Cooling process observed just after filling the cavity with hot glycerol: **left**, geometry of the cavity with two partial divisions and flow inlet at the bottom; **center**, PIV evaluated velocity field; **right**, temperature field visualized by TLC. The cavity was inclined 45°, initial fluid temperature 50°C. Both isothermal walls (upper and lower) are kept at 10°C. Due to the sudden cooling of the fluid, the TLC tracers change color from dark blue (hot) to red (cold regions).

for the mould-filling problem. To compare with the experimental results, several numerical simulations of transient and steady states were performed. Our scope is to understand and explain the observed discrepancies between the measured and calculated flow patterns.

The solidification process of water, the working fluid, is observed as it sets in at both isothermal walls. Initially, almost uniformly and parallel to the wall, a layer of solid builds at both walls. However, as time progresses, the remaining effects of natural convection start to modify the heat transfer, resulting in a faster solidification at lower parts and an increase of asymmetry between upper and lower wall. The entire process depends on the inclination angle of the cavity. FIGURE 4 shows the experimental result obtained for water freezing in the cavity tilted at 11°. The initial water temperature was 9°C, the isothermal walls temperature −7°C.

Vapor Bubble Growth

Modeling heat transfer during nucleate boiling is essential for many industrial processes. Despite a large number of studies devoted to this phenomenon during the past fifty years, the problem is still far from being resolved. In fact, numerous

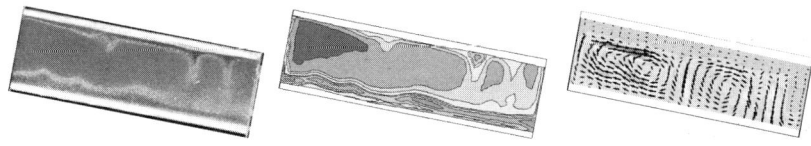

FIGURE 4. Solidification of water observed in the mould model without divisions. Ice layer at both isothermal walls 5 min after the filling process was finished: **left**, TLC tracer image; **center**, PIT evaluated isotherms; **right**, PIV evaluated velocity vectors and velocity magnitude contours. (COLOR PLATE 7.)

attempts have been undertaken to develop a general correlation for nucleate boiling heat transfer, but none has led to a satisfactory result in a broad range of governing parameters. One of the still unsolved problems is the proper description of the process of formation, growth, and detachment of a single vapor bubble. This fundamental problem for the boiling process phenomenon appears to be very difficult, both for experimental investigations and for theoretical or numerical modeling.

Microscopic observations supported by a high-speed illumination system, CCD camera, and frame grabbers were used to obtain a quantitative description of the vapor bubble interface dynamics. The seeding of the fluid with thermochromic liquid crystals is used to visualize the temperature and velocity fields surrounding the bubble. The experiments were performed for water boiling in a low-pressure environment inside a small cube-shaped cavity. The six walls of the cavity are equipped with several internal passages for water circulation from the thermostat. This allows us to maintain the cavity and the fluid inside at constant temperature (to within ±0.1°C). In the experiments the bulk temperature of the liquid T_l and the wall temperature T_b were varied over the range 30°C–70°C.

The bubbles and flow in the cavity could be observed through five glass windows. In the present experiment, a bubble was observed through one of the side windows using a 3CCD-color camera.[11] To obtain images of a well-defined bubble interface, back light illumination was applied using the opposite window. For this purpose, both a strobe light and a halogen spot lamp were used. The second light source, equipped with the halogen lamp, was located perpendicular to the optical axis of the camera. It was used to produce a 1 mm light sheet for flow visualization around the bubble.

The bubbles were observed through a 50 mm lens using an extension tube. A typical bubble diameter was about 2 mm and the mean velocity of the interface exceeds 0.1 m/sec. This means that the relative velocity in the plane of the CCD sensor was very high, requiring a short illumination time and high-speed imaging. To obtain sharp images of the interface when a halogen light was used, we activated the electronic shutter of the camera with a typical opening time of 4 msec. The strobe illumination from a standard stroboscopic lamp was used to study bubble dynamics.

The flow and temperature field surrounding the departing bubble turns out to be very complex and difficult to analyze. The wide range of velocities, the sudden change of the flow direction and the generation of local vortices were typical features of all our experiments. FIGURE 5 shows a vapor bubble departing from the hot surface. Liquid crystal tracers visualize the surrounding temperature and velocity fields.

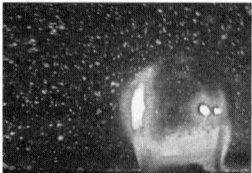 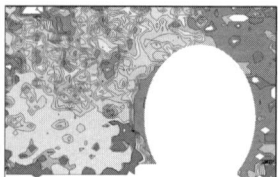 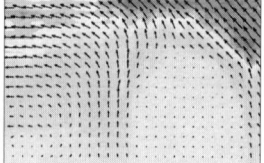

FIGURE 5. Flow field surrounding a departing vapor bubble. Water at $P = 5.3$ kPa, $T_l = 35.5$°C, $T_b = 66.5$°C. **Left**, TLC tracers; **center**, PIT evaluated temperature field; **right**, PIV evaluated velocity vectors and velocity magnitude contours. (COLOR PLATE 8.)

Within the surface layer they exhibit blue color, indicating an increase of temperature in the fluid. At the right side of the bubble, a relatively cold stream of water is visualized by the predominating red color of the TLCs. One can realize that the presence of the bubble strongly deforms the symmetry of the temperature and velocity field (FIG. 5 center and right).

The main experimental problem with visualizing the flow surrounding a growing bubble was due to strong light reflections from its surface. This noticeably diminished the resolution of the PIV and PIT evaluation in the vicinity of the bubble surface. One of the possibilities for minimizing this effect was to mask part of the images obtained from the light sheet illumination by the second image of the bubble alone, obtained by backlight illumination. To acquire two separate images of the same object (backlight and light sheet illumination), we used two different colors. A blue filter was used for the light sheet and a red LED diode was used for the backlight illumination. With an RGB color camera, two different images were obtained for the red and blue channel, easy to use for the masking procedure. This technique was successfully applied to improve the PIV evaluation of single bubbles. To preserve the tracer color information, essential for the PIT evaluation, a second greyscale camera was necessary to create the appropriate mask for the bubble.

CONCLUSIONS

An extension of the PIV technique, by using thermochromic liquid crystals as seeding, opens a new possibility to study thermally driven flows. Image-processed data makes available quantitative full-field information about the temperature and velocity fields, which will undoubtedly encourage the study of situations that have been, until now, too complex to consider. The non-invasive character of the method and its relative simplicity offers a valuable tool for the full field verification and validation of numerical results.

ACKNOWLEDGMENTS

This is a summary of a study that I was fortunate to share with colleagues and students, first at the Max-Planck-Institut and presently at my home institution. In particular, I would like to acknowledge the contribution of W. Hiller, St. Koch, A. Cybulski, M. Rebow, and J. Pakleza (CNRS), to what has been a team effort over a number of years. The Polish Scientific Committee (KBN Grant No. 8T09A00820) partly supported this work.

REFERENCES

1. HILLER, W. & T.A. KOWALEWSKI. 1987. Simultaneous measurement of the temperature and velocity fields in thermal convective flows. *In* Flow Visualization IV. C. Veret, Ed.: 617–622. Hemisphere, Paris.
2. DABIRI, D. & M. GHARIB. 1991. Digital particle image thermometry: the method and implementation. Exp. Fluids **97:** 77–86.

3. HILLER, W.J., ST. KOCH, T.A. KOWALEWSKI & F. STELLA. 1993. Onset of natural convection in a cube. Int. J. Heat Mass Transfer **36**: 3251–3263.
4. PARK, H.G., D. DABIRI & M. GHARIB. 2001. Digital particle image velocimetry/thermometry and application to the wake of a heated circular cylinder. Exp. Fluids **30**: 327–338.
5. HILLER, W.J., ST. KOCH & T.A. KOWALEWSKI. 1988. Simultane Erfassung von Temperatur- und Geschwindigkeitsfeldern in einer thermischen Konvektionsströmung mit ungekapselten Flüssigkristalltracern. 2D-Meßtechnik DGLR-Workshop, Markdorf, DGLR-Bericht 88-04, DGLR Bonn. 31–39.
6. HAY, J.K. & D.K. HOLLINGSWORTH. 1996. A comparison of trichromic systems for use in the calibration of polymer-dispersed thermochromic liquid crystals. Exp. Thermal Fluid Scs. **12**: 1–12.
7. QUÉNOT, G., J. PAKLEZA & T.A. KOWALEWSKI. 1998. Particle image velocimetry with optical flow. Exp. Fluids **25**: 177–189.
8. KOWALEWSKI, T.A. & M. REBOW. 1999. Freezing of water in the differentially heated cubic cavity. Int. J. Comp. Fluid Dyn. **11**: 193–210.
9. ABEGG, C., G. DE VAHL DAVIS, W.J. HILLER, *et al.* 1994. Experimental and numerical study of three-dimensional natural convection and freezing in water. Proc. 10th Intl. Heat Transfer Conf., Brighton. **4**: 1–6.
10. KOWALEWSKI, T.A., A. CYBULSKI & T. SOBIECKI. 2001. Experimental model for casting problems. *In* Computational Methods and Experimental Measurements. Y.V. Esteve, G.M. Carlomagno & C.A. Brebia, Eds.: 179–188. WIT Press, Southampton.
11. KOWALEWSKI, T.A., J. PAKLEZA, J.B. CHALFEN, *et al.* 2000. Visualization of vapor bubble growth. 9th Int Symp. on Flow Visualization, Edinburgh. I. Grant & G.M. Carlomagno, Eds. CD ROM Proceedings. ISBN 0953399117, Edinburgh 176.1–9.

Multiphase Bubbly Flow Visualization Using Particle Image Velocimetry

YASSIN A. HASSAN

Department of Nuclear Engineering, Texas A & M University, College Station, Texas, USA

> ABSTRACT: This article describes advances made in using particle image velocimetry (PIV) techniques in the study of multiphase bubbly flow. One of the fundamental issues in bubbly flow is the prediction of the velocity field. A methodology that allows for velocity field measurements of both components of a two-phase bubbly flow is presented. The bubble shape is also constructed via a shadow imaging technique combined with PIV.
>
> KEYWORDS: bubbly flow; particle image velocimetry; tracking algorithms; turbulence

INTRODUCTION

Two-phase bubbly flow is widely applied in engineering and environmental processes. The interaction of the dispersed phase with the continuous phase has a great effect on transfer processes between the phases. The relative velocities between the phases, the interfacial area, and the shape of the dispersed phase are the key dependent parameters in the drag, heat, and mass transfer between the phases. Although physical understanding of bubbles rising in a liquid is of significant practical importance in many areas of engineering and in numerous medical, chemical, and biological processes, neither the interactions between bubbles in clusters nor the bubble-induced pseudoturbulence (i.e., the generation of velocity fluctuations by bubbles and their wakes in laminar flow) are fully understood. The more complex the flow, as in bubble columns with high void fraction, the more difficult is measurement and computation. Modeling bubbly flow with computational fluid dynamics (CFD) codes requires detailed information about the full field velocity close to the bubble and its wake. Such information is not easily available.

Many measurement systems that use image-processing techniques have recently been developed and applied to bubbly flow. These approaches have helped in clarification of the flow structure around the bubbles that appears to act locally, but influences global (channel scale) turbulence. The local and instantaneous description of such flows is generally a difficult task due to the tremendous number of initial and boundary conditions that need to be specified. It is, therefore, necessary to adopt statistical description methods. Volume, area average, time or ensemble average, and conditional sampling have been employed in an attempt to understand bubbly flow

Address for correspondence: Yassin A. Hassan, Ph.D., Department of Nuclear Engineering, Texas A & M University, College Station, Texas 77843-3133, USA.
hassan@cedar.tamu.edu

behavior. These average processes exhibit shortcomings: the averaging operator smooths out the fluctuations existing in the flow phases and, consequently, vortex shedding in the wake of bubbles and deformation of the interfaces are missed. These fluctuations play a significant part in the magnitude of the Reynolds stress tensors needed in CFD code simulations. For the range of bubble diameters (2–5 mm) considered in the present study, the wakes are significant. Here we present the velocity field results for bubbly flow using state of the art particle image velocimetry (PIV) technique.[1,2] The bubble shapes were obtained by using a shadow imaging technique in conjunction with the PIV method.[3]

METHODOLOGY

Most methods developed for velocity measurements fall into two categories: optical methods and opaque methods. Optical methods, such as laser doppler velocimetry (LDV) and pulsed laser velocimetry (PLV) can be applied to two-phase flow when the particle sizes are small, with low concentrations. For opaque materials, magnetic resonance imaging (MRI), neutron radiography imaging (NRI), gamma ray, and X-ray methods are examples of non-invasive measurement tools that can be used in certain flow situations. Ultrasonic resonance is also used for flow regime and velocity determination. Modern developments in image processing and advances in power computing have been responsible for the use of flow visualization to obtain quantitative velocity data with accuracy. PIV is one of several tools that were developed during the past decade and applied for various fluid applications. Here we employ a PIV setup for capturing the liquid velocity vector field and bubble trajectories simultaneously, in circular ducts with laminar and turbulent flow.

The application of PIV has been increasingly successful during the past decade. PIV provides instantaneous full field velocity distributions in a two-dimensional plane.[1] It can be applied to study two-phase flow if the component phases can be separated during analysis.[2–8] With the recent improvement of digital imaging technology, PIV measurement techniques are now capable of capturing high-resolution digital images of gas–liquid two-phase flow, in which the continuous liquid phase and the dispersed gas phase are unsteady and multidimensional.

EXPERIMENTAL SETUP

Bubbly flow in the test facility is generated in a sintered metal cylinder placed at the bottom of a glass pipe with internal diameter 5 cm. Connected to the pipe are the bubble generator, two pumps, and one reservoir tank. The main liquid flow enters above the bubble generator at the bottom end of the pipe. A secondary liquid flow is injected through the lower side of the bubble generator. This secondary flow is used to control the air bubble sizes. Air is injected into the air chamber surrounding the sintered metal tube. The secondary pump draws the water from the reservoir tank, passes it through a flow gauge, and then into the centre of the bubble generator, where the air passes through the sintered metal tube. The main flow pump draws water from the reservoir, passes it through a flow gauge, and then injects it below the test section.

A clear transparent plastic box encloses a portion of the pipe where the PIV measurement volume is located. This box is filled with mineral oil to reduce pipe refraction effects. The pipe is about 200 cm long and the test section has an approximate L/D ratio of 30, where L is the length from the bottom of the pipe and D the pipe inlet diameter. The test facility is illustrated in FIGURE 1.

A twin Nd:YAG high energy (400 mJ) pulsed laser (9.0 nsec pulse width) is used as illumination source. Typical high-energy optics are used to manipulate the laser beam and form the laser sheet necessary for PIV measurements. The synchronization signals originate from the cameras and are passed through a pulse generator to trigger the laser. The high-resolution ($1,016 \times 1,016$ pixels) digital camera and associated frame grabber boards have two modes of operation. The normal, continuous, mode uses 30 frames per second. The *triggered* mode enables the cameras to capture two consecutive frames with a very small time delay controlled by the user; the system has a capture rate of 15 Hz or less. As each camera captures frames, they are transferred to a personal computer for temporary storage in computer RAM. Each computer is capable of holding about 400 sequential images in RAM, resulting in 13.3 seconds of continuous run time before the frame data is transmitted to a hard disk. Two other digital cameras are used for shadow PIV. The cameras are capable of capturing frames at 30 Hz with a resolution of 640×480. Illumination is supplied by red light emitting diodes (LEDs) that are opposite each of the cameras around the measurement volume. A screen diffuses the LED light and filters are attached to each

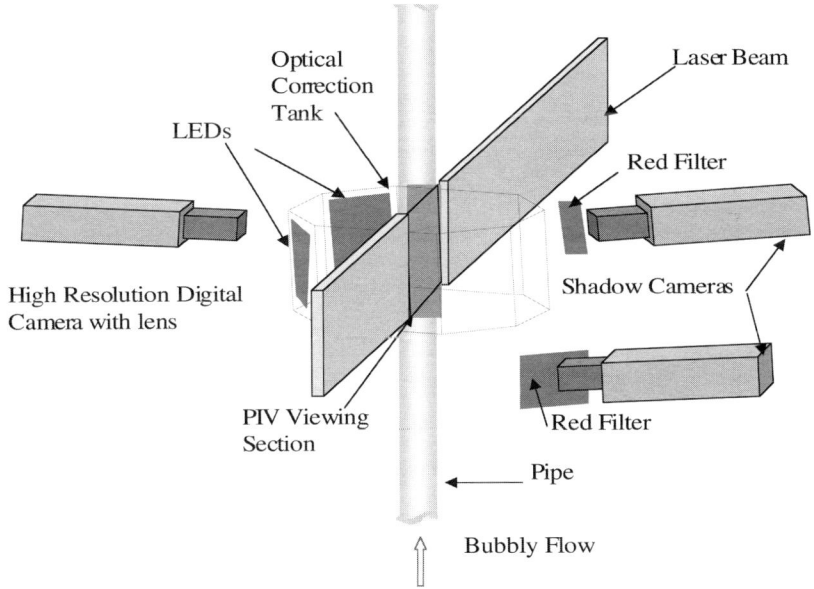

FIGURE 1. Test facility for bubbly flow.

camera to remove any reflected laser light. A tracer size of 6µm was found to be very suitable for this system.

The tracer particles are neutrally buoyant and chemically compatible. Particle tracking velocimetry (PTV) is used to determine the velocity vectors at certain instants. PTV is the preferred dilute phase analysis method. Three different algorithms available to us are used for the tracking process: a binary cross-correlation,[2] an ART2 neural network using a Hough transform,[4] and the spring model.[5]

RESULTS

Measurements can be performed to capture the effects of rising air bubbles in the continuous liquid phase. For two-phase flow, the upper scale for velocity is approximately the velocity of the gas phase. As the gas passes through the continuous liquid, it drags liquid in the vicinity of the bubble and accelerates the liquid to approximately the velocity of the bubble. The first series of the tests were for a single bubble rising in a stagnant liquid. FIGURE 2 represents the results of the vorticity ω_y on the X–Z plane, position $y = 0$, at $t = 33.3$ msec after the bubble has entered the viewing volume and for bubble trajectory within the bubble center. The subscript, y, refers to the vorticity direction. The plot of the vorticity ω_z on the X–Y plane, position $z = 0.61$, at $t = 33.3$ msec after the bubble has entered the viewing volume and for bubble trajectory through the pipe center is shown in FIGURE 3. It is interesting to note the complex structure of the flow around the bubble and its influence on the stagnant liquid. Large values of positive and negative vorticities are delineated. This demonstrates the complex interaction between the bubble wake and surrounding

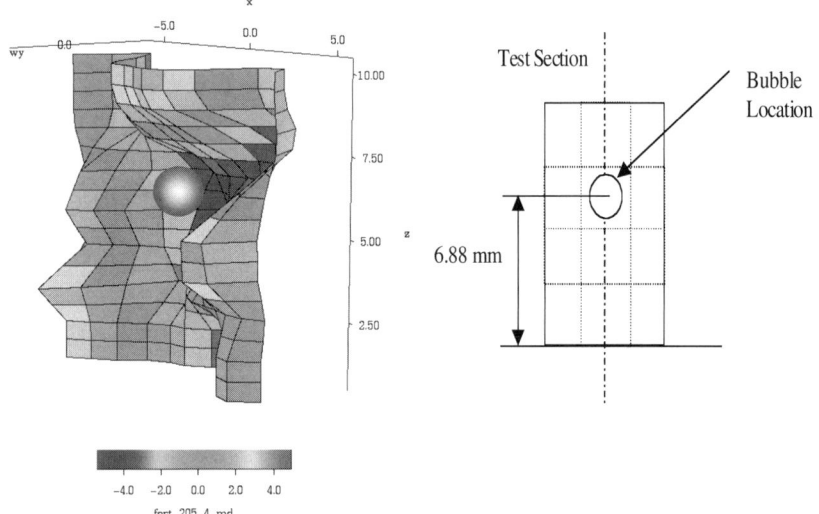

FIGURE 2. Surface plot of ω_y on the X–Z plane, position $y = 0$, at $t = 33.3$ msec after the bubble has entered the viewing volume.

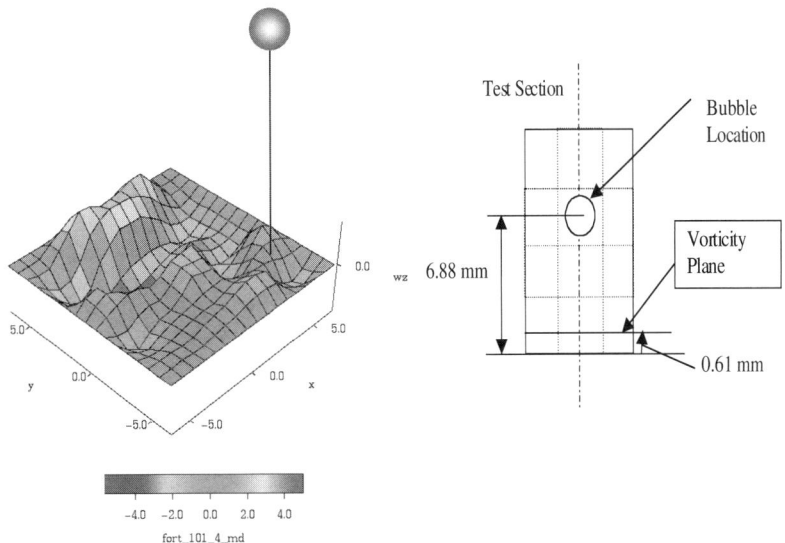

FIGURE 3. Surface plot of ω_z on the X–Y plane, position $z = 0.61$, at $t = 33.3$ msec after the bubble has entered the viewing volume.

fluid and its momentum exchange. The vorticity ω_z in FIGURE 3 presents the one of the snapshots of the upstream flow field behavior at a distance of 6.17 mm from the bubble. With the three-dimensional measurements of the velocity field around the bubble, we are able to compute the vorticity components $(\omega_x, \omega_y, \omega_z)$ simultaneously. Due the availability of the quantitative flow measurements using PIV, the wake structure behind the bubble can be accurately identified.

FIGURE 4 presents an example of the simultaneous two-shadow camera images. From these two images the three-dimensional positions (x, y, z) of the bubbles can be determined utilizing the calibration process. It is also clear that the bubble size varies from one bubble to another, and the bubble shape is not spherical. The bubble sizes were between 0.8 to 2 mm. Enhanced images are obtained via filtering. The relative

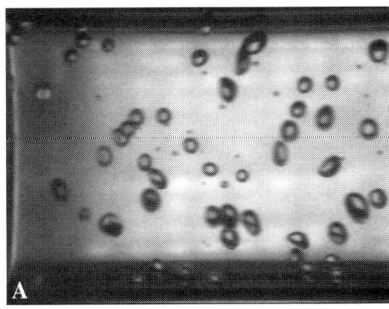 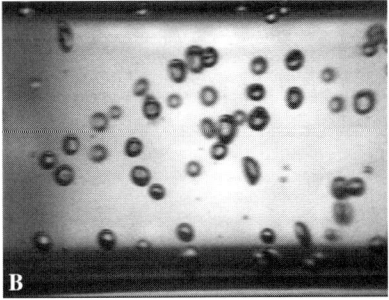

FIGURE 4. Example of the simultaneous two-shadow camera **A** and camera **B** images.

velocity between the bubble velocity and the liquid phase can be determined by subtracting the liquid velocity from the bubble velocity obtained using the PTV algorithm. The view areas of the shadow cameras are 48.3 mm × 68.4 mm and the area of the PIV image is 48.3 mm × 48.3 mm. About 3,000 images were obtained from the PIV camera.

CONCLUSIONS

PIV is a powerful tool that provides better understanding of the complex physical phenomena of multiphase flow. Instantaneous full-field velocity components can be estimated. A key advantage of this scheme is the ability to obtain simultaneous data for velocities of the continuous and dispersed phases. These data can be used to examine and validate computational fluid dynamic models. However, a large number of images is needed to achieve a reasonable turbulent average quantities.

REFERENCES

1. ADRIAN, R.J. 1991. Particle-imaging techniques for experimental fluid mechanics. Annu. Rev. Fluid Mech. **23:** 261–304.
2. HASSAN, Y.A., T.K. BLANCHAT, C.H. SEELEY, JR. & R.E. CANAAN. 1992. Simultaneous velocity measurements of both components of a two-phase flow using particle image velocimetry. Int. J. Multiphase Flow **18:** 371–395.
3. HASSAN, Y.A., W. SCHMIDL & J. ORTIZ-VILLAFUERTE. 1998. Investigation of three-dimensional two-phase flow structure in a bubbly pipe flow. Meas. Sci. Technol. **9:** 309–326.
4. HASSAN, Y.A. & O.G. PHILIP. 1997. A new artificial neural network tracking technique for particle image velocimetry. Exp. Fluids **19:** 342–347.
5. OKAMOTO, K., Y.A. HASSAN & W.D. SCHMIDL. 1995. New tracking algorithm for particle image velocimetry. Exp. Fluids **19:** 342–347.
6. GUI, L. & W. MERZKIRCH. 1996. Phase separation of PIV measurements in two-phase flow by applying a digital mask technique. ERCOFTAC Bull. **30:** 45–48.
7. MURAI, Y., S. WATANABE, F. YAMAMOTO & Y. MATSUMOTO. 1997. Three-dimensional of bubble motions in bubble plume using stereo image processing. Proc. 2nd Int. Workshop, PIV'97-Fukui, Japan. 13–18.
8. SONG, X., L. SHEN, Y. MURAI & F. YAMAMOTO. 1999. Separation of particle-bubble images in multiphase flow. Proc. 3rd Int. Workshop on PIV'99. Santa Barbara, USA. 15–20.

Velocity Measurements by Particle Image Velocimetry Using a Direct Intercorrelation Algorithm

Application to the Interaction between a Water Mist and a Liquid Pool Fire

JEROME RICHARD, ARNAUD SUSSET, AND JEAN-PIERRE VANTELON

*Laboratoire de Combustion et de Détonique - UPR 9028 au CNRS,
ENSMA – University of Poitiers, Futuroscope Chasseneuil, France*

ABSTRACT: The aim of this study was to obtain better characterization of the interaction between a water mist and a diffusion flame of small-scale size. A PIV algorithm was used to determine the flow pattern with and without a water mist and its relation to the velocity of water droplets.

KEYWORDS: PIV; intercorrelation algorithm; pool fire; water mist

INTRODUCTION

Gaseous agents (e.g., halons) were widely used to extinguish fire during the past few decades. However, for environmental reasons, the use of halons was progressively withdrawn and research for an alternative has become an important challenge. In this context, fire suppression systems based on water mist are now considered to offer a reliable technique. Studies[1,2] of the extinguishing properties of water mists have identified three dominant mechanisms acting simultaneously: heat extraction or gas phase cooling, oxygen displacement or dilution, and attenuation of radiant heat fluxes. It is of great interest to study the dynamic interaction between a cloud of water droplets and flame. The objective of the present paper is to describe the analysis by tomographic video recording of the flow pattern, using a PIV algorithm. An original algorithm for separating spots due to the diffusion of water droplets and soot particles in the flow pictures was also developed.

EXPERIMENTAL

Experiments were performed with small scale heptane pool fires. The heptane was contained in a circular steel pan, 10 cm deep, with a diameter of 23 cm. Before each test the pan was filled to 5 mm below the pan lip and during the test, the level

---

Address for correspondence: J. Richard, Ph.D., Laboratoire de Combustion et de Détonique UPR 9028 au CNRS, ENSMA—University of Poitiers, Téléport 2-1, Avenue Clément Ader, BP 40109, France 86961 Futuroscope Chasseneuil, Cedex.
jerome_richard@caramail.com

was kept constant by using a gravity liquid feeding system. The water mist system consisted of three bifluid nozzles placed symmetrically with respect to the flame axis (120°) and directed toward the fuel surface. The applied air pressure (1 bar) and water flow (3.5 g·sec^{-1}/nozzle) were chosen in such a way that the fire did not extinguish and could be maintained long enough for experimental purposes. Under these conditions, the mean drop size (characterized by PDA) was 30 µm.

The laser was a double pulse Yag laser (532 nm) giving 25 mJ/pulse in 10 nsec. The laser sheet, 2 mm deep, was directed along a vertical cross section. Pictures were obtained with a high resolution CCD video camera (1,300 × 1,030 pixel^2) fitted with a 50 mm (f/1.2) lens, permitting visualization of a field of 25 × 30 cm. The combination of an interferential filter (532 nm ± 1 nm) and a liquid crystal shutter transmitting 30% provided the means to eliminate part of the intense radiation due to soot particles. The time between two successive pictures was 2 msec.

IMAGE PROCESSING

Spots from water droplets and filaments due to the diffusion of soot particles are evident in FIGURE 1 (left). The information concerning the dynamics of the droplets alone was obtained by using an original algorithm designed to isolate the spots. This preprocessing step was applied to all the pictures corresponding to the application of water mist. FIGURE 1 (middle) shows the filaments of soot, and FIGURE 1 (right) shows only the spots due to water droplets.

The PIV algorithm was applied to pictures obtained from the flame without a water mist (reference pictures), as well as to pictures showing filaments of soot (I_S pictures), and finally to pictures showing spots due to the water droplets (I_G pictures). The PIV calculation is based on a direct intercorrelation algorithm that was developed in our laboratory.[3]

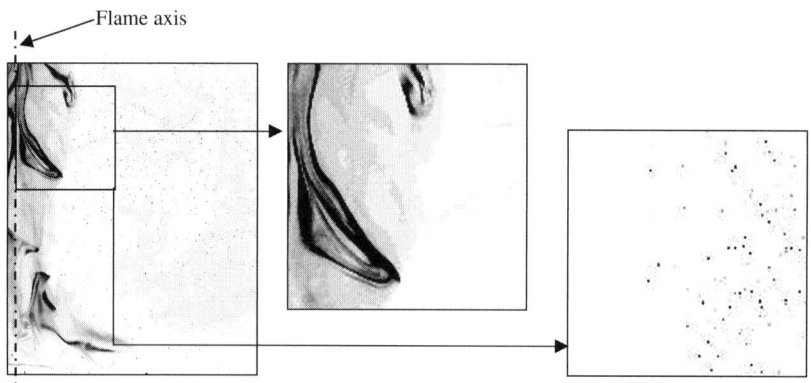

FIGURE 1. Spots from water droplets and filaments due to soot particles: **left**, initial image; **middle**, part of the image showing only the soot, (I_S image); **right**, part of the image showing only the spots due to water droplets, (I_G image).

RESULTS AND DISCUSSION

Because the flame characteristics (with and without water mist addition) are statistically axisymmetric, the measurements are also considered to be axisymmetric and, consequently, representative of those at the same radial distance in a given plane.

Flame without Water Mist Addition (Reference Flame)

Previously,[4,5] pool fire studies of flames without water mist addition allowed us to validate our image processing procedure. The overall flow velocity field is shown in FIGURE 2. The soot aggregates are small enough to allow the assumption that the calculated velocities are representative of the gaseous flow velocity. Therefore, from the observed trend of this field, the major characteristics of the flame base structure are evident: chemical reaction and air entrainment near the fuel surface and upward movement of the hot gases created by buoyancy as a result of temperature gradients.

The variable showed by color levels in FIGURE 2 is the validation criterion. It is defined as the local ratio of the number of vectors validated by the PIV algorithm, to calculate the soot mean velocity, and the total number of vectors in the field. This variable may be viewed as a soot presence probability rate. This can be linked with the soot volumetric fraction field which we verified by measuring the monochromatic absorption coefficient using a classical laser attenuation method.[6]

Flame with Water Mist Addition

Soot Correlation

During the application of water mist, the flame loses its diffusion flame structure. The result of the interaction is a cyclic behavior involving three periods. In the first period, the overall trend is that of a classical diffusion flame. However, due to the momentum generated by the mist delivered by the three nozzles, the tip of the flame

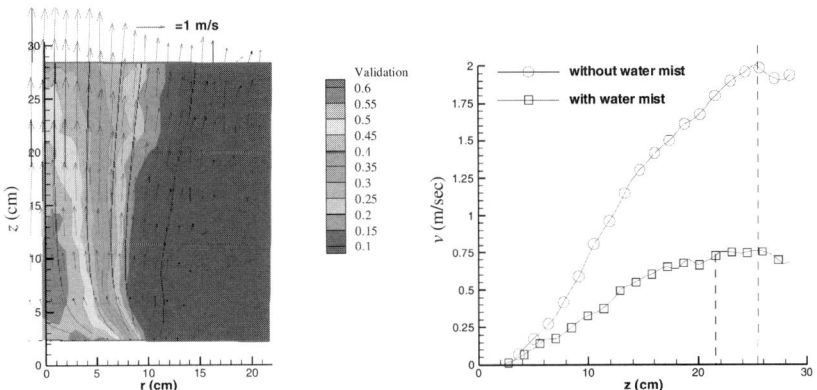

FIGURE 2. Left, soot mean velocity field for the flame without water mist. **Right**, soot mean velocity versus the distance above the fuel surface, on the flame axis.

is flattened. In the second period, the flame seems to "explode" with random lateral projections of pockets of warm unburned fuel and products. During this phase of lateral expansion, there are clearances from part or even the entire liquid surface. The flame looks like an annular flame anchored to the pan edge. It is during this period of expansion that extinguishment can occur: the generation of flammable vapors and the level of temperature decrease enough to reduce the vapor–air mixture temperature near the surface to below the flammability limit. If it is not the case, heptane vapor is reignited and the flame is restored in the third period.

It can be seen in FIGURE 2 (right) that the soot is clearly decreased by the momentum of the water spray. The cyclic expansion of the flame gives rise to an increase in the radial component of the velocity and, therefore, to an inflection of the corresponding lines that turn outward in the upper region (see FIGURE 3 left). It can also be seen that the presence of soot given by the validation criterion field is correlated with the temperature field shown in FIGURE 4. For example, the highest temperature zone is located on the axis, in the bottom region of the flame, between 2 and 10 cm (FIG. 4). This corresponds to the zone where the values of the validation criterion (i.e., the soot presence probability) are also high.

Droplet Correlation

The overall flow velocity field of water droplets is shown in FIGURE 3 (right). It can be seen that the direction of the velocity lines of the droplets is inflected, because of the conjunction of two phenomena: upward movement of the flame gases created by buoyancy that deflects the water droplets from their initial direction and periodic expansion, described above, that adds an important positive radial component that partially compensates for the negative radial component of the initial velocity.

FIGURE 3 (right) depicts a region wherein the droplets velocity is very low. This zone is located on the axis at approximately 15 cm above the fuel surface and may correspond to an area where the initial momentum of the droplets is diminished by the upward movement of the hot gases and soot, and by the radial component added by the expansion effect. Also shown in FIGURE 3, by color level, is the local droplet

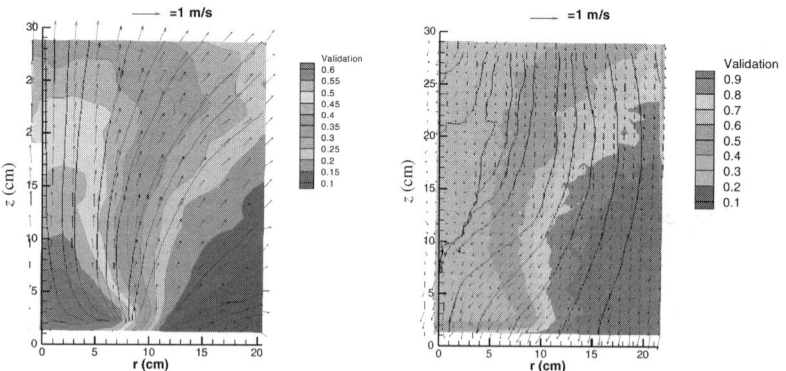

FIGURE 3. Right, soot mean velocity field for the flame with water mist. **Left**, droplet mean velocity field and relative concentration for the flame with water mist addition. (COLOR PLATE 9.)

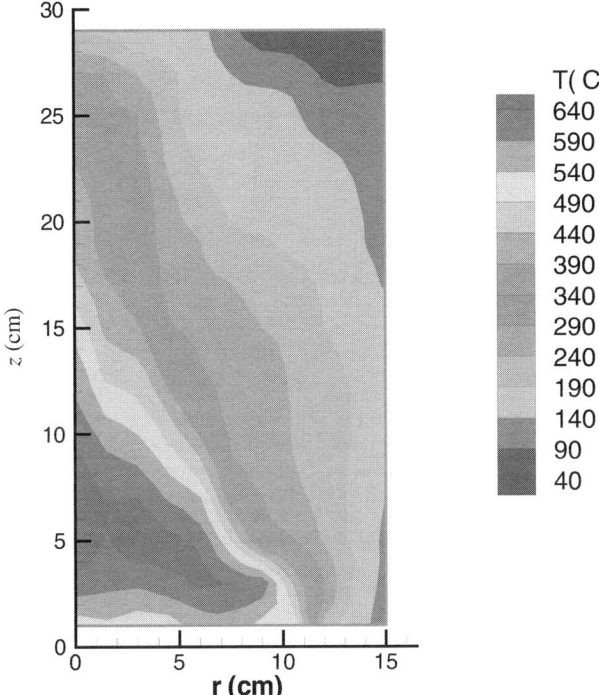

FIGURE 4. Mean temperature field for the flame with water mist.

concentration ratio, $\alpha = C/C_{max}$, where C is the droplet concentration and C_{max} is the maximum droplet concentration. Three distinct zones of α are evident. The first zone is in the bottom of the studied area, outside of the flame. In this zone, the droplet concentration is maximum. The second zone corresponds to the axis of the flame, where the droplet concentration is minimum. The third zone, between these two zones, is a transition zone wherein the droplet concentration decreases as α decreases.

The field of the drop concentration ratio α is affected by two phenomena:

1. The mean fluid movement determined by a combination of the momentum of the spray, the upward movement of the flame gases and the expansion effect due to water vaporization.

2. The effect of the temperature field (FIG. 4). The temperature gradient between the periphery of the flame and its axial part creates a force that repels the droplets out of the flame zone.

CONCLUSION

The aim of this study was to obtain a better characterization of the interaction between a water mist and a diffusion flame of small-scale size. A PIV algorithm permitted the determination of the flow pattern with and without a water mist and the velocity of water droplets. An original algorithm separating spots due to the diffusion of water droplets from the filaments due to diffusion of the soot particles in the flow pictures was also developed.

Future plans call for us to determine the value of C_{max}, to obtain quantitative information about the concentration of droplets, and to develop an algorithm to obtain the diameters of the droplets.

REFERENCES

1. RABASH, D.J., Z.W. ROGOWSKI & G.W.V. STARK. 1960. Mechanism of extinction of liquid fuel fires with water spray. Combust. Flame **4**: 223.
2. MAWHINNEY, J.R., B.Z. DLUGOGORSKI & A.K. KIM. 1994. A closer look at the fire extinguishing properties of water mist. Fire Safety Science. Proc. 4th Int. Symp., Ottawa, Canada. 47.
3. SUSSET, A., J.M. MOST, D. HONORÉ & M. PERRIN. 2000. Développement d'un traitement itératif par corrélation directe pour l'application de la PIV aux écoulements à forts gradients de vitesse. 7ième Congrès de l'A.F.V.L., Marseille, France.
4. MCCAFFREY, B.J. 1979. Purely buoyant diffusion flames: some experimental results. Natioal Bureau of Standards, NBSIR 79-1910.
5. BOUHAFID, A., J.P. VANTELON, P. JOULAIN & A.C. FERNADEZ-PELLO. 1989. On the flame structure at the base of a pool fire. 22nd Int. Symp. on Combustion. The Combustion Institute. 1291.
6. RICHARD, J., J.P. GARO, J.M. SOUIL & J.P. VANTELON. 2002. On the flame structure at the base of a pool fire interacting with a water mist. Second Mediterranean Int. Combustion Symp., Sharm El-Sheikh, Egypt. 443.

Visualization of Bubble–Fluid Interaction by a Moving Object Flow Image Analyzer System

H.M. CHOI,[a,b] T. TERAUCHI,[a] H. MONJI,[a] AND G. MATSUI[a]

[a]*Institute of Engineering Mechanics and Systems, University of Tsukuba, Tsukuba, Japan*

[b]*Fluid Flow Group, Korea Research Institute of Standards and Science, Yuseong, Daejeon, Korea*

ABSTRACT: This paper deals with interaction between a bubble and fluid around it, visualized by a moving object flow image analyzer (MOFIA) consisting of a three-dimensional (3D) moving object image analyzer (MOIA) and two-dimensional particle image velocimetry (PIV). The experiments were carried out for rising bubbles of various sizes and shapes in stagnant water in a vertical pipe. In the MOFIA employed, 3D-MOIA was used to measure bubble motion and PIV to measure fluid flow. The 3D position and shape of a bubble and the velocity field were measured simultaneously. The experimental results showed that the interaction was characterized by the shape, size and density of a bubble. Concretely, they showed the characteristics of bubble motion, wake shedding, and flow field.

KEYWORDS: bubble motion; trajectory; flow field; PIV; 3D-MOIA; MOFIA

INTRODUCTION

Dispersed two-phase flow of solid particles or gas bubbles in a continuous fluid phase appears frequently in energy related facilities, such as pipelines for minerals and charcoals, chemical plants, and nuclear and steam power stations. Flow characteristics of the dispersed two-phase flows have been extensively investigated in order to enhance understanding of the mechanisms involved and to improve the efficiency and safety of the relevant facilities.

Particles or bubbles in dispersed two-phase flow are characterized by size, shape, buoyancy (or density), deformation (bubbles), and so forth that simultaneously affect the characteristics of the dispersed two phase fluid. Zun[1] studied the transverse lift force on a bubble, generated by velocity gradient of the liquid phase, where the relative velocity varies from near the wall to the core of a pipe. Matsui and Monji[2] studied the effect of the size, density, shape, and buoyancy of a particle/bubble on its motion by using 3D-MOIA. Fujiwara *et al.*[3] measured the velocity distributions around a single bubble by using PIV. Furthermore, Matsui and Monji[4] visualized the two-dimensional (2D) water velocity field around a rising single bubble/particle in a pipe by a hybrid PIV and 2D-MOIA.

Address for correspondence: H.M. Choi, Fluid Flow Group, Korea Research Institute of Standards and Science, P.O. Box 102, Yuseong, Daejeon 305-600, Korea.
hmchoi@kriss.re.kr

The basic characteristics of dispersed two-phase flow, including the velocity field and bubble motion, are better understood by investigating the interactions between a bubble and the surrounding fluid, between the bubbles, and between a bubble and a pipe wall. The hybrid PIV-MOIA serves to understand the interaction between bubble motion and the velocity field around the bubble. The hybrid PIV consists of PIV, measuring the velocity field around the bubble, and 2D-MOIA measuring the bubble motion. Monji et al.[5] and Choi et al.[6] measured bubble motion and ambient flow by the hybrid PIV and investigated the bubble–fluid interaction. However, these 2D-MOIA studies were unable to determine the 3D location and shape relating to the motion of a bubble.

The spatial relation between a bubble and a velocity field is important when bubble–fluid interaction is discussed. To investigate the effects of bubble size, shape, deformation, and buoyancy on bubble motion and flow, a new system, MOFIA, previously called Hybrid PIV,[7] consisting of 3D-MOIA[7,8] and PIV was applied. MOFIA revealed the 3D spatial relation between a bubble and the velocity field, as well as the motion and shape of the bubble.

EXPERIMENTAL APPARATUS AND VISUALIZATION METHODS

The experimental apparatus consisted of a bubble injector and a vertical pipe with inner diameter $D = 0.04$ m, with a test section $7.5D$ long. The test section was located at $30D$ downstream from the bubble injector. The vertical pipe was made of transparent acrylic pipe 2 m long.

The vertical pipe was filled with stagnant water. An air bubble was injected by a syringe into the water to make a single bubble. The syringe was pushed slowly to assure a zero or very small bubble velocity. The experiments were done with four bubble sizes: 2.0, 5.7, 7.3, and 9.1 mm equivalent diameter (de). The equivalent diameter of the bubbles was calculated based on the displacement of the syringe. The motion of the injected bubble and the velocity field of the water were measured simultaneously by the MOFIA system at the test section.

FIGURE 1 shows a schematic diagram of the test section and the MOFIA system. MOFIA mainly consists of a CCD camera and a double pulse Nd–YAG laser (25 mJ) for PIV, two high-speed CCD cameras and two strobe lights for 3D-MOIA, and a pulse generator. A personal computer controlled both the PIV and the MOIA, and recorded the images. A water jacket prevented the distortion of the visual images.

A light sheet of the double pulse Nd–YAG laser illuminated the test section twice at an interval of 2 msec and two flow images were taken for water velocity measurement. The water velocity was obtained by the cross correlation of the two flow images. The time interval between the water velocity measurements was 70 msec. A fluorescent polymer particle was used as a tracer. Its size and density were 10 μm diameter and 1,500 kg/m^3, respectively.

Two stroboscope and the high-speed CCD camera pairs were used to take an image of the bubble. The back light system, or the stroboscope located opposite the CCD camera, was used to take clear contours of the particle image. Each bubble image was approximated by an ellipsoid.[9] After correcting the images, the 3D position and shape of the bubble was reconstructed based on the two images, assuming

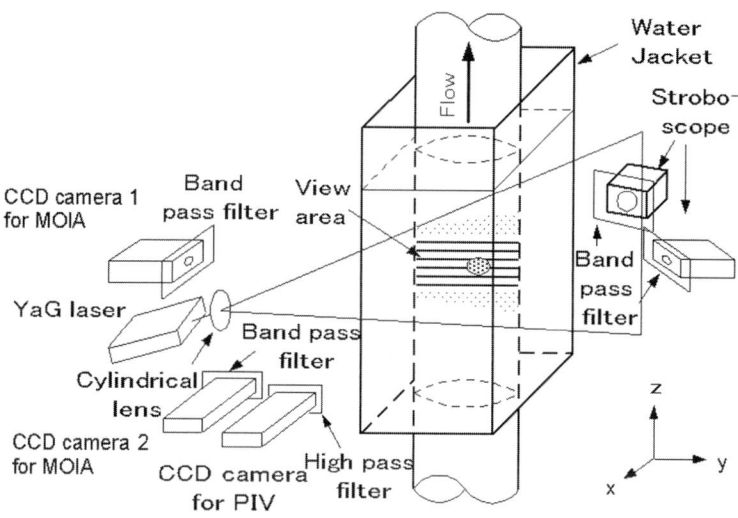

FIGURE 1. Schematic of test section and MOFIA.

an oblate spheroidal bubble.[10] This assumption was reasonable for small bubbles, and was better than assuming a spherical shape. The time interval between images for MOIA was 5 msec and the resolution of the bubble images was 0.05 mm/pixel.

PIV and the MOIA were synchronized, based on the pulses generated by the pulse generator. The velocity field and the reconstructed bubble shape were combined and the flow field, including the bubble, was obtained as a result of MOFIA. To avoid interference between the PIV and the MOIA, optical filters with different ranges of wavelength were installed in front of the three CCD cameras and the two stroboscopes. Furthermore, by using an optical filter, it was possible to eliminate reflection of the laser light on the pipe wall and the bubble.[5]

RESULTS

FIGURE 2 shows a time series of visualization results by MOFIA, in which the spatial relation between a 3D bubble and a velocity field is shown. The bubble size was $de = 9.1$ mm based on the volume of the injected air. In these pictures the arrows indicate the velocity vectors of the water. The gray levels in the velocity field show a component of vorticity along the perpendicular axis to the velocity filed. The vorticity in a white region is clockwise and counterclockwise in a black region. The time interval between results, 70 msec, was not sufficient to analyze the bubble motion. Therefore, the detailed bubble motion obtained by MOFIA every 5 msec is discussed separately. FIGURE 2A shows the wake behind the bubble and the flow pushed in front of the bubble. The bubble motion was spiral and wake shedding can be observed in FIGURE 2B and C—a well known phenomenon from a previous study.[4] The vortex in the wake separated at the lower edge of the inclined major axis of the bubble (FIG. 2B) and changed the direction of motion of the bubble.[4] FIGURE 3 shows

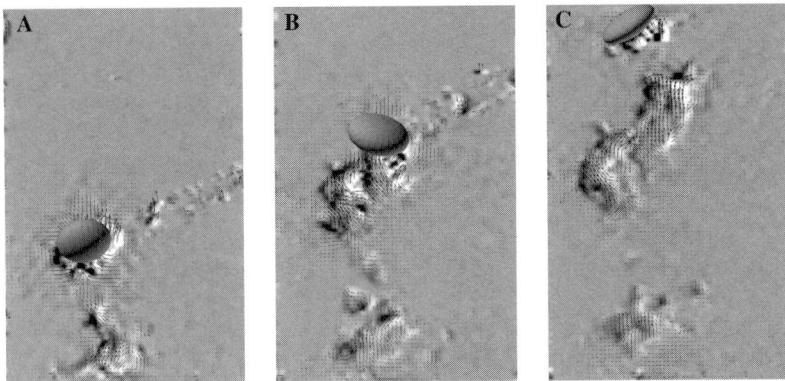

FIGURE 2. Typical example of the motion of a bubble (de = 9.1 mm) and the flow around it: **(A)** t = 0 msec; **(B)** t = 70 msec; **(C)** t = 140 msec.

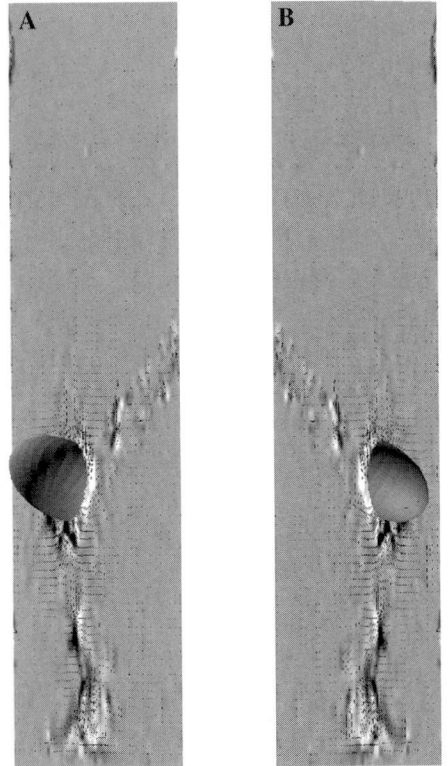

FIGURE 3. Inclined velocity fields: **(A)** +20°; **(B)** −20°.

the same flow field as FIGURE 2A, but seen from different directions or angles. The visual axis or angle in FIGURE 2A is perpendicular or $+90°$ to the plane of the velocity field, but each (visual) angle between the visual axis and the flow field in FIGURE 3 is $+20°$ or $-20°$. The sign + represents the clockwise direction. Based on FIGURE 3, it is clear that the center of the bubble was located almost on the plane of the velocity field and that the flow in FIGURE 2A occurred near the bubble interface. FIGURE 3 demonstrates that the 3D spatial relation between the bubble and the velocity field can be obtained by the MOFIA and that the flow field can be seen from any direction.

The bubble motion is shown in FIGURES 4 and 5. To investigate the bubble motion in detail, the bubble motion is shown at every 25 msec in FIGURE 4. The bubble images were taken by MOIA. The rising velocity of the bubble was 0.194 m/sec. The characteristic point of the bubble motion was that the direction of the moving bubble at all times was the direction of the minor axis of the oblate spheroidal bubble. The drag force due to the water acted in the opposite direction to the bubble motion and the bubble deformed. Thus, bubble deformation maintains the minor axis parallel to the direction of motion.

FIGURE 5 shows the trajectories of the five kinds of bubbles, projected on the cross section. The equivalent diameters of bubbles were (a) $de = 2.0$ mm, (b) $de = 5.7$ mm, (c) $de = 7.3$ mm, and (d) $de = 9.1$ mm. There are five trajectories of a single bubble in each figure. A solid dot represents the starting point, when the bubble entered the test section, and the trajectory was drawn by tracing the gravitational center of the bubble. When bubble motion is rectilinear or zigzag, the motion is 2D and the trajectory projected on the cross section is a dot or a line; that is, the dimension of the

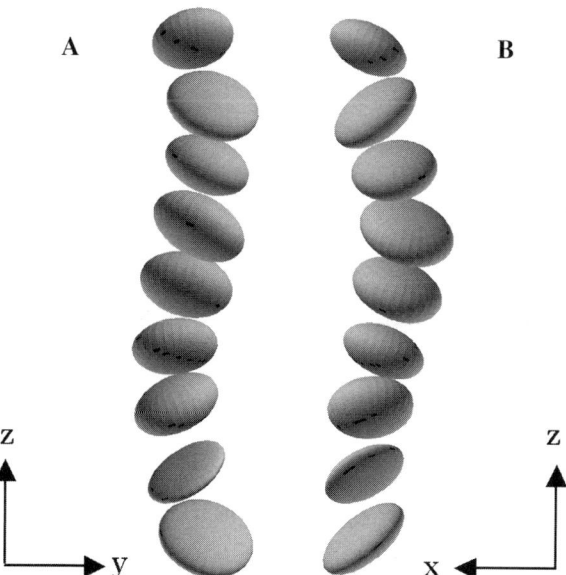

FIGURE 4. Bubble ($de = 9.1$ mm) motions at every 25 msec: (**A**) y–z plane; (**B**) x–z plane.

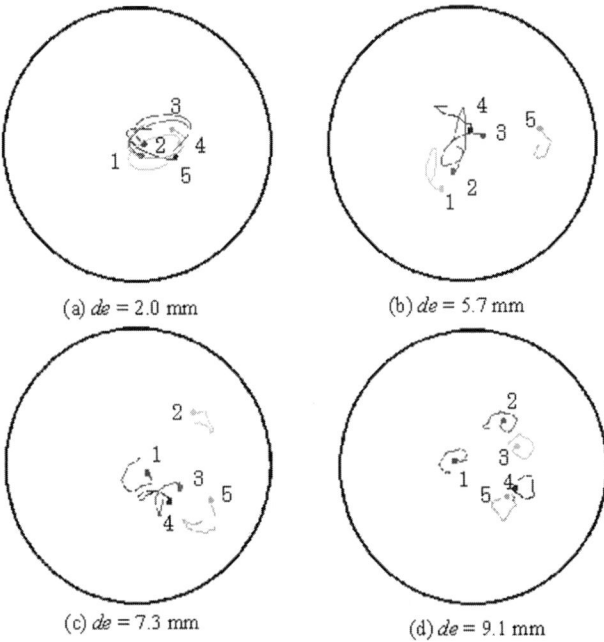

FIGURE 5. Trajectories of bubbles projected on the cross section.

trajectory is less than two. On the other hand, when the bubble motion is spiral, the projected trajectory is a circle or an ellipsoid. Most trajectories in (a) are located in the core part of the pipe and their shapes look ellipsoidal. The trajectories in (b) and (c) are curved and straight lines. Therefore, the bubble motions were rectilinear, zigzag, or spiral, depending on the bubble. The trajectories in (d) were almost closed circles, corresponding to spiral motion. The motion of the small and large bubbles was spiral and that of the middle sized bubbles was spiral, zigzag, or rectilinear.

FIGURES 2, 4, and 5 suggest that the spiral motion was caused by the deformation effect of the bubble and the successive change of its attitude or position due to 3D wake shedding. Also, the minor axis of the bubble approximated an oblate spheroid and always kept the same direction as the moving bubble.

SUMMARY

The new MOFIA measuring and analyzing system was developed in order to obtain the spatial relation between a bubble and a flow field around it. Using the system, the 3D motion of the bubble and the 2D flow field around it were measured simultaneously. The main results were: (1) Based on the instantaneous velocity field and 3D position and shape of the bubble obtained, the spatial relation between the bubble and the velocity field was clear. (2) The minor axis of an oblate spheroidal bubble always stayed parallel to the direction of motion of the bubble due to the deformation of the bubble and the drag force by the fluid. (3) The 3D-MOIA allowed us to discriminate whether bubble motion was zigzag or spiral. In general small and large bubbles moved spirally.

REFERENCES

1. ZUN, I. 1980. The transverse migration of bubbles influenced by walls in vertical bubby flow. Int. J. Multiphase Flow **6**(6): 583–588.
2. MATSUI, G. & H. MONJI. 1995. Behavior of a single particle/droplet/bubble in vertical liquid flow. *In* Two-Phase Flow Modeling and Experimentation. Edizioni ETS. 789–795.
3. FUJIWARA, A., *et al.* 1998. Investigation of oscillatory bubble motion using a dual shadow technique and its surrounding flow field by LIF-PIV. Proc. 3rd Int. Conf. Multiphase Flow, ICMF'98, Lyon, France. (CD-ROM.)
4. MATSUI, G. & H. MONJI. 1998. An image processing study of bubble behavior in upward bubble flow. Proc. 3rd Int. Conf. Multiphase Flow, ICMF'98, Lyon, France. (CD-ROM.)
5. MONJI, H., *et al.* 2000. Measurement of velocity field and a particle/bubble motion by PTV and PIV. Proc. 9th. Int. Sym. on Flow Visualization. (CD-ROM.)
6. CHOI, H.M., *et al.* 2000. Measurement of particle/bubble motion and turbulence around it by hybrid PIV. Proc. Second Japanese-European Two-Phase Flow Group Meeting. (CD-ROM.)
7. MATSUI, G., *et al.* 2001. Bubble/particle motion and ambient flow by PIV and 3D-PTV. Proc. 4th Int. Conf. Multiphase Flow, ICMF 2001.
8. MATSUI, G., *et al.* 2001. Visualization techniques in gas–liquid two-phase flow field. Proc. Third Int. Symp. on Measurement Techniques for Multiphase Flows.
9. LUNDE, K. & J. PERKINS. 1995. A method for the detailed study of bubble motion and deformation. *In* Advances in Multiphase Flow. Elsevier Science.
10. MONJI, H. & G. MATSUI. 1996. Gas-phase characteristics in sliding bubble flow obtained by image processing. Proc. Japan–U.S. Seminar on Two-Phase Flow Dynamics (Supplement). 155–162.

Effect of Electrolytes on Bubble Coalescence in Columns Observed with Visualization Techniques

MARÍA EUGENIA AGUILERA, ANTONIETA OJEDA, CAROLINA RONDÓN, AND AURA LÓPEZ DE RAMOS

Departamento de Termodinámica y Fenómenos de Transferencia, Universidad Simón Bolívar, Caracas, Venezuela

ABSTRACT: Bubble coalescence and the effect of electrolytes on this phenomenon have been previously studied. This interfacial phenomenon has attracted attention for reactor design/operation and enhanced oil recovery. Predicting bubble coalescence may help prevent low yields in reactors and predict crude oil recovery. Because of the importance of bubble coalescence, the objectives of this work were to improve the accuracy of measuring the percentage of coalescing bubbles and to observe the interfacial gas–liquid behavior. An experimental setup was designed and constructed. Bubble interactions were monitored with a visualization setup. The percentage of air bubble coalescence was 100% in distilled water, about 50% in 0.1 M sodium chloride (NaCl) aqueous solution, and 0% in 0.145 M NaCl aqueous solution. A reduction of the contact gas–liquid area was observed in distillate water. The volume of the resulting bubble was the sum of the original bubble volumes. Repulsion of bubbles was observed in NaCl solutions exceeding 0.07 M. The percentage of bubble coalescence diminishes as the concentration of NaCl chloride increases. High-speed video recording is an accurate technique to measure the percentage of bubble coalescence, and represents an important advance in gas–liquid interfacial studies.

KEYWORDS: bubble coalescence; electrolyte solutions; gas–liquid interface

BACKGROUND

Bubble coalescence has attracted considerable attention in past few decades for the purposes of reactor design and operation and aerated beds in bubble column and stirred tank reactors, mainly because the gas–liquid contact area is one of the key parameters in mass transfer. When bubbles coalesce the contact area diminishes and mass transfer decreases. Bubble size distribution is a result of bubble formation, bubble coalescence, and breakup. For these reasons it is important to predict the bubble coalescence in order to prevent low yields in reactors and it is also important to enhanced crude oil recovery.

Address for correspondence: María Eugenia Aguilera, M.Sc., Departamento de Termodinámica y Fenómenos de Transferencia, Universidad Simón Bolívar, Apartado Postal 89000, Caracas 1080-A, Venezuela.
meaguile@usb.ve

Studies have been made in an attempt to explain bubble behavior in electrolyte solutions. Prince and Blanch[1,2] examined the amount of salt required to immobilize the gas–liquid interface of the film between coalescing bubbles, proposing a phenomenologic model for rates of bubble coalescence and bubble breakup in turbulent gas–liquid dispersions. They developed a measurement technique to test the validity of the model. Craig et al.[2] performed experiments using N_2, He, Ar, or SF_6 as the gas phase, and $NaNO_3$, KBr, $CaCl_2$, $MgSO_4$, or NaCl as electrolyte solutions. They found that temperature, viscosity and surface tension have no influence on bubble coalescence. They concluded that some electrolytes and some mineral acids have no effect at all on bubble coalescence. They assigned a property (α or β) to each anion or cation, the combinations $\alpha\alpha$ or $\beta\beta$ inhibit coalescence (e.g., NaCl is $\alpha\alpha$), and the combinations $\beta\alpha$ or $\alpha\beta$ have no effect (e.g., HCl is $\beta\alpha$). No exceptions have yet been found. Craig et al.[3] pointed out that these observations could be explained only by the local influence of the ions on water structure, possibly in a way related to the hydrophobic interaction. The data obtained by Craig et al.[3] show that electrolytes affect the hydrophobic interaction (e.g., KBr).

Zahradník et al.[4] performed experiments with NaCl, KCl, KI, NaSCN, NaOH, $CaCl_2$, $BaCl_2$, Na_2SO_4, and $MgSO_4$ in order to include electrolytes with typical coalescent, noncoalescent, and transient behavior. They conclude that the transition electrolyte concentrations are linked up with the hydrodynamics of bubble beds containing solutions of respective electrolytes and can be employed for estimation of the effect of electrolytes on gas holdup in bubble column reactors.[5]

Liendo[5] designed a coalescence cell with basis on work of Zahradník et al.[4] The set up consisted of a plexiglass cell with two glass capillaries in opposite positions. Liendo[5] used air as gas phase and NaCl, $NaNO_3$, HCl, HNO_3, CH_3COOH, $CuSO_4$, $MgSO_4$, and NH_4OH in electrolyte solutions. Liendo[5] obtained results in agreement with those of Craig et al.[3]

EXPERIMENTAL SETUP

Ueyama et al.[6] developed a system for measuring bubble coalescence time and bubble age. Coalescence was observed with the help of a laser system and a high-speed video camera, but the setup and the measuring system were complicated to construct and to use. Zahradník et al.4 developed a simpler setup but did not use a high speed video camera.

In this study, a new setup was designed and constructed, based on the recent work of Liendo.[5] The gas circuit was improved and a digital high-speed video camera (4,500 fps) used to visualize the bubble coalescence experiments with a magnifying lens. A fiber optic lamp was used to improve the focus on the interfaces. All the experiments were recorded on the computer. The bubble coalescence visualizations were carried out in a vertical glass column with two glass capillaries (I.D. 2 mm) 5 mm apart. Coalescence was observed in stagnant distilled water, and electrolyte solutions of various concentrations. The electrolyte solutions were prepared with distilled water and NaCl of analytical grade in concentrations from 0.001 M to 0.145 M. The selection of NaCl as the electrolyte was made based on previously reported work demonstrating that NaCl is one of the electrolytes that inhibits coalescence.[3,5] The gas phase was air at a constant feed rate. The average air velocity as measured from video films was 0.008 m/sec. All experiments were carried out at room temperature and atmospheric pressure. Pairs of bubbles were formed simultaneously in capillaries, the frequency of bubble formation was 28 pairs per minute.

Note that the new experimental device allows for the substance in the capillaries to be changed, including changing the gas phase to study liquid–liquid interactions in water–crude oil and surfactants.

RESULTS AND DISCUSSION

The percentages of bubble coalescence were measured, and the results were more precise than those obtained with previous techniques. FIGURE 1 shows the percentage of bubble coalescence versus the concentration of NaCl compared with other studies.[5] The difference observed is due to the speed of the video cameras used: Liendo[5] used a slow video camera and the interface at the moment of coalescence was not observed clearly, so it was difficult to determine if coalescence actually occurred.

As can be seen in FIGURE 1, bubble coalescence was 0% in 0.145 M NaCl solution, about 50% in 0.1 M solution, and 100% in distilled water. In some experiments, the interactions in the interface showed that the repulsive forces increase as concentration of NaCl increases. Some authors demonstrated that such behavior is independent of surface tension.[1]

In FIGURE 2 A, the air bubbles are formed simultaneously in distilled water. As can be seen, there is sufficiently strong attractive force between the bubbles to cause coalescence in water. In these experiments, the attractive van der Waals forces overcome the repulsive forces and the total free energy diminishes, reducing the contact gas–liquid area (FIG. 2 B and C). Finally, a new elliptical bubble arises (FIG. 2 D), and the volume of this bubble is the sum of the volume of the two original bubbles.[5]

FIGURE 3 shows the bubble interactions in 0.125 M NaCl solution. In FIGURE 3 A bubbles are formed. It can be shown that NaCl acts to prevent coalescence, because bubbles approach and seem to coalesce (FIG. 3 B). However, a substantial hydrodynamic repulsion force acts and bubbles are unable to coalesce. Therefore, the

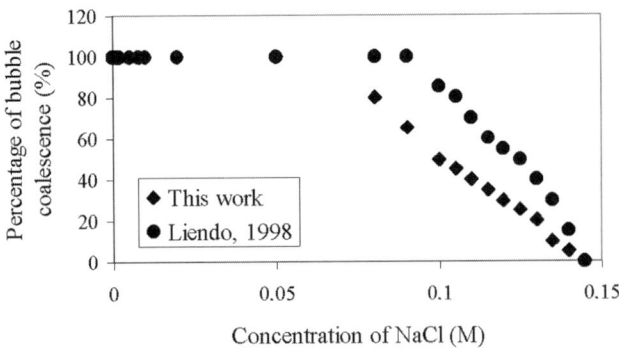

FIGURE 1. Percentage of bubble coalescence versus concentration of aqueous NaCl solution.

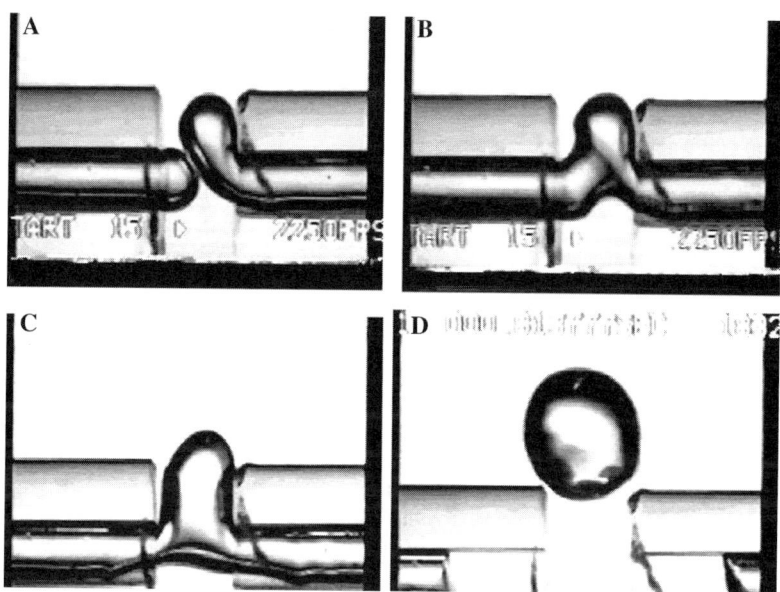

FIGURE 2. Bubble interactions in distilled water: **(A)** formation of bubbles, **(B)** coalescence of bubbles, **(C)** formation of a new bubble, **(D)** a new elliptical bubble is formed.

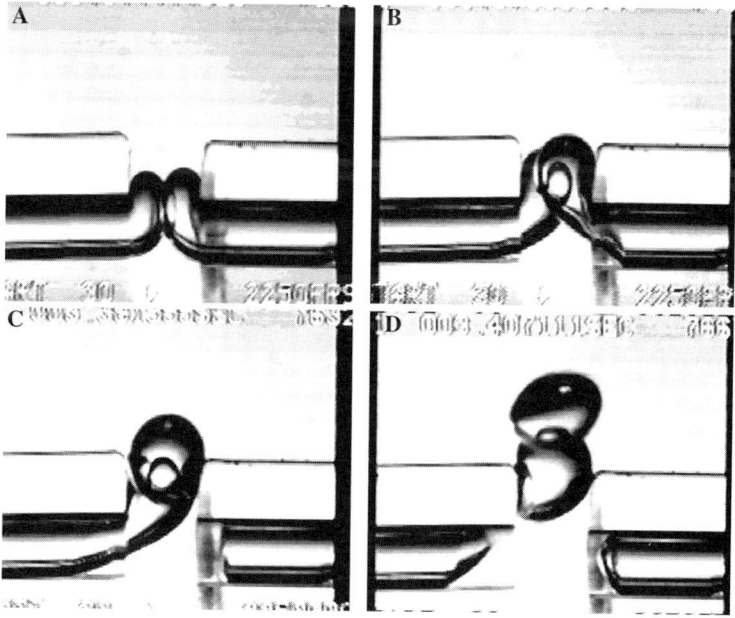

FIGURE 3. Bubble interactions in a 0.125 M NaCl solution: **(A)** formation of bubbles, **(B)** superposition of bubbles, **(C)** separation of bubbles, **(D)** two spherical bubbles are formed.

bubbles separate completely (FIG. 3C) and two spherical bubbles arise (FIG. 3D). In these experiments coalescence was not observed, but at concentrations below 0.08 M some bubbles coalesced. As can be seen, the percentage of coalescence decreases as NaCl concentration increases. This behavior was also observed by Craig et al., who defined NaCl as an αα combination.[3]

SUMMARY

An experimental setup was designed and constructed, and bubble coalescence was observed with a digital high-speed video camera using NaCl aqueous solutions as the liquid phase and air as the gas phase. It was observed that concentrations of NaCl exceeding 0.08 M inhibit air bubble coalescence, consistent with previous studies.[3,4] The prediction of bubble coalescence, and the accuracy of these measurements are also important in some research fields. The contribution of this work is in the visualization of the coalescence phenomenon. Since attraction and union of bubbles could be clearly observed, bubble volume, contact gas–liquid area, and percentage of bubble coalescence could also be accurately measured with a simple setup. Finally, high speed video techniques and adequate illumination allow bubble volumes and contact gas–liquid area to be studied in detail.

ACKNOWLEDGMENT

The authors thank Agenda Petróleo, Project N° 7003788, PDVSA, for the high speed video camera and Universidad Simón Bolívar and CONICIT for financial support.

REFERENCES

1. PRINCE, M. & H. BLANCH. 1990. Transition electrolyte concentrations for bubble coalescence. AIChE J. **36**(9): 1425–1429.
2. PRINCE, M. & H. BLANCH. 1990. Bubble coalescence and break-up in air-sparged bubble columns. AIChE J. **36**(10): 1485–1499.
3. CRAIG, V., B. NINHAM & R. PASHLEY. 1993. Effect of electrolytes on bubble coalescence. Nature Lett. **364**: 317–319.
4. ZAHRADNÍK, J., M. FIALOVA, F. KASTÁNEK, et al. 1995. The effect of electrolytes on bubble coalescence and gas holdup in bubble column reactors. Trans. IChemE **73**: 341–346.
5. LIENDO, L. 1998. Estudio de la Coalescencia de Burbujas en Sistemas Electrolíticos. Thesis, Universidad Simón Bolívar, Chemical Engineering Department, Caracas, Venezuela.
6. UEYAMA, K., M. SAEKI & M. MATSUKATA. 1993. Development of system for measuring bubble coalescence time by using a laser. J. Chem. Eng. Japan **26**(3): 308–314.

X-ray-Based Flow Visualization and Measurement

Application in Multiphase Flows

AXEL SEEGER,[a] KLAUS AFFELD,[a] LEONID GOUBERGRITS,[a]
ULRICH KERTZSCHER,[a] ERNST WELLNHOFER,[b] AND RENE DELFOS[c]

[a]*Biofluidmechanics Laboratory, Charite, Berlin, Germany*

[b]*German Heart Institute Berlin, Berlin, Germany*

[c]*Laboratory for Aero and Hydrodynamics, Delft University of Technology, Delft, the Netherlands*

ABSTRACT: Information concerning continuous or discreet phase flow in multiphase systems is desired for various practical and analytical applications. The potential of X-ray-based flow visualization and measurement of multiphase flow is demonstrated here by two non-intrusive methods: (1) Measurement of the three-dimensional (3D) velocity field of the continuous liquid phase in a bubble column by X-ray-based particle tracking velocimetry (PTV) of seeded particles. (2) Liquid flow visualization in a bubble column by injecting an X-ray absorbing liquid into the bubble column. X-rays have the advantage that they are not affected by the various refraction indices of the multiphase system and penetrate the multiphase flow in undistorted straight lines. Hence, in contrast to optical methods, both of these X-ray-based methods are independent of the void fraction and are applicable to opaque liquids.

KEYWORDS: X-ray; particle tracking velocimetry; multiphase flow; bubble column

INTRODUCTION

Multiphase flows are used in many industrial applications. Examples of this are bubble columns and loop reactors, both of which are widely used in biotechnology and in chemical process engineering, for instance, in waste water treatment, fermentation, and the oxidation of various substances.[1] Despite their widespread use, flow phenomena in these systems are not yet clearly understood due to lack of adequate flow visualization and measurement techniques. Moreover, flow simulation results obtained by computational fluid dynamics (CFD) for these systems cannot be validated for lack of reliable experimental data. Of particular difficulty is the measurement of the velocities of liquid and solid phases in the presence of a gaseous phase. These velocities are commonly measured by optical methods, such as laser doppler velocimetry (LDV) and particle image velocimetry (PIV). However, these methods

Address for correspondence: A. Seeger, Dipl.Ing., Biofluidmechanics Laboratory, Charite, Spandauer Damm 130, D-14050 Berlin, Germany.
  axel.seeger@charite.de

are unsatisfactory in multiphase flows with large voids or large solid fractions.[2,3] According to Larue de Tournemine et al.,[2] the maximum void fraction for which optical methods can be applied in a bubble column, is about 5%. The value depends on the length of the optical paths and on the bubble diameters. Furthermore, in a flow with solid particles, the solid phase is usually opaque and does not allow the light to penetrate the medium. Also, the light is distorted by reflections and refractions from the phase boundaries.

Some specific X-ray-based flow visualization and flow measurement methods are well known. The medical application of X-ray angiography is a good example. This is a commonly used technique to visualize the flow in coronary vessels, where optical access is impossible.[4] An X-ray absorbing liquid is injected into the coronary vessels to be visualized and the motion of the liquid is recorded. The cardiologist can, therefore, detect and locate a stenosis. X-rays were also applied to multiphase flows, and used to measure the local time-averaged void fraction in bubble columns by X-ray tomography,[5] and to determine the overall void fraction dynamically.[6] Both methods are independent of the void fraction. For the measurement of the local time-averaged void fraction in bubble columns, the X-ray absorption in the bubble column is measured from many different angles.[5] This enables one to calculate a local time-averaged X-ray absorption coefficient. Since the X-ray absorption coefficients of liquid and gas differ enormously, the local time-averaged void fraction can be calculated from the local absorption coefficients. The results are accurate, since X-rays penetrate a multiphase flow in straight lines, with no reflection and dispersion on the phase boundary. However, this procedure is very time consuming.

X-rays can also be used to measure the local velocities of the liquid phase in bubble columns and the solid phase in multiphase flows. The various methods[2,3,7–14] available to measure local liquid velocities in a bubble column are listed in TABLE 1.

All the optical methods are limited to flow with a low density of bubbles and a small void fraction. The reason for this is the reflection and refraction on phase boundaries. Other methods, such as the electrodiffusion method and hot film velocimetry (HFV) can be used in bubble columns with a high void fraction. However, these are intrusive single point methods that disturb the flow and, therefore, flow structures are difficult to obtain. A single point non-intrusive method that works independently of the void fraction is computer aided radioactive particle tracking (CARPT).[8]

TABLE 1. Available methods for measuring the velocity of the liquid phase in a bubble column

	Small void fraction (less than 1%) optical methods	Larger void fraction (greater than 1%)
Point measurements	laser Doppler velocimetry (LDV)[7,10]	hot-film velocimetry (HFV)[12] computer aided radioactive particle tracking (CARPT)[8] electrodiffusion method[13]
2D or 3D methods	particle image velocimetry (PIV)[8,9] laser induced fluorescence (LIF)[11] particle tracking velocimetry (PTV)[7]	X-ray based PTV (XPTV)[14]

Here we present liquid flow visualization by injection of X-ray absorbing liquid in the bubble column, as well as by measurement of the local liquid velocity in particle seeded bubble columns by X-ray-based PTV. This non-intrusive multipoint method is independent of void fraction. It allows the assessment of three-dimensional vector fields in a volume.[14,15]

METHODS

The Experimental Bubble Column

The bubble column had an inner diameter of 110mm and a fluid height of 100mm. Sixty-nine injection needles with an inner diameter of 0.34mm were used as gas dispergers. The use of the needles makes the gas distribution more uniform than the use of a perforated plate.

X-ray-Based PTV in Bubble Columns

The X-ray-based PTV was used to measure the local liquid velocity in the bubble column. The liquid was seeded with X-ray absorbing particles. The particles, which had the same density as the liquid, moved with the liquid and were as small as the visibility on the screen permitted. The velocity of the particles was measured by detecting their motion.

A typical experimental setup is shown in FIGURE 1.

Two X-ray-sources, S1 and S2, generate X-rays that are directed through the bubble column onto image intensifiers. A point P describing the particle was mapped on the two image intensifiers I1 and I2 generating the points P1 and P2. The point P can be reconstructed from P1 and P2. By observing the motion of the particle, its velocity can be obtained by its displacement. The image intensifier converts the X-rays into visible light and magnifies their intensity. Digital cameras directed at the image

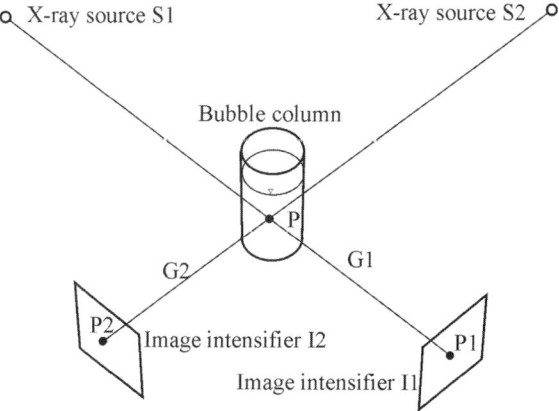

FIGURE 1. Experimental setup.

intensifier took the images. About 50 X-ray absorbing particles were added to the liquid simultaneously. Details of this method are described elsewhere.[14]

A biplane X-ray unit (Philips Integris BH 3000), developed for visualizing blood flow in human coronary vessels, was used in these experiments. The German Heart Institute in Berlin (DHZB) provided the unit. The grayscale images had a resolution of 512×512 pixels. An X-ray flash was generated and an image was taken every 20 ms alternating on each X-ray system and yielding 25 image pairs per second. An image series of up to 1,000 images (500 image pairs) could be obtained. The investigated volume had a size of about $12 \times 12 \times 12 \, cm^3$. The method was first validated in a bubble column with a small void fraction and compared with standard velocity measurement methods, such as laser doppler velocimetry and optical PTV. The results for the velocity and the standard deviation of the velocity were found to be similar.[14,15] X-ray PTV was then applied to the experimental bubble column. The results are presented below.

Flow Visualization with X-ray Absorbing Liquid Injection in an Air–Water Bubble Column

The X-ray absorbing liquid was injected in the center of the bubble column. The injection needle had an inner diameter of 1.5 mm to ensure a fast injection. Images were taken during the injection and the fluid dispersion could be clearly visualized. Ultravist Iopromide, (Schering AG, Germany) 300 mg/ml, was used as the X-ray absorbing liquid, water as liquid, and air as gaseous phase.

RESULTS

Local Liquid Velocity Measurement in an Air–Glycerin Bubble Column

An example of flow in a bubble column is shown in FIGURE 2, as measured by X-ray PTV. Glycerin with a viscosity of $850 \times 10^{-6} \, m^2/sec$ was used as liquid, resulting in a slower flow and, thus, easier particle tracking. The superficial gas velocity was set to 6 mm/sec and the resulting void fraction above the injection needles was about 8%. The vector field is the mean value of 465 image pairs having 17,552 3D-velocity vectors. The recording time was about 19 sec. The flow was assumed to be stationary during this period.

Flow Visualization with X-ray Absorbing Liquid Injection in an Air–Water Bubble Column

The superficial gas velocity was set at 10 mm/sec. FIGURE 3 shows a time sequence of images obtained during the injection of X-ray absorbing fluid. Clearly, this kind of data can be used to study the flow and dispersion characteristics of bubble columns with various void fractions and bubble distributors.

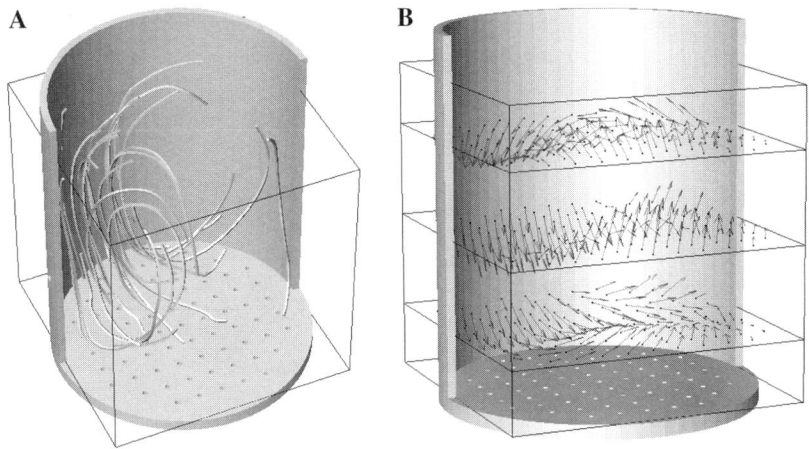

FIGURE 2. (**A**) Particle trajectories in a glycerin–air bubble column. (**B**) The vector field calculated from the particle trajectories. The vector field is the mean value of 465 image pairs having 17,552 3D-velocity vectors. The recording time was about 19 sec. The flow was assumed to be stationary during this period. The injection needles are not shown. The visualization was performed with AMIRA (Indeed—Visual Concepts GmbH, Berlin).

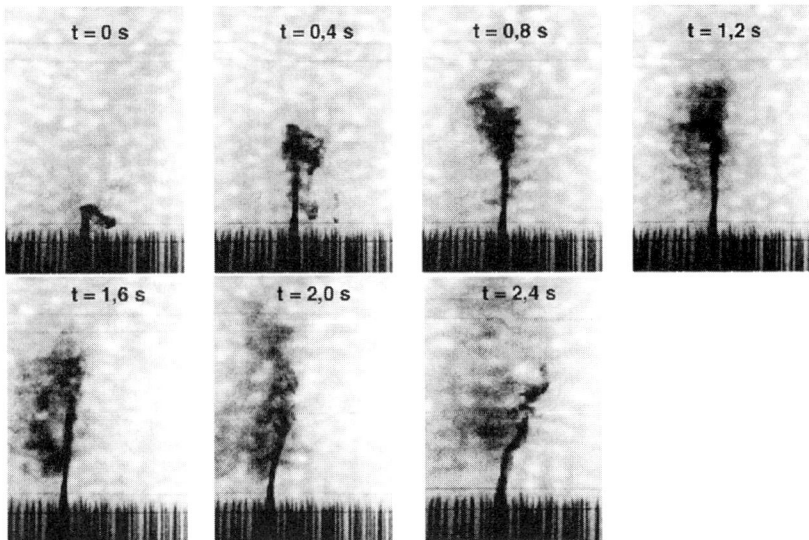

FIGURE 3. Images of injection of an X-ray absorbing liquid into water. The images are contrast and brightness enhanced.

SUMMARY

An X-ray-based technique for flow visualization and velocity measurements is shown here to be a powerful technique where optical methods prove insufficient. The X-ray-based PTV method is a non-intrusive technique that can be used to measure the liquid or solid velocity in multiphase flows at many points in the volume and in 3D. The measurement time to get a 3D-velocity field is small. Using faster technical devices (X-ray sources, cameras, and image intensifiers) may allow up to 500 image pairs per second to be captured. This permits the measurement of bubble columns with higher velocities (e.g., a bubble column with water as liquid). Furthermore, it reduces the measurement time.

ACKNOWLEDGMENT

Axel Seeger was supported by a fellowship from the "Studienstiftung des deutschen Volkes", Bonn, Germany.

REFERENCES

1. DUDUKOVIC, M.P., F. LARACHI & P.L. MILLS. 1999. Multiphase reactors—revisited. Chem. Eng. Sci. **54:** 1975–1995.
2. LARUE DE TOURNEMINE, A., V. ROIG & C. SUZANNE. 2001. Experimental study of the turbulence in bubbly flows at high void fraction. Proc. 4th Int. Conf Multiphase Flow, New Orleans, USA.
3. MUDDE, R.F., D.J. LEE, J. REESE & L.S. FAN. 1997. Role of coherent structures on Reynolds stresses in a 2D bubble column. AIChE J. **43:** 913–926.
4. WEBSTER, J.G. 1988. Encyclopedia of medical devices and instrumentation. John Wiley, New York.
5. MEWES, D. & D. SCHMITZ. 1999. Tomographic methods for the analysis of flow patterns in steady and transient flows. Proc. 2nd Int. Symp on Two Phase Flow Modeling and Experimentation. 29–42.
6. BEINHAUER, R. 1971. Dynamische Messung des relativen Gasgehalts in Blasensäulen mittels Absorption von Röntgenstrahlen. Dissertation, TU Berlin.
7. BORCHERS, O. & G. EIGENBERGER. 2000. Particle tracking velocimetry for simultaneous investigation of liquid and gas phase in bubbly flow. 9th Int. Symp of Flow Visualization, Edinburgh.
8. CHEN, J., A.K. AL-DAHHAN, M.H. DUDUKOVIC, et al. 1999, Comparative hydrodynamics study in a bubble column using computer-automated radioactive particle tracking (CARPT)/computed tomography (CT) and particle image velocimetry (PIV). Chem. Eng. Sci. **54:** 2199–2207.
9. CHEN, R.C., J. REESE & L.S. FAN. 1994. Flow structures in a three-dimensional bubble column and three-phase fluidized bed. AIChE J. **40:** 1093–1104.
10. MUDDE, R.F., J.S. GROEN & H.E.A. VAN DEN AKKER. 1997. Liquid velocity field in a bubble column: LDA experiments, Chem. Eng. Sci. **52:** 4217–4224.
11. BRÖDER, D. & M. SOMMERFELD. 2000. A PIV/PTV system for analysing turbulent bubbly flows. Proc. 10th Int. Symp. Application of Laser Techniques to Fluid Mechanics, Lisbon. Paper 10.1.
12. FRANZ, K., T. BORNER, H.J. KANTOREK & R. BUCHHOLZ. 1984. Flow structures in bubble columns. Germ. Chem. Eng. **7:** 365–374.
13. PAULI, J. 1991. Einsatz der Elektrodiffusionsmeßtechnik in Gas-Flüssigkeitsströmungen mit Sauerstoff als Polarisator, VDI-Fortschrittsberichte, Reihe 3, Nr. 278.

14. SEEGER, A., K. AFFELD, L. GOUBERGRITS, et al. 2001, X-ray-based assessment of the three-dimensional velocity of the liquid phase in a bubble column. Exp. Fluids **31**(2): 193–201.
15. SEEGER, A., K. AFFELD, U. KERTZSCHER, et al. 2001, Assessment of flow structures in bubble columns by X-ray based particle tracking velocimetry. Proc. 4th Int. Sym. Particle Image Velocimetry, Göttingen, Germany.

Simultaneous Velocity and Concentration Measurements of a Turbulent Jet Mixing Flow

HUI HU,[a] TETSUO SAGA,[b] TOSHIO KOBAYASHI,[b] AND NOBUYUKI TANIGUCHI[b]

[a]*Department of Mechanical Engineering, Michigan State University, East Lansing, Michigan, USA*

[b]*Institute of Industrial Science, University of Tokyo, Tokyo, Japan*

> ABSTRACT: A method for the simultaneous measurement of velocity and passive scalar concentration fields by means of particle image velocimetry (PIV) and planar laser induced florescence (PLIF) techniques is described here. An application of the combined PIV-PLIF system is demonstrated by performing simultaneous velocity and concentration measurements in the near field of a turbulent jet mixing flow. The distributions of the ensemble-averaged velocity and concentration, turbulent velocity fluctuation, concentration standard deviation, and the correlation terms between the fluctuating velocities and concentration in the near field of the turbulent jet flow are presented as the measurement results of the simultaneous PIV-PLIF system.
>
> KEYWORDS: PIV; PLIF; PIV-PLIF system; jet flow; Reynolds stress; turbulence; flux correlation; concentration

INTRODUCTION

Simultaneous information concerning a passive scalar property and a velocity field is required in many fluid flow investigations, including mixing in combustion chambers and distributions of drugs in biomedical applications. The possibility of measuring velocity and a scalar at the same time, with a high spatial and/or temporal resolution is also of fundamental importance for the validation and development of models for turbulence and turbulent mixing. For example, in a turbulent jet mixing flow, the species concentration field is determined by molecular diffusion and transported by the turbulent flow field. When considering the Reynolds-averaged scalar conservation equation, the effects of turbulent transport appear in terms of the correlation between the concentration and velocity fluctuations; that is, expressions, such as $\overline{u'c'}$ and $\overline{v'c'}$. Experimental characterization of these correlation terms is needed for the development and validation of physical models. This requires the simultaneous measurement of velocity and concentration fields.

With the rapid development of modern optical techniques and digital image processing techniques, whole-field optical diagnostics, such as PIV and PLIF techniques, are assuming ever-expanding roles in the diagnostic probing of fluid mechanics. The

---

Address for correspondence: Hui Hu, Ph.D., A22, RCE Building, Department of Mechanical Engineering, Michigan State University, East Lansing, MI 48824, USA.
huhui@egr.msu.edu

advances of PIV and PLIF techniques in recent years have lead them to be mature techniques for whole-field measurement of velocity and concentration and/or temperature in an objective plane or over the volume of an objective fluid flow. The development of a PIV-PLIF combined system is described here for the simultaneous measurement of velocity and concentration in the near field of a turbulent jet mixing flow and the determination of the correlation terms between the fluctuating velocities and concentration.

EXPERIMENTAL SETUP AND TECHNIQUES

The experimental set up contained a circular test nozzle ($D = 30$ mm) installed in the middle of a water tank ($600 \times 600 \times 1,000$ mm^3). Fluorescent dye (rhodamine B, concentration of about 0.3 mg/liter) for PLIF or PIV tracers (hollow glass particles $d = 8$–12μm) were premixed with water in a jet supply tank, and jet flow was supplied by a pump. The flow rate of the jet flow, which was used to calculate the representative velocity and Reynolds number (Re) values, and was measured by a flow meter. A cylindrical plenum chamber with comb structures was installed upstream of the test nozzle to insure the jet flow had fully developed. The turbulent levels of the core jet flows at the exit of test nozzles were about 3%. An overflow system was used to keep the water level in the test tank constant during the experiment. The investigation region is at the near field of the jet flow ($Y/D < 5.0$). The distance between the exit of the test nozzle and the free surface of the water in the test tank was about 30D. Therefore, the effect of the water free surface in the test tank on the vortical and turbulent structures in the near field of the jet flow was negligible, and the jet flow exhausted from the test nozzle was considered to be a free jet. During the experiment, the core jet velocities (U_0) at the exit of the test nozzle was set to be about 0.20 m/sec. The Re value of the jet flow, based on the nozzle exit diameter and the core jet velocity, was about 6,000.

Pulsed illumination laser sheets (thickness about 1.5 mm) were generated by a double-pulsed Nd:YAG laser system (Quantel Inc.). The frequency of the double-pulsed illumination was 10 Hz. The pulsed illumination duration was 4 nsec, and power 200 mJ/pulse. The time interval between the two pulses was adjustable, set at 3 msec in the present study.

A simultaneous image recording system was designed by using optics and two high-resolution CCD cameras (TSI PIVCAM10-30, 1K by 1K resolution). The emission peak of rhodamine B is about 590 nm, and the wavelength of the illuminating laser light scattered by the PIV tracer particles is 532 nm. Two kinds of optical filters were used to separate LIF lights from scattered laser light, and then recorded separately to obtain PLIF and PIV image simultaneously. A bandpass optical filter (532 ± 5 nm) was installed at the head of camera #1: only the scattered laser light was transmitted to form the PIV image on the CCD sensor of camera #1, and LIF light was blocked out. A high pass filter (greater than 580 nm pass) was installed in the head of the camera #2 to filter out the scattered laser light (wavelength 532 nm). The LIF light (peak at 590 nm) passed through the optical filter to generate the LIF image on the CCD censor of camera #2.

Rather than tracking individual particles, an improved spatial correlation analysis method, named the hierarchical recursive PIV method,[1] was used for PIV image processing. The hierarchical recursive PIV method is actually a recursive operation process of conventional spatial correlation. The recursive operation starts with a large interrogation window size and search distance, the same as the conventional correlation analysis based PIV image processing methods. By using the results of the former iteration step as the approximate offset values in the next iteration step, the interrogation window size and search distance were reduced hierarchically. Whereas the conventional correlation method used 64 by 64 pixels or 32 by 32 interrogation windows, the hierarchical recursive PIV method could reduce the final interrogation window to 8 by 8 pixels with spurious vectors being less than 2%.

In order to obtain a whole field quantitative concentration distribution in the objective flow from PLIF images, an improved whole-field calibration procedure[2] was conducted to account for the non-uniformity of the laser sheet. To improve the accuracy level of the PLIF measurements, the averaged background was subtracted from the LIF images. All the PLIF images were also normalized to account for the laser sheet intensity variations. A general mapping method[3] was used to obtain the spatial correlation between the PIV and PLIF images. The spatial resolution of PIV results was determined by the sizes of the interrogation window used for the correlation operation. The final interrogation window size for PIV image processing was 8 by 8 pixel; therefore, the concentration data were also averaged over 8 by 8 sub-windows during the PLIF image processing.

Once the velocity and concentration fields were calculated, it was relatively straightforward to calculate the various ensemble-averaged velocity (U, V), turbulent velocity fluctuations ($\sqrt{(\overline{uu})}, \sqrt{(\overline{vv})}$), mean concentration (c), concentration standard deviation ($\sqrt{(\overline{cc})}$), and the turbulent flux correlation terms ($\overline{uc}, \overline{vc}$) between the velocity and concentration.

RESULTS AND DISCUSSIONS

FIGURE 1 represents a typical pair of the instantaneous PIV and PLIF measurement results. Since the final interrogation window size was 8 by 8 pixels for PIV image processing, about 50,000 vectors could be obtained for every instantaneous

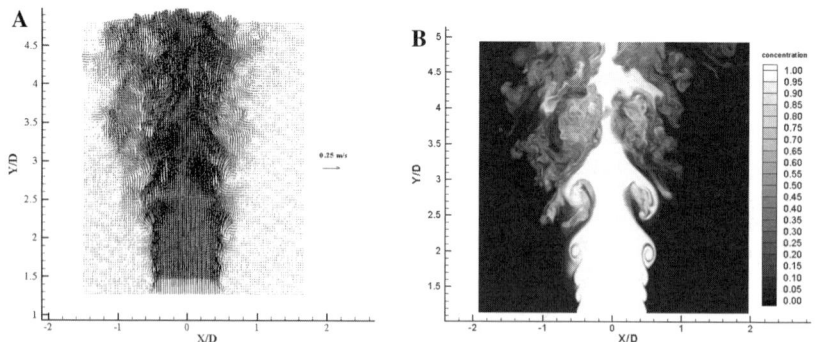

FIGURE 1. Typical instantaneous measurements from the combined PIV–PLIF system: **(A)** PIV measurements; **(B)** simultaneous PLIF measurements.

PIV frame. The velocity vectors shown in FIGURE 1A display only 25% of the PIV velocity vectors. FIGURE 1B shows the instantaneous concentration field obtained by PLIF image processing, which is the simultaneous measurement result of the PIV results shown in FIGURE 1A. The contour levels given in the figure represent rhodamine B concentration levels normalized by the jet source concentration $\xi_0 = 0.3$ mg/l. It is well known that the shear layers in jet flows are unstable via Kelvin–Helmholtz instability. The instability grows downstream and rolls into coherent vortex rings. The vortex ring structures merge as they move downstream and then break down into small vortex structures. The transition of the jet flow into turbulence occurs when the large vortex rings break down into small-scale vortices. All of these processes can clearly be seen from the PIV-PLIF simultaneous measurement results shown in FIGURE 1.

Two hundred and fifty PIV and PLIF image pairs, captured simultaneously at a frame rate of 10 Hz, were used to calculate the ensemble-averaged values. The ensemble-averaged values were also normalized with the core jet velocity $U_0 = 0.20$ m/sec and jet source concentration $\xi_0 = 0.3$ mg/l.

FIGURE 2 shows the profiles of ensemble-averaged velocity and mean concentration profiles in the three downstream locations of the turbulent jet flow. The measurements of Lemoine et al.,[4] who used single point measurement techniques (LDV and LIF) downstream of a circular jet flow at the same Re value are also given in the figures. It can be seen that the present PIV and PLIF simultaneous measurements agree with the results of Lemoine et al. reasonably well.

FIGURE 3 presents the distributions of various ensemble-averaged terms, which include ensemble-averaged velocity, ensemble-averaged concentration, turbulence intensity, concentration standard deviation and the radial and axial turbulent flux terms ($\overline{uc}, \overline{vc}$). The ensemble-averaged velocity and mean concentration distributions indicate that there is a high speed and high concentration region in the center of the jet turbulent flow, also called the potential core region. High turbulence intensity and high concentration fluctuation regions exist in the shear layers between the jet flow and ambient flows, whereas both the turbulence intensity and concentration fluctuation values are low in the potential core region. The potential core region extended to $Y/D > 4.0$ downstream, which is consistent with the finding that the length of the potential core region of a conventional circular jet flow ranges between

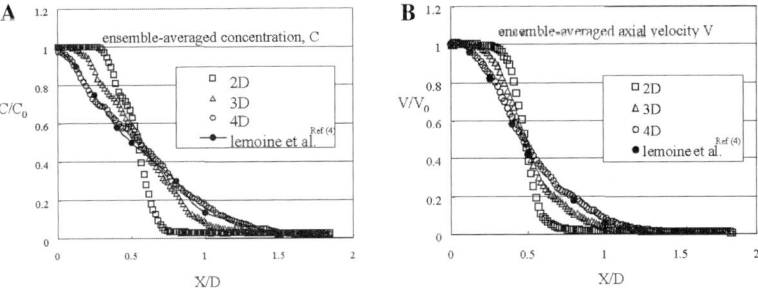

FIGURE 2. Ensemble-averaged concentration and velocity profiles.

$4D$ and $6D$.[5] Although the ensemble-averaged axial velocity component is much bigger than the radial velocity component in the near field of the turbulent jet flow, the axial turbulent flux $\overline{vc}$ and radial turbulent flux $\overline{uc}$ were found to be almost of the same order as the present measured results.

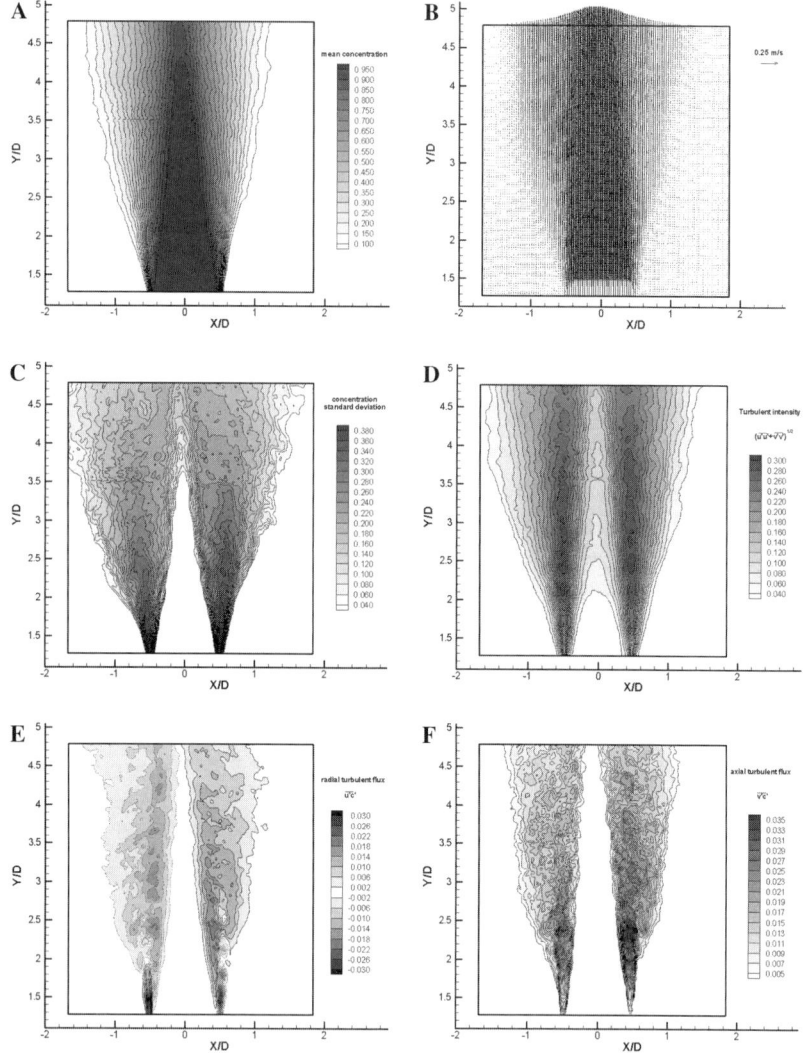

FIGURE 3. Ensemble-averaged measurements from the combined PIV–PLIF system: (**A**) mean concentration distribution; (**B**) mean velocity distribution; (**C**) concentration standard deviation distribution; (**D**) tubulence intensity; (**E**) radial turbulent flux term distribution; (**F**) axial turbulent flux term distribution.

SUMMARY

The development of a high resolution PIV-PLIF combined system that could provide simultaneous measurement of velocity and concentration in fluid flow, is described. The PIV tracer particles and fluorescent dye (rhodamine B) were premixed in an objective fluid flow and the objective fluid flow was illuminated by a double pulsed Nd:YAG laser. The LIF light and scattered illuminating laser light were separated successfully by using two kinds of optical filters, and recorded simultaneously by two high-resolution CCD cameras. The system was applied to measure the velocity and concentration distributions in the near field of a circular jet flow simultaneously. The distributions of the ensemble-averaged velocity and concentration, turbulent velocity fluctuation, concentration standard deviation, and the correlation terms between the fluctuating velocities and concentration were presented as the measured results of the simultaneous PIV-PLIF system.

REFERENCES

1. HU, H., et al. 2000. Improve the spatial resolution of PIV results by using hierarchical recursive operation. Proc. of 9th Int. Symp.on Flow Visualization, Edinburgh, Scotland, UK.
2. HU, H., et al. 2000. Simultaneous velocity and concentration measurements in a turbulent jet flow by using PIV–PLIF combined system. Proc. of 4th JSME-KSME Joined Thermal Eng. Conf., Kobe, Japan.
3. SOLOFF, S.M., et al. 1997. Distortion compensation for generalized stereoscopic particle image velocimetry. Meas. Sci. Tech. **8:** 1441–1454.
4. LEMOINE, F., et al. 1996. Simultaneous concentration and velocity measurements using combined laser-induced fluorescence and laser doppler velocimetry: application to turbulent transport. Exp. Fluids **20:** 341–327.
5. HINZE, J.O. 1959. Turbulence. McGraw-Hill, New York.

PIV Measurement and Numerical Analysis of a New Refrigeration Compartment of a Refrigerator

SEONG-HO CHO, IN-SEOP LEE, JAY-HO CHOI, AND YOUNG-SOK NAM

Core Technology Team, Digital Appliance Research Laboratory, LG Electronics Inc., Gasan-Dong, Keumchen-Gu, Seoul, Korea

> ABSTRACT: A new cooling duct system was developed in order to increase the refrigerator performance in terms of uniform temperature distribution particle image velocimetry (PIV) and the commercial software FLUENT was used to compare the performance of the conventional cooling system with the new cooling system. The new cooling system effectively cooled the door basket region, thus increasing the uniformity of the temperature distribution in the refrigeration compartment.
>
> KEYWORDS: PIV; temperature; numerical analysis; refrigerator

INTRODUCTION

Recent studies on refrigerators focus on increasing the performance in terms of energy consumption and noise reduction. A thorough investigation of the flow inside the refrigerator is needed in order to achieve such goals.[1–4] Experimental and numerical analysis methods were employed to investigate the flow inside a refrigerator. The side by side type refrigerators use a single evaporator, and a single fan to cool the freezer and refrigerator compartments. The high-temperature, high-pressure refrigerant flows through the evaporator, cooling the surroundings in the process, and the cool air is distributed into the freezer and refrigerator compartments according to a prescribed volume ratio. The cool air flowing into the refrigeration compartment passes through a damper insulation and out of front and rear outlets. Once the refrigeration compartment is cooled to a prescribed temperature, the temperature is controlled by the opening and closing of the damper door, located inside the damper insulation. In this refrigeration cooling system, the upper shelf in the refrigeration compartment is over-cooled and the lower shelf and door basket regions are under-cooled, resulting in non-uniform cooling of the refrigeration compartment. Consequently, to induce a uniform temperature distribution inside the refrigeration compartment, ducts were installed from the damper insulation leading to outlets installed on the left and right walls of the compartment, near the lower shelf and lower door baskets. Two-dimensional PIV measurements[5] were carried out at the shelf regions and door basket regions inside the refrigeration compartment. To analyze the veloc-

---

Address for correspondence: S.H. Cho, Ph.D., Core Technology Team, Digital Appliance Research Laboratory, LG Electronics Inc., 327-23, Gasan-Dong, Keumchen-Gu, Seoul, Korea.

sunnyhus@lge.co.kr

ity and temperature distribution inside the entire refrigeration compartment, the commercial software FLUENT was used for numerical analysis of the conventional duct system and the new duct system of the refrigerator.

PIV MEASUREMENTS

To compare of the conventional duct system and the new duct system of the refrigerator, the PIV system shown in FIGURE 1 was used to measure the velocity distribution in each shelf and door basket. The PIV system consists of a two-head Nd:YAG laser (maximum 300 mJ/7 nsec pulse), light sheet forming optics, a laser arm to direct the light source, a pulse synchronizer, and a 1 K × 1 K CCD camera (30 frames/sec). Lubricating oil (glycerin, SG-α 10G) was used with an atomizer to supply the seeding particles. Fields-of-view of 120 mm by 120 mm were used to capture 50 particle images and calculate the mean velocity field.

The mean velocity field results for the upper shelf of the refrigeration compartment with the conventional duct system and with the new duct system are compared in FIGURE 2. It can be seen that the flow induced by the conventional duct system and the new duct system show similar patterns. The flow in the upper shelf region of the two cooling systems was found to wrap around the tip of the upper shelf to reach the lower shelf regions, forming a vortex flow in the rear wall region with the flow originating from the rear outlet in the refrigeration compartment. The flow in the lower shelf regions showed a significantly lower velocity and no noticeable flow characteristics were found. In the lower-most shelf region, the cooling flow was found to flow along the rear wall in the downward direction and, after reaching the lowest horizontal surface, changing direction and flowing toward the door baskets. From these

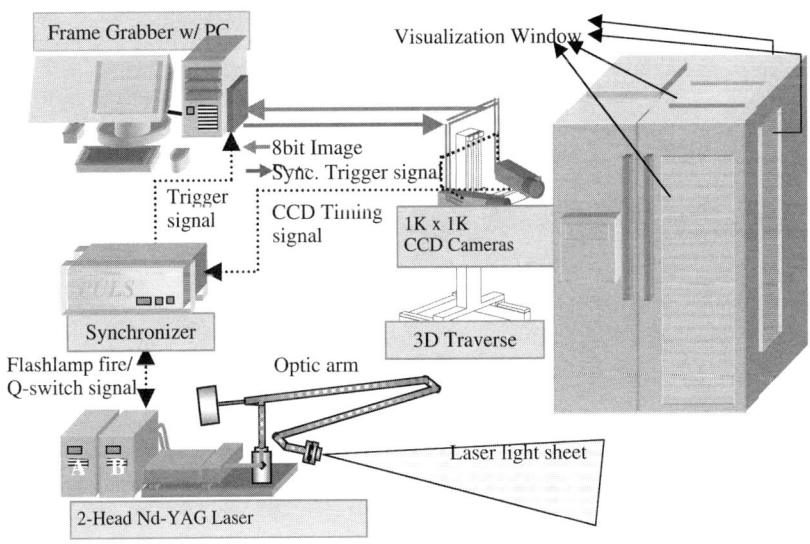

FIGURE 1. Schematic of the PIV measurement system.

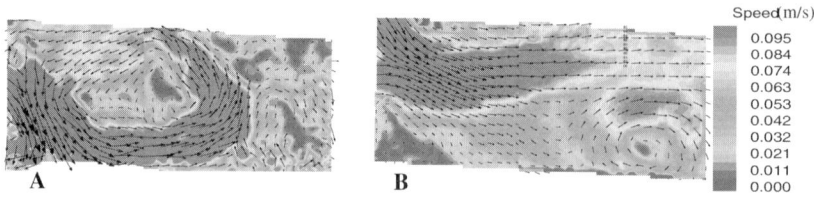

FIGURE 2. Mean velocity distribution of the upper shelf of the refrigeration compartment: **(A)** conventional cooling system; **(B)** the new cooling system. (COLOR PLATE 10.)

similar flow characteristics the two duct systems could be anticipated to show similar performance in the cooling of the shelf region.

The mean velocity field results of the basket regions with the conventional duct system and with the new duct system are shown in FIGURE 3. Due to the new duct system, flow into the lower basket could be observed, and this was found to efficiently cool the region; effectively leading to a more uniform temperature distribution inside the refrigerator compartment. The relatively high-speed flow from the right outlet was found to reach the center region of the door basket, and the relatively low-speed flow from the left outlet was found to flow along the left wall downward, with no significant effect on the door basket regions.

NUMERICAL SIMULATION

Numerical analysis of the velocity and temperature distribution inside the refrigeration compartment was carried out using the commercial software FLUENT. A structured grid system (grid number, 320,000) was made with the body fitted coordinate, with the boundary condition taken as the average heat transfer coefficient calculated with a 30°C surrounding temperature. The left wall was adjacent to the −18°C freezer compartment and the boundary condition was set according to its

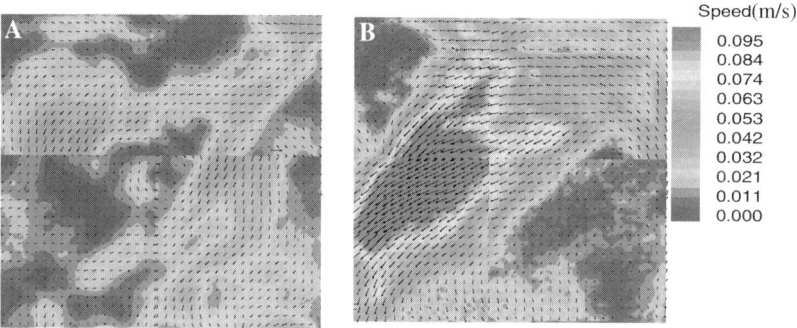

FIGURE 3. Mean velocity distribution of the door basket region of the refrigeration compartment: **(A)** conventional cooling system; **(B)** the new cooling system.

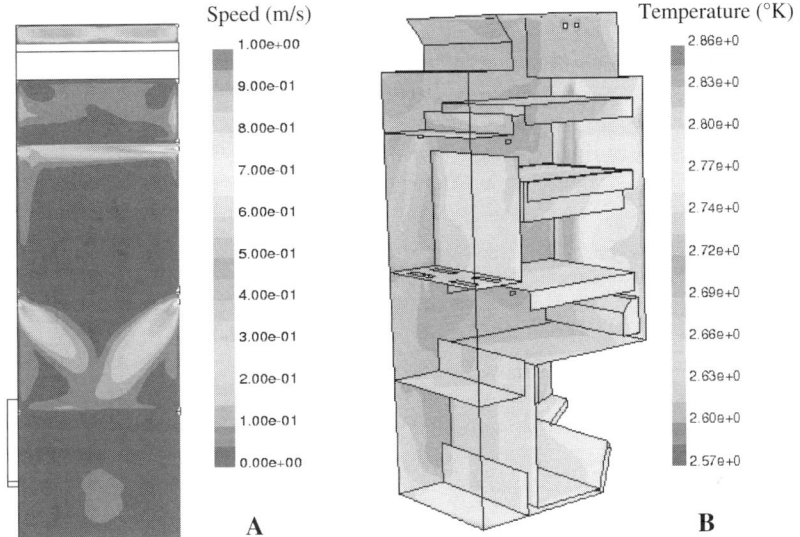

FIGURE 4. (**A**) Velocity distribution at the outlet plane of the refrigeration compartment. (**B**) Temperature distribution at the wall of the refrigeration compartment.

average heat transfer coefficient. The outlet velocity and temperature at each outlet were set as found by experiments. The standard k–ε model was used as the turbulence model.[6] FIGURE 4 shows the velocity distribution at the center plane of the refrigeration compartment and temperature distribution at the wall of the refrigeration compartment with the new cooling duct system. A cooling flow can be observed in FIGURE 4A, only in the home-bar and lower door basket region. The other door basket regions do not show any significant cooling flow. The upper wall outlets are directed perpendicular to the wall surface whereas the lower wall outlets are directed 45° downward compared to the wall surface. The lower wall outlets are tilted 45° downward in order to increase the flow rate into the door baskets, and improve the cooling rate of the region, thus inducing a more uniform temperature distribution. As can be seen in FIGURE 4B the heat is transferred out of the refrigeration compartment through the wall located adjacent to the freezer compartment, and transferred into the compartment through the other walls. Also the refrigeration compartment is cooled mainly by the outlets at various locations. This result suggests that the new duct system exhibits an improved uniform cooling performance of the refrigerator.

SUMMARY

The PIV technique and FLUENT were used to analyze the flows inside the refrigeration compartment in a conventional cooling duct system and in the new cooling duct system. An overall uniform temperature distribution was successfully achieved by installing additional ducts inside the left and right walls of the compartment, leading cooled air to flow toward the lower door basket.

REFERENCES

1. ATWOOD, T. & H.M. HUGHES. 1990. Refrigerants and energy efficiency. Intl. J. Refrig. **13:** 270–273.
2. JANSSEN, M.J.P., J.A. DE WIT & L.J.M. KUIJPERS. 1992. Cycling losses in domestic appliances and experimental and theoretical analysis. Intl. J. Refrig. **15**(3): 152–158.
3. TASSOU, S.A. & T.Q. QURESHI. 1998. Comparative performance evaluation of positive displacement compressors in variable-speed refrigeration applications. Intl. J. Refrig. **21**(1): 29–41.
4. VINEYARD, E.A., J.R. SAND & R.H. BOHMAN. 1995. Evaluation of design options for improving the energy efficiency of an environmentally safe domestic refrigerator-freezer. ASHRAE Trans. **101**(1): 1422–1430.
5. LEE, I.-S., S.-J. BAEK, M.-K. CHUNG, et al. 1999. A study of air flow characteristics in the refrigerator using PIV and computational simulation. J. Flow Vis. Image Process. **6:** 333–342.
6. LAUNDER, B.E. & D.B. SPALDING. 1974. The numerical computation of turbulent flow, Comp. Meth. Appl. Mech. Eng. **3:** 269–289.

Visualization of Jet Flows over a Plate by Pressure-Sensitive Paint Experiments and Comparison with CFD

KOZO FUJII, NOBUYUKI TSUBOI, AND NOBUYOSHI FUJIMATSU

The Institute of Space and Astronautical Science, Yoshinodai, Sagamihara, Kanagawa, Japan

ABSTRACT: Flow fields created by underexpanded sonic jets impinging on an inclined flat plate were studied experimentally using the pressure sensitive paint (PSP) measurement technique. The measurement system and some representative results are presented here. Two binders, thin-layer chromatography (TLC) plates and anodized aluminum (A-A) plates were tested with bathophen ruthenium chloride as a luminophore. The results show that both the binders can be used. TLC plates are preferable because their luminescent intensity is almost twice that of the A-A plates. Quantitative measurements require accurate temperature calibration. A preliminary effort to elucidate the flow structure by combining the PSP results with a computer simulation of the same flow field is presented. Although good agreement is obtained between the experimental and numerical results, future quantitative comparisons are necessary to yield a useful tool in the analysis of the jet-plate interaction flows.

KEYWORDS: jet; pressure-sensitive paint; CFD

INTRODUCTION

Problems created by the impingement of supersonic jets on solid objects appear in a wide variety of situations in the area of space transportation systems. Examples can be found in plume-wall interactions during the rocket launchings, multistage rocket separations, and attitude-control thruster operations. The flow fields are, in general, extremely complex. They contain mixed subsonic and supersonic regions, shock/shock or shock/expansion interactions, contact discontinuities, and instabilities inside the turbulent shear layer. Although one can find some literature and understanding of the flow mechanism is gradually increasing,[1,2] a systematic study of the flow field is still missing.

The present paper describes recent efforts at the Institute of Space and Astronautical Science (ISAS) to better understand the flow fields created by jet impingement on an inclined flat plate. The experimental and computational analysis were conducted in parallel. The research was separated into three steps. The first step was to conduct the PSP experiment and assure a quantitatively reliable experiment by choosing

Address for correspondence: Kozo Fujii, Ph.D., High Speed Aerodynamics Space Transportation Research Division, The Institute of Space and Astronautical Science, 3-1-1, Yoshinodai, Sagamihara, Kanagawa, 229-8510, Japan.
fujii@flab.eng.isas.ac.jp

adequate molecular sensors and binders. The second step was the evaluation of experimental results. The third step involved the CFD simulation and comparison of the PSP data, pressure sensor data, and Schlieren pictures with the CFD simulation results. By validating the CFD simulation based on a comparison with the PSP experiments in the second step, the detailed flow structure can be discussed. These three steps show the flow mechanism of the jet impingement over a flat plate, resulting in the estimation of the location of maximum pressure and heating, which is important for many engineering problems. The present paper focuses on the experiment and a qualitative comparison with the CFD simulations.

The mechanism of PSP is well known[3–5] and is only briefly described here. Pressure-sensitive paints consist of a luminescent compound, or dye, that is quenched by oxygen and dispersed in an oxygen permeable polymeric binder. The luminescence is induced by the excitation of the dye at its absorption wave length. The emitted intensity of the PSP is inversely proportional to the partial pressure of oxygen. The relationship between the intensity of the luminescence and the partial pressure of oxygen can be expressed in terms of the Stern-Volmer relation, from which the pressure data is obtained by measurement of luminescence intensity, namely, the picture images taken by a CCD camera.

THE PSP EXPERIMENT

The experiments were carried out in a small induction-type wind tunnel. The test section was connected to the large low-pressure chamber and air was sucked into the test section when the valve was opened. The duration was roughly 10–15 seconds for a typical supersonic external-flow experiment, but can be much longer, as in the present experiment wherein a small sonic nozzle with diameter 5 mm was allowed to induce smaller mass flows into the test section. The ratio of the ambient pressure and the jet total pressure was varied by the control of a valve vane. The flat plate was placed in the test section. The location and the angle of the flat plate could be changed, but the only results shown here are for a 45 degree angle with

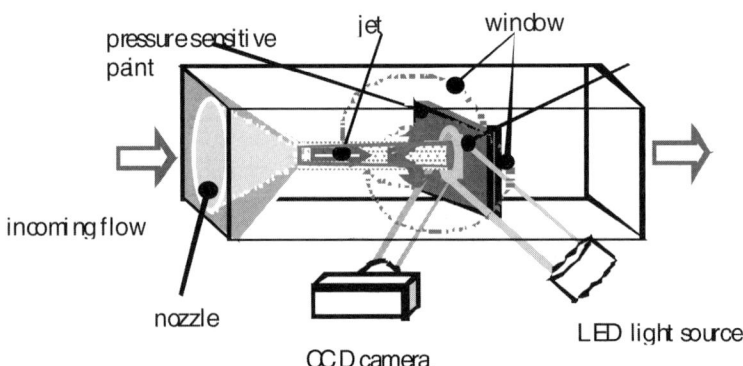

FIGURE 1. Schematic of the experimental setup.

distances of 10mm (distance/diameter ratio, 2.0) at a pressure ratio 125.0, and 20mm (distance/diameter ratio, 4.0) at a pressure ratio 3. FIGURE 1 shows schematically the experimental setup for the PSP measurement. The LED light source could be located inside the test section, depending on the flow conditions.

The PSP data for a flat plate was acquired using a 12-bit CCD camera system illuminated with 779 blue LED arrays. The blue LED has a narrow wavelength range (about 460nm) which was adequate for paint of ruthenium complexes having an excitation wavelength of about 460nm. The ruthenium complexes emit 620nm luminescence and the images were acquired using a CCD camera. Two types of binders were tried; one was a silica-gel thin-layer chromatography (TLC) plate, 0.5mm thick, and the other was an anodized aluminum (A-A) plate. Both showed clear images of jet impingement.

EXPERIMENTAL RESULTS

The results obtained with the TLC plate showed strong temperature dependence, but almost double the emission of the anodized aluminum plate. When TLC plates were used, line-by-line temperature calibration using the thermocouples, was required before achieving a quantitative description of the pressure distributions[6] (not shown here). Temperature calibration for the entire image using thermography was also attempted with TLC plates as a binder (not shown here). The final images were obtained by subtracting the non-flow images from the original flow images, thus eliminating the effects of the luminescent intensity dependence on the light source location, angle, and other factors affecting the luminescent intensity. FIGURE 2A shows the calibration curve that relates the final image intensity to the quantitative pressure level. The curve is almost linear, leading to an image of the pressure distributions over a flat plate shown in FIGURE 2B. Again, the typical ring-like structure is observed. Five test cases, with much smaller pressure ratios, in which complex flow patterns appear, are presently underway. Some of the preliminary results for the small pressure ratios are used here for the comparison with the CFD results.

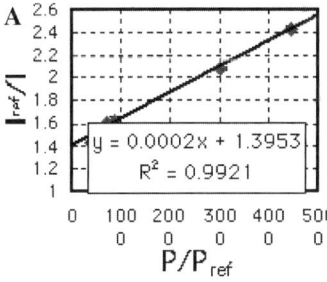
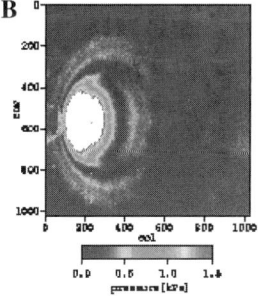

FIGURE 2. Calibration curve (**A**) and pressure distributions (**B**) over an inclined flat plate.

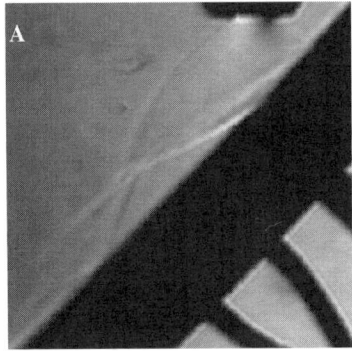
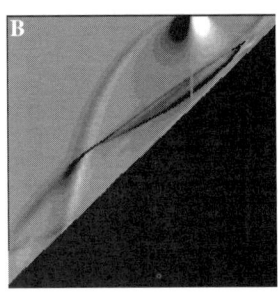

FIGURE 3. Comparison of a Schlieren picture with a simulated Schlieren picture of the CFD result for a pressure ratio 125.0: **(A)** experiment, **(B)** CFD result.

COMPARISON BETWEEN CFD AND THE EXPERIMENTAL RESULTS

The three-dimensional (3D) compressible Euler and Navier–Stokes equations are solved by using computer code developed in house. Because some of the cases showed flow unsteadiness, two computer programs, using explicit and implicit time integration, were used to identify the unsteadiness of the flow fields. Only the inviscid solutions are presented in the discussion below.

The convective terms were discretized, by the so-called TVD upwind scheme. The viscous terms, when necessary, were evaluated by central differences. For the implicit time-integration method, the LU-ADI factorization algorithm[7] was used. The Yee-Harten upwind scheme was used for explicit time-integration. The computer

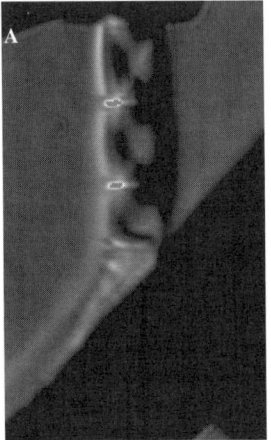
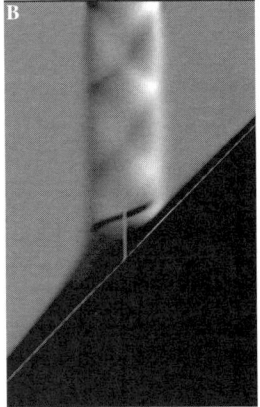

FIGURE 4. Comparison of a Schlieren picture with a simulated Schlieren picture of the CFD result for a pressure ratio 3.0: **(A)** experiment, **(B)** CFD result.

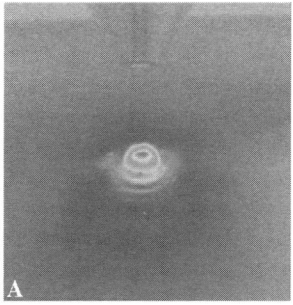 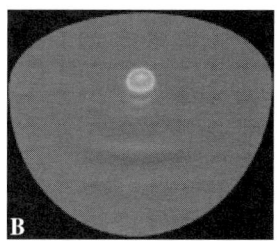

FIGURE 5. Comparison of the PSP pressure map with pressure computed by the CFD method, pressure ratio 3.0: (**A**) experiment, (**B**) CFD result.

programs were used for a wide variety of CFD applications and validated by the comparison with experiments.[2,7–9] Only two results from explicit time-integration are shown here. FIGURE 3A shows the Schlieren image taken in an experiment at pressure ratio 125. The angle of the flat plate was 45 degrees and the distance was 10 mm (distance/diameter ratio, 2.0). The same flow field was computed and the image simulating the Schlieren picture was obtained with a postprocessing technique. The result is shown in FIGURE 3B. The flow pattern is well captured by the CFD simulation. FIGURE 4 presents a similar comparison for a pressure ratio 3.0. The angle of the flat plate was 45° and the distance was 20 mm (distance/diameter ratio, 4.0). In this case, flow unsteadiness occurs in the computation and the image based on the result, averaged for a given time span, is presented. The cell structure is clearly captured by the CFD simulation, but the flow field near the wall shows a difference. For instance, the plate shock wave is almost normal to the jet, but is at an angle in the computation. This is probably due to the viscous effect and therefore the flow simulations with the Navier–Stokes equations are currently underway. FIGURE 5 presents the comparison of the PSP pressure map with the pressure map computed by the CFD simulation. Future quantitative analysis is necessary, but the results are promising, since the wavy flow pattern created by the shock reflection in the jet near the wall is well captured by the CFD simulation. The results indicate that the combination of the PSP measurement, Schlieren images, and CFD simulations can offer a useful tool for the analysis of the mechanism of the jet impingement flow field.

SUMMARY

Experimental and computational efforts at the ISAS for the analysis of the flow fields created by underexpanded sonic jets impinging on the inclined flat plate are considered here. Only limited preliminary results are presented. Plates with TLC and A-A binders were successfully tested with bathophen ruthenium chloride as a luminophore and LED arrays as a light source. The experiments were conducted using an induction-type test section attached to the low-pressure chamber. Schlieren pictures and the PSP images were compared with the CFD results and showed good qualitative agreement. The essential feature of the flow field was well captured in the CFD simulations and the result indicates that the combination of the PSP measurement, Schlieren images, and CFD simulations offers a useful tool for the analysis of the mechanism of the jet impingement flow field.

REFERENCES

1. LAMONT, P.J. & B.L. HUNT. 1980. The impingement of underexpanded, axisymmetric jets on perpendicular and inclined flat plate. J. Fluid Mech. **100**(3): 471–511.
2. TSUBOI, N., A.K. HAYASHI, T. FUJIWARA, et al. 1991. Numerical simulation of a supersonic jet impingement on a ground. SAE Trans. J. Aerospace Section 1, part 2. **10:** 2168–2180.
3. BENCIC, T.J. 2001. Calibration of detection angle for full field pressure-sensitive paint measurement. AIAA Paper 2001-0307.
4. AMER, T.R., T. LIU & D.M. OGLESBY. 2001. Characterization of pressure sensitive paint intrusiveness effects on aerodynamic data. AIAA Paper, 2001-0556.
5. SHIMBO, Y., K. ASAI, H. KANDA, et al. 1998. Evaluation of several calibration techniques for pressure sensitive paint in transonic testing. AIAA Paper 98-2502.
6. FUJIMATSU, N., N. TSUBOI, K. FUJII & M. MATSUMOTO. 2001. PSP measurement for underexpanded axisymmetric jets impinging on an inclined flat plates. Proc. First Int. Symp. on Advanced Flow Information.
7. OBAYASHI, S., K. MATSUSHIMA, K. FUJII & K. KUWAHARA. 1986. Improvement in efficiency and reliability for Navier–Stokes computations using the LU-ADI factorization algorithm. AIAA Paper 86-0338.
8. FUJII, K. & S. OBAYASHI. 1987. Navier–Stokes simulations of transonic flows over a wing fuselage combination. AIAA J. **25**(12): 587–1596.
9. FUJII, K. & Y. TAMURA. 1997. Effect of engine integration on the aerodynamic characteristics of a spaceplane. Int. J. Comput. Fluid Dynam. **8**(4): 235–246.

Transonic Injection in Interaction with Transverse Compressible Flow

R. DIZENE,[a,b] J.M. CHARBONNIER,[a,c] E. DORIGNAC,[b,d] AND R. LABLANC[a,e]

[a]*Laboratoire d'Etudes Aerodynamiques, Poitiers, France*

[b]*Department of Mechanical Engineering, USTHBV University, El Alia, Bab Ezzouar, Alger, Algeria*

[c]*CNES, Toulouse, France*

[d]*Laboratoire d'Etudes Thermiques, Poitiers, France*

[e]*ENSMA, Chasseneuil du Poitou, Futuroscope, France*

ABSTRACT: An extensive study devoted to modelling blade cooling was undertaken at CEAT a few years ago in collaboration with SNECMA. For the turbomachinery applications, an experimental configuration of a turbulent boundary layer with heat transfer was studied for compressible and incompressible flows. The research presented here is a part of that study and this paper reports on the experimental results of an investigation concerned with a row of transonic jets interacting with a transverse flow. In many applications, the cooling layer does not emerge onto the surface from a tangential slot but comes from a slot normal to or inclined to what is otherwise a flush surface. In this case the freestream interacts with the coolant flow. The secondary (jet) flow is introduced at an angle of 45° to the mainstream flow direction. Visualization studies using the surface flow patterns and surface temperature flow patterns are reported and discussed.

KEYWORDS: imaging; visualization; transonic injection; jet; compressible flow

INTRODUCTION

The interaction of a transonic circular jet with a transonic deflecting stream has, during the past few years, been the subject of our interest. This type of flow configuration exhibits many features, such as production of large-scale eddies and high levels of turbulence, including flow reversal. In addition, there are a number of applications of this type of flow, such as film cooling, mixing of two fluid streams, and impingement cooling. Each application has its own specification. Thus, in film cooling it is of interest that the jet remains as close as possible to the wall from which it is issued and spreads itself out on the surface. With impingement cooling, the jet should penetrate across the flow and spread itself out on the surface opposite to that from which it is injected. In mixing two streams, it is usually desirable that the jet mixes fully and uniformly with the main flow.

Address for correspondence: R. Dizene, Ph.D., Department of Mechanical Engineering, USTHBV University, P.O. Box 32, El Alia, Bab Ezzouar, Algiers, Algeria.
   dizene@yahoo.fr

This paper reports experimental results of an investigation that concerns film cooling. The overall aim is to contribute to better understanding of the interactions in film cooling, with the intention of highlighting physical characteristics that develop in compressible interaction. To accomplish this, it is necessary to study the fundamental process by which a fluid jets interact with mainstreams in the neighborhood of solid surfaces. To understand the basic process of this interaction, the configuration studied[1] concerns the interaction of a row of jets and a transverse flow stream the results reported here are flow visualizations and surface flow visualizations using an oil-powder mixture to observe the wall streamlines. In addition, results are reported of temperature visualizations using an infrared camera technique.[2]

SURFACE FLOW AND TEMPERATURE PATTERNS

The visualization of flow patterns plays a singularly important role in the advancement of our physical understanding of the mechanics of fluids. Flow visualization has led to the discovery of flow phenomena, has helped in the development of mathematical models for complex flow problems and in the verification of existing principles, and has been an important tool in the development of complicated engineering systems. This section deals with the observation of certain flow characteristics close to the wall of a flat plate. The plate wall is coated either with a highly emissive black paint that indicates the wall temperature distribution, or layered pigments showing the streamline pattern of the flow adjacent to the wall.

Surface Oil Flow

This technique enables a picture of the flow pattern at the surface of the flat plate model placed in a wind tunnel flow to be obtained quickly and easily. The surface is coated with a specially prepared mixture of appropriate oil and a fine pigment powder. Oil dark smoke is used in the case of a perpendicular jet and oil of magnesia in the case of a row of inclined jets. The air stream causes the mixture to flow along the plate surface. The consistency of the mixture is suitably chosen in order to observe streaky deposits of the pigment powder indicating the direction of flow near the surface. It is the art of the experimenter to find a paint of suitable consistency by mixing oil and a pigment of his choice. There are several requirements to meet in mixing the proper paint.[1] Two of these are the oil viscosity and surface skin friction, so that the paint leaves a clearly defined pattern of streaks, which is also a requirement for good photographic quality. The technique allows one to observe, in particular, the separation and reattachment lines, if the flow field contains separated regions. Still photographs and movie pictures, taken from outside of the wind tunnel, were studied after a tunnel run. Note that the surface oil flow technique yields qualitative information but it is obviously unsuitable for showing correctly the pattern of unsteady flow.

Surface Temperature Visualization

Aerodynamic test models that are investigated in a high speed flow facility, are exposed to considerable heat transfer from the fluid to the body under study. In our case, the surface temperature distribution and heat transfer between the mainstream and the film-cooled wall had to be known for the analysis of the cooling process. The surface temperature or the local heat transfer rate could be measured by means of thermocouples inserted in the plate. However, like all probe type measurements, this technique yielded the desired parameters only at discrete points on the model surface. An alternative method was the use of visualization techniques, such as the scanning camera technique developed by Dorignac[2] in the Thermal Study Laboratory of CEAT to obtain detailed[3] film effectiveness data. The present data was obtained in the same open circuit transonic wind tunnel. The surface temperature and heat flux on the flat plate model were determined using an AGEMA infrared scanning camera SWB 880. The infrared beam, viewed the test surface through a scuttle of vinyl chloride that was used as an infrared window in the test section upper wall. The camera was calibrated and the emissivity coefficient obtained was 0.87. The test surface of the model was coated with high emissivity flat-black paint. More details of this technique are presented below.

EXPERIMENTAL

Wind Tunnel and Test Model

The experiments were conducted in the University of Poitiers Aerodynamic Study Laboratory. The experiments, with the row of jets configuration, were performed in a small transonic wind tunnel; 600 mm long with 40×80 mm square cross section. The crossflow travelled horizontally and the jets were injected vertically upward. Injection of the secondary air was through a 5 mm-ID nozzle and introduced at angle of 45° to the mainstream flow direction. The normal Mach number was 0.8 and the Reynolds number (Re) based on the orifice diameter was 0.7×10^5. The test apparatus is described in detail elsewhere.[1,2] All visualizations were made in a rectangular cross section transonic wind tunnel. The sidewalls were set to diverge slightly to maintain a uniform free stream speed of the air in the tunnel.

The mainstream velocity and the boundary layer profiles were determined using a total and static pressure probe, in addition to static wall taps. With the normal free stream speed of 235 m/sec, the boundary layer displacement thickness at the upstream edge of the injection holes was about 15 mm. The Re value, based on this velocity and diameter $D = 5$ mm was 0.78×10^5. Air was injected through five tubes spaced three-diameters apart across the span. The tubes were long enough to assure fully developed turbulent flow at the exit in the absence of a mainstream flow. The flow inside the tunnel was assured by a supersonic ejector placed downstream of the sonic throat.[1,2] The overall flow was determined by measuring the pressure drop across an orifice plate. The injection abscissa ($X = 0$) was located 380 mm from the tunnel test section inlet. The dimensions of the measurement surface were $X = 20D$ long and $Z = -1.5D$ wide. The thermal test conditions were inverted by using a wall temperature of $T_W = 313$ K, which was lower than the jet temperature, $T_{gj} = 333$ K, but higher than the mainstream temperature, $T_{ge} = 286$ K. Hence, the temperature difference between main and jets flow was 40°C.

Scanning Camera Technique

The surface temperature and heat flux on the flat plate model were determined by using an AGEMA infrared scanning camera SWB 880. The camera had a depth of field of approximately 5 cm when focused on the test section and 35 cm from the camera. The thermocouple measurements provided accurate temperature measurements at seven locations and the infrared scanning camera provided isothermal contours at selected temperature levels, which would have been possible only with a multitude of thermocouples. Note that the use of the test model sheet limited the view field to observe only three holes out of five.

For some coolant flow rates, four separate isothermal levels were mapped along the X direction with 250×250-pixel resolution. The span that could be covered was about $X = 40D$. The flux density was deduced from an application of numerical method developed by Dorignac et al.,[3] based on stationary conduction model. The validation method was obtained by comparing the isothermal contour value that resulted from a thermocouple measurement at the same position.

The visualized and measured temperatures were affected by numerous parameters, such as the emissivity coefficient and some radiation from the surroundings. The data were presented in a plane view of the flat plate with the coolant holes. Realization of the position of the hole on the isothermal display was difficult. This position was realized when using the high temperature variations, whereby the isotherm contours were positioned with an uncertainty of ± 0.26 mm. The uncertainties in the measured temperatures indicated by the scanning camera were approximately 10%.[2]

RESULTS

Surface Shear Flow Visualization

The surface shear flow visualized by the oil flow shows flow features, such as a separation region just ahead of the jet exit, an entertainment flow in the vicinity of the jet exit, and a wake flow downstream of the exit. One oil flow picture at the same blowing rate ratio is presented in FIGURE 1. The surface flow is seen to be quite symmetric to the centre line of the flat plate (x-axis), indicating satisfactory test conditions. The overall flow characteristics indicate two flow regions with different shear flows. When the crossflow approaches the jet exit, a deceleration of the crossflow induces a boundary layer separation immediately upstream of the jet exit. The upstream separation occurs in a crescent area just ahead of the jet exit. Inside the separation region, the flow is highly affected by jet entrainment.[4,5] However, the resolution of the oil flow is not fine enough to show details. The jet exit is followed by a wake downstream of the jet exit. Inside the wake, the shear flow is weak. On the

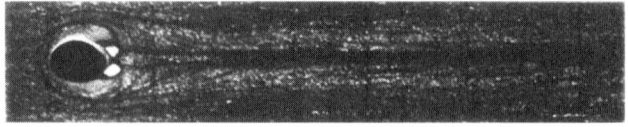

FIGURE 1. Surface oil flow visualization; single row of inclined jets (center jet): 5 jets, $M = 0.72$, $R = 0.5$, $T_w = 40°C$, $T_g = 40°C$.

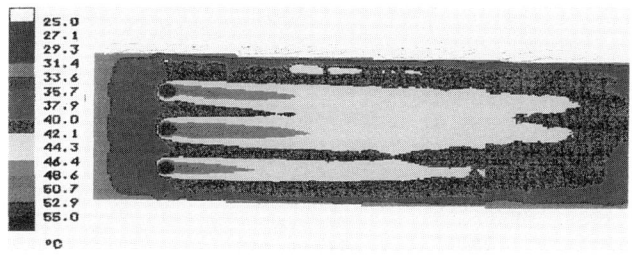

FIGURE 2. Surface temperature visualization. (COLOR PLATE 11.)

boundary of the wake, the shear flow is quite strong, indicating large scale turbulent mixing. Downstream of the jet for a distance of about 3 to $5D$, the surface flow is strongly influenced by jet entertainment. The lateral scale of the wake increases downstream of the jet exit.

Surface Temperature Visualization

The temperature distribution results are presented in FIGURE 2. The principal features of these results are the streamwise and spanwise pronounced thermal markers. The jet thermal marker is noted for a distance of about 25 to 30 holes diameter along the x-axis. The individual character of the jets[6,7] appears clearly for $x = 15D$, although a light coalescence of jets is visible from a distance of about 16 to $20D$. The differences observed for each jet are due to a slight difference in the hole flow rate injections and are estimated to be about $\pm 2°C$. Successful employment requires balanced injection conditions.[3,4] The low temperature differences visualized and measured (-0.7%) demonstrate the potential of this technique for the acquisition of highly detailed film effectiveness data.

SUMMARY

The results presented here validate the utility of experimental techniques using visualization and imaging to elucidate fundamental fluid mechanics of the jet in crossflow. A counter-rotating vortex pair could be developed immediately from the jet exit. The data confirm the well separated jet patterns on the surface and the low coalescence that develops far downstream in the streamwise direction. We plan to extend these results by modifying the transonic tunnel flow dimensions and to investigate, by visualization, the instantaneous jet/cross-flow interaction in more detail so as to clarify the jet/wake flow characteristics.

ACKNOWLEDGMENT

This study was sponsored by the DRET as part of agreements 88/042 and 89/273. The authors wish to thank Mr. Henry Garem for his help and technical advice during the experiment investigations.

REFERENCES

1. DIZENE, R. 1993. Etude d'interaction de jets avec un ecoulement transversal compressible en vue des applications au refroidissement de parois. Deuxième partie: rangée de jets obliques. These de Doctorat, Ceat Université de Poitiers, France.
2. DORIGNAC, E. 1990. Contribution à l'étude de la convection forcée sur une plaque plane en présence de jets pariétaux dans un écoulement subsonique. Thése de Doctorat, Ceat Université de Poitiers, France.
3. DORIGNAC, E., J.J. VUILLERME, J.L. BOUSGARBIES & P. DENIBOIRE. 1991. Heat transfer and film cooling following injection through inclined circular tubes of little dimension compared with the boundary layer thickness. Eurotherm 25, Heat Transfer in Single Phase Flow, Pau, France.
4. DIZENE, R., J.M. CHARBONNIER, E. DORIGNAC & R. LEBLANC. 2000. Etude expérimentale d'une interaction de jets obliques avec un ecoulement transversal compressible. I. Effets de la compressibilité en régime subsonique sur les champs aérothermiques. Int. J. Thermal Sci. **39**(3): 390–403.
5. LOZANO, A., B. YIP & R.K. HANSON. 1992. A tracer for concentration measurements in gaseous flows by planar laser-induced fluorescence. Exp. Fluids **13**: 369–376.
6. DIZENE, R., J.M. DORIGNAC, E. CHARBONNIER & R. LEBLANC. 2000. Etude expérimentale d'une interaction de jets obliques avec un ecoulement transversal compressible. II. Effets du taux d'injection sur les transferts thermiques en surface. Int. J. Thermal Sci. **39**(5): 571–581.
7. BLAIR, M.F. & R.D. LANDER. 1975. New techniques for measuring film cooling effectiveness. ASME J. Heat Transfer **Nov.**: 539–543.

Visualization and Phase Doppler Particle Analysis Measurements of Oscillating Spray Propagation of an Airblast Atomizer under Typical Engine Conditions

PETER SCHOBER, ROBERT MEIER, OLAF SCHÄFER, AND SIGMAR WITTIG

Institut für Thermische Strömungsmaschinen,
University of Karlsruhe, Karlsruhe, Germany

> ABSTRACT: Propagation of a kerosene spray formed by a prefilming airblast atomizer within a pulsating flame has been investigated. The measurements were performed in a 400 kW single combustor rig at typical engine conditions of up to 0.8 MPa inlet pressure and 673 K inlet temperature. The homogeneity of the fuel spray propagation in the reacting zone substantially influences the local temperature distribution in the reaction zone and, therefore, the formation of thermal NO_x. Phase locked visualization of the spray propagation and the flame luminescence was applied to the pulsating combustion process in order to analyze the complex interaction between the time dependent flow field, the spray propagation, and the combustion process. The spray characteristic was additionally investigated by means of time resolved PDPA measurements in the reacting flow.
>
> KEYWORDS: airblast atomizer; doppler; PDPA; combustion; gas turbine

INTRODUCTION

New concepts for improving gas turbine combustion focus on a significant reduction of NO_x emission. The formation of NO_x in a combustion process is mainly determined by the local combustion temperature. Low emission combustion, therefore, requires a homogeneous lean mixture of fuel and air resulting in low combustion temperatures and NO_x-emissions.[1] However, various difficulties appear in well mixed lean combustion, including flashback, auto-ignition, and serious combustion oscillation. Complex interactions occur between the atomization process, spray propagation, droplet evaporation, fuel dispersion, and chemical reaction in lean liquid fuelled gas turbines. The development of new CFD design tools for a liquid fuelled combustor requires more investigation so as to predict these interactions properly.

One approach for lean combustion in liquid fuelled gas turbines is based on prefilming airblast atomizers, such as the type investigated in this study. Prefilming airblast atomizers are now employed in a wide range of aircraft, industrial, and marine

---

Address for correspondence: Peter Schober, Dipl.Ing., Institut für Thermische Strömungsmaschinen, University of Karlsruhe, 76128 Karlsruhe, Germany.
peter.schober@ait.uni-karlsruhe.de

gas turbines. A key feature of this type of atomizer is that the liquid fuel is exposed at an atomization edge to high velocity air streams, disintegrating the fuel to a dispersed spray. Due to combustion oscillations, the flow field inside the combustor, as well as the velocity of the air streams around the atomization edge, are transient, thus affecting both the primary atomization[2] and spray propagation. The optimization of the atomizer, therefore, requires time-resolved investigation of the two-phase flow with respect to the interaction between the combustion oscillation and the atomization process.[3]

The investigation of the various unsteady processes in gas turbine applications requires new approaches to the measurement techniques.[4–6] The present study focuses on the application of phase locked visualization techniques and time resolved PDPA on the oscillating fuel spray propagation flame structure of a prefilming airblast atomizer. The experiments were performed at various elevated pressures, inlet temperatures, and air/fuel ratios (AFR), using conditions that can be found in aircraft jet-engines at cruise.

THE EXPERIMENTAL SETUP

The Test Facility

The experiments were performed at the high pressure, high temperature test facility of the Institut für Thermische Strömungsmaschinen (ITS) at the University of Karlsruhe, which is capable of reaching a maximum pressure of 1 MPa and a hot air temperature of up to 1,100 K. A maximum airflow of up to 1.4 kg/sec could be directed to a hot or cold air supply system that could be controlled independently. Non-vitiated hot air was supplied via a heat exchanger fired by a 2 MW natural gas burner. The high pressure laboratory offered four slots for various test sections.

Combustor Test Section

To investigate spray propagation under realistic conditions, an annular model combustor (see FIGURE 1) consisting of two quartz glass tubes was employed.[7] The model combustor was confined by a pressure vessel with two plain quartz glass

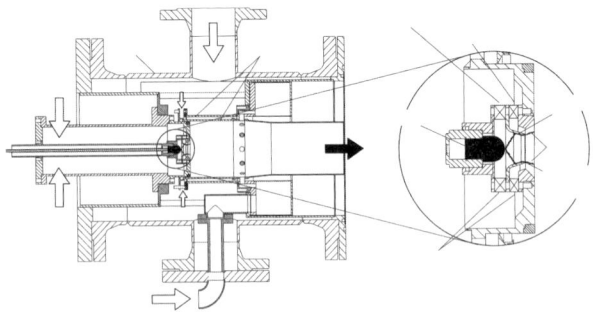

FIGURE 1. Cross-section of the high pressure test rig, including the model combustion unit.

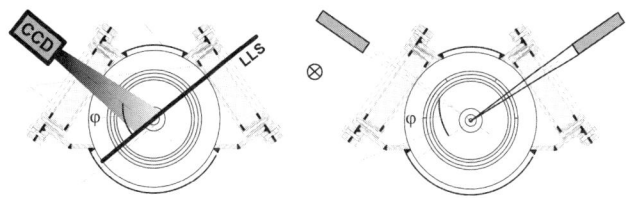

FIGURE 2. Optical access to the combustor with laser light sheet and PDPA setup.

windows with a size of $224 \times 133\,\text{mm}^2$ (see FIGURE 2). The hot air supply from the high pressure facility was guided separately to the combustion and dilution air inlets. The fully independent dilution air supply enabled various combustion/dilution air ratios with the same combustor set up. Additionally, a cooling air flow scavenged the outer pressure vessel. The cold air inside the pressure vessel flowed through the gap between the two quartz class cylinders of the model combustor to the exhaust duct in order to increase the heat transfer from the inner cylinder and avoid glowing.

Various combustion modes exist depending on the test conditions. The test conditions used in this study are summarized in TABLE 1.

The investigated prefilming airblast atomizer produced a vortex flow which stabilized the flame via a co-rotating pair of radial swirlers. To gain a uniform fuel sheet on the inner surface of the prefilmer, the fuel was spread via a simplex hollow cone atomizer onto the prefilmer. Driven by the shear stress from the gaseous flow onto the fuel film, the thin sheet atomized at the edge of the prefilmer, where the liquid sheet was exposed to high velocity air streams. The fuel was supplied via a water cooled tube to a simplex atomizer to ensure constant fuel temperatures.

MEASUREMENT TECHNIQUES

Phase Locked Flame and Spray Visualization

To analyze the interaction of the flame and the spray in the presence of combustion oscillations, phase locked visualization of the flame and the fuel spray were performed. The flame luminescence images were captured with a 3-chip color CCD camera and a frame grabber PC card. To investigate spray propagation, the laser light sheet (LLS) technique was employed. FIGURE 2 shows a cross section through the combustor test rig and the optical test set up. The LLS technique permits a detailed and very effective 2D characterization of the liquid fuel spray. Single shot images of the spray gave a good impression of the shape and the droplet distribution. Thus, in addition to this qualitative information, quantitative information on the spray cone angle, penetration depth, and spray flow direction could be obtained.

The statistical distribution of droplets was investigated by capturing a series of images. A pulsed Nd:YAG Laser was used to illuminate the spray. The laser provided pulses with a duration of 15 nsec at a frequency of 10 Hz. The pulse energy could be raised to 140 mJ at a wavelength of 532 nm. The laser beam was directed via mirrors to the combustor rig. A lens system formed a light sheet with a thickness of 0.5 mm that was directed through the optical access radially to the combustor axis.

TABLE 1. Test conditions used in this study

Condition	Non-oscillating		Oscillating	
mass flow inlet pressure (MPa)	0.2	0.4	0.4	0.8
inlet temperature (K)		573		
pressure drop Δp (%)		3		
combustion air (g/sec)	42	65	65	140
dilution air ratio		1/3		
mass flow cold air (g/sec)	80	100	100	200
λ (stoichiometric air/fuel ratio)	2.8–3.5		1.0–2.7	
fuel		kerosene Jet A		

The 2D images were captured with a 12-bit black and white fast shutter CCD-camera with a resolution of $1{,}280 \times 1{,}024$ pixels. The strong illumination of the droplets during the high energy laser pulse permitted the shutter time of the CCD-camera to be reduced to 100 nsec. Since the light intensity of the flame luminescence was relatively low compared to the illuminated droplets, only scattered light from the droplets was detected. The CCD camera was positioned at an off-axis angle of 72° given by the design of the optical access to the model combustor. The distortion caused by the tilted camera position and the glass cylinders is compensated by the image acquisition software, based on a reference image placed in the light-sheet plane. 150 single 12-bit grey scale images were captured and averaged for each measurement. Applying a high- and low- pass filter to erase background light, as well as side wall reflections, the averaged images were reduced to 8 bit for the post processing software. Finally, the files were transformed to a false color representation.

To obtain a trigger signal, the pressure oscillations of the combustion process were measured via a microphone close to the pressure vessel. The dominant oscillating frequency was determined via a real-time FFT analysis. The pressure signal was then band pass filtered with respect to the dominant frequency. The intercept of the rising slope of the filtered, sinusoidal microphone signal was used to produce a trigger signal for the measurement systems. The trigger signal could be shifted with an adjustable time delay via a special trigger unit. A PC based timing unit released the laser pulse and controlled the camera image acquisition.

Time Resolved PDPA Measurements

The time and spatial resolved analysis of droplet size and velocity distribution was performed with a DANTEC phase doppler particle analyzer (Dual-PDPA). This technique permitted the detection of size and velocity of individual spherical particles in liquid or gaseous flows.[8,9] The particle concentration and the volume flux could be calculated based on these two quantities. The main features of the technique were non-intrusive, on-line, *in situ*, and calibration free measurements with very high accuracy and spatial resolution. The measurement technique required optical access to the measurement area usually from two directions. The refractive indices of the particle and the continuous medium must usually be known. Due to the changing refractive index of the remaining fractions in an evaporating droplet, an off-axis angle of 72° (FIG. 2) was chosen. This arrangement proved to minimize the uncertainties caused by this effect.[10]

Contrary to the visualization, the PDPA measurements were not phase locked. To resolve the phase relationship a time stamping procedure was applied. Time stamping of the measured droplet velocities and sizes was performed on the base of the 0° trigger signal of the microphone. In a postprocessing step the measured values were assigned, relative to the trigger signal in the phase space, in steps of 10°. To identify phasing with no significant droplet appearance, a threshold minimum of 100 droplets was set for each averaged phase step of 10°. The number of droplets measured was limited to 20,000 for each measurement position. However, due to the extremely difficult measurement conditions the acquisition was also timed-out after 240 seconds, since the overall measurement time was limited by the optical quality of the glass cylinders, which suffered from the high thermal stress. Density fluctuations in the reacting turbulent flow caused strong beam steering, which lowered the coincident validation rate significantly. The validation also declined as a result of low signal-to-noise ratio due to the strong flame luminescence at elevated pressure.

RESULTS AND DISCUSSION

FIGURE 3A shows a stationary situation at 0.4 MPa inlet pressure, 573 K inlet temperature, and $\lambda = 3.0$. The predominantly bluish flame luminescence covering most of the combustor volume indicates an overall lean combustion at low temperature, resulting in low NO_x emission caused by a high recirculation of the exhaust gas. The weak blue flame luminescence also indicates low soot formation. Depending on the fuel–air ratio, the inlet temperature, and the inlet pressure, the model combustor exhibits large pressure oscillations. Increasing the fuel mass flow or the pressure slightly causes the combustion process to start oscillating with a frequency of approximately 620 Hz.

FIGURE 3B and C shows two typical situations (0.8 MPa, $T_{inlet} = 573$ K, $\lambda = 2.0$), captured at two different phase logs in the oscillating cycle for the same time integral conditions. The combustion varies in one oscillation cycle between extinction, lean (FIG. 3C) and rich (FIG. 3B) combustion with stoichiometric flame temperatures and a high soot and NO_x formation. CFD calculations of the reacting two-phase flow would, therefore, lead to erroneous results with a stationary approach. The images in

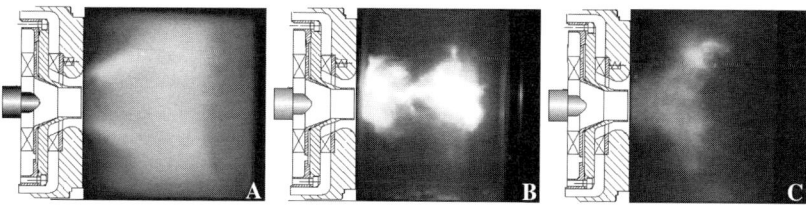

FIGURE 3. Flame luminescence images at various operating conditions: **(A)** 0.4 MPa, $\lambda = 3.0$, non-oscillating; **(B)** 0.8 MPa, $\lambda = 2.0$, oscillating; **(C)** 0.8 MPa, $\lambda = 2.0$, oscillating.

FIGURE 4. Droplet frequency distribution at various operating conditions: **(A)** 0.4 MPa, λ = 3.0, non-oscillating; **(B)** 0.8 MPa, λ = 2.5, dt = 300 nsec; **(C)** 0.8 MPa, λ = 2.5, dt = 500 nsec.

FIGURE 4 show a droplet frequency distribution of 150 images. FIGURE 4A shows the non-oscillating case at 0.4 MPa inlet pressure. The droplets propagate as a dense spray. No fuel droplets were found in the center of the atomizer. FIGURE 4B and C show the droplet frequency distribution at 0.8 MPa inlet pressure and 573 K inlet temperature at two different phasings relative to the pressure oscillation. The overall time duration of one cycle was about 1,600 nsec. Typical for the oscillation was the presence of fuel close to the axis of the combustor. This enabled the flame to travel in the recirculation zone upstream to the atomizer until the fuel close to the axis was consumed (FIG. 3). The increase of the temperature inside the prefilmer evaporates the present fuel and blocks the flow from the radial swirls through the airblast atomizer.

FIGURE 5 shows typical results of the PDPA measurements. The droplets could be detected only in the phase range between 180° and 360°. The droplet velocities on the axis ($r = 0$, $z = 6$ mm) downstream the back plane of the combustion unit show strong oscillation between 100 m/sec and −20 m/sec.

The highly reduced oscillation velocities at the outer radius of $R = 8$ mm indicate that substantial mechanisms for flashback into the prefilmer were taking place in a small central recirculation zone close to the axis. The combustion oscillations also showed a strong effect on the droplet sizes. In the non-oscillating cases, the average

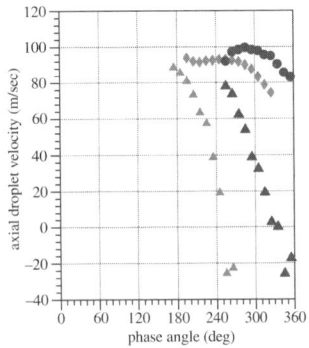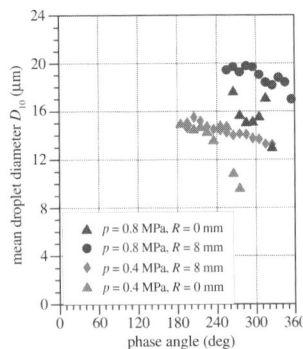

FIGURE 5. Axial droplet velocities and mean diameter D_{10} (λ = 2.0, T_{inlet} = 573 K, $z = 6$ mm).

droplet diameter decreased with an increase of air pressure due to better atomization, caused by higher Weber numbers. The measurements of the time-resolved droplet diameter in the oscillating cases show that increasing pressure leads to increased droplet diameters close to the axis. The increase of the air density associated with the high velocity peaks of the oscillating air flow enhanced the constriction of the cone angle of the simplex atomizer. This forced more and larger droplets at the side of the prefilmer toward the axis, enabling flashback inside the vortex recirculation zone due to the presence of fuel.

CONCLUSIONS

The results presented herein demonstrate that the combination of phase-locked laser light sheet visualization and phase resolved PDPA measurements provide excellent tools for the investigation of two-phase flow in the presence of combustion oscillations. The results clarify the importance of knowing the interaction between the combustion oscillation and two-phase flow needed to validate transient CFD calculations.

The periodic flashback into the prefilmer resulted in serious combustion oscillations, driven by the transport of fuel in regions close to the combustor axis that were extremely lean in the non-oscillating case. The combustion oscillations substantially affect the droplet size and velocity distribution, with corresponding effects on droplet evaporation, dispersion, and chemical reaction.

ACKNOWLEDGMENT

The investigation of the MTU airblast atomizer was funded by MTU Motorenund Turbinen Union, Munich, Germany.

REFERENCES

1. SANTAVICCA, D.A., R.L. STEINBERGER, K.A. GIBBONS, et al. 1993. The effect of incomplete fuel–air mixing on the lean limit and emissions characteristics of a lean prevaporized premixed (LPP) combustor. AGARD Conf. Proc. No. 536, AGARD Meeting on Fuels and Combustion Technology for Advanced Aircraft Engines.
2. MEIER, R., G. MAIER & S. WITTIG. Highspeed Imaging of the Instationary Break-up of thin Liquid Sheets. 16th Annual Conf. on Liquid Atomization and Spray Systems. Darmstadt, Germany.
3. MCDONELL, V.G., M. ADACHI & G.S. SAMUELSEN. 1992. Structure of reacting and nonreacting swirling air-assisted sprays. Combust. Sci. Tech. **82:** 225–248.
4. WILLMANN, M., R. MEIER & S. WITTIG. 1997. Visualization of the instationary atomization and spray propagation inside a reacting modell gasturbine combustor. 13th annual Conf. on Liquid Atomization and Spray Systems ILASS-Europe. Florence, Italy.
5. KREUTZ-IHLI, T., D. FILSINGER, A. SCHULTZ & S. WITTIG. 2000. Numerical and experimental study of unsteady flow field and vibration in radial inflow turbines. ASME J. Turbomachin. **122:** 1–8.

6. SCHÄFER, O., R. KOCH & S. WITTIG. 2000. Measurement of the periodic flow of an enclosed lean premixed prevaporized stagnation flame. 10th Int. Symp. on Application of Laser Techniques to Fluid Dynamics, Lisbon, Portugal.
7. MEIER, R., K. MERKLE, G. MAIER, et al. 1999. Development of an improved prefilming airblast atomizer for gas turbine application. Fifth Annual Conf. on Liquid Atomization and Spray Systems, Toulouse, France.
8. DURST, F. & M. ZARÉ. 1975. Laser Doppler measurements in two-phase flows. Proc. LDA-Symp., Copenhagen, Denmark.
9. BACHALO, W.D. & S.V. SANKAR. 1990. Analysis of the light scattering interferometry for spheres larger than light wavelength. Int. Symp. on Application of Laser Techniques to Fluid Dynamics, Lisbon, Portugal.
10. WILLMANN, M., A. GLAHN & S. WITTIG. 1996. Phase-Doppler particle sizing with off-axis angles in Alexander's darkband—a promising approach for complex technical spray systems. Eighth Int. Symp. on Application of Laser Techniques to Fluid Dynamics, Lisbon, Portugal.

Numerical Visualization of Two-Phase Plume Formation in a Stratification Flow Environment

BAIXIN CHEN,[a] MASAHIRO NISHIO,[b] YONGCHEN SONG,[a] SATOSHI SOMEYA,[b] AND MAKOTO AKAI[b]

[a]*Research Institute of Innovative Technology for the Earth, RITE-Tsukuba Division, Ibaraki, Japan*

[b]*National Institute of Advanced Industrial Science and Technology, Research Institute for Energy Utilization, Tsukuba-shi, Japan*

ABSTRACT: Evolution of two-phase plumes driven by air bubble buoyancy in a stratification ambient in a rectangular tank is visualized numerically by means of two-phase flow theory and large-eddy simulation technology. With a focus on the discrete nature of the buoyant dispersed phase and the role of momentum exchange between two phases in plume formation, we investigated the phenomena of mass *entraining-in* and *peeling-out* for continuous phase plume, which may result from a complicated and intricate interplay with phase interaction and dynamic stability of the stratification ambient, respectively. Numerical simulations show that although mass *entraining-in* and *peeling-out* appear to be distinguished entirely in the vertical direction, they interact or couple locally within inner of the plume and present a discontinuity in nature. The numerically visualized three-dimensional density field also indicates the same plume characteristics.

KEYWORDS: two-phase plume; numerical visualization; LES; stratification

INTRODUCTION

Two-phase plumes are common buoyancy flow phenomena involved in hydraulic engineering and some environmental applications. Air bubble plumes applied to reservoir destratification,[1] CO_2 droplet plume used in CO_2 ocean sequestration[2,3] which is one of the acceptable options of mitigating green house gases, and deep-sea blowout of oil and natural gas,[4] are some of the classic examples. Experimental results[5] have shown that the overall characteristics of two-phase plumes are different from those of single-phase plumes for which self-similarity can be assumed, because the interactions between discrete bubbles or droplets and the continuous ambient produce a multilayered peeling structure. This is due to the slip velocity between two phases/fluids and the dynamic stability of the stratification ambient. Based on these experimental results, theoretical models[6,7] were developed to study the mean flow

Address for correspondence: Baixin Chen, Ph.D., Research Institute of Innovative Technology for the Earth (RITE), RITE-Tsukuba Division, AIST-EAST, 1-2-1 Namiki, Tsukuba-shi, Ibaraki 305-8564, Japan.
 b.chen@aist.go.jp

characteristics of the continuous-phase plume. The models were established by a set of ordinary differential equations with respect to the central direction of the plume by assuming that the plume is in steady flow and ignoring the details of turbulence. Experimental constants, (e.g., entrainment coefficient and constants in the equation for mean velocity distribution along the plume section) were calibrated from experimental data. The models predicted overall characteristics of the continuous plume, such as the peeling height, the plume height, and the plume efficiency as a function of bubble/droplet releasing rate. However, the details of the inner physics of plume formation and ambient mass entraining-in and peeling-out were rarely investigated, especially by means of numerical simulations. This paper describes our initial efforts to simulate and visualize this two-phase plume formation and inner behavior of ambient mass entraining and peeling by applying two-phase flow theory and large-eddy simulation technology. First, the numerical results are validated by existing experimental data[6] for mean section density and the multilayer outline of the continuous plume for various evolution times. Second, typical three-dimensional fields of density, ambient mass entraining and peeling, and velocity vectors are *visualized* in detail. Finally, the mechanism and the inner interaction of plume formation are discussed.

GOVERNING EQUATIONS

Assuming that bubbles or droplets with a mean droplet diameter d_0 are released at a height H from the bottom of a rectangular tank that is filled with stratification ambient fluid, the fluid element selected from the system contains a continuous phase fluid and a discontinuous phase fluid (i.e., bubbles or droplets). If the density of bubbles/droplets is different from that of the continuous fluid, buoyancy will cause them to rise (positive buoyancy) or fall (negative buoyancy). The droplet velocity and void fraction, as discontinuous fluid, depend on the intertranslation of momentum with the continuous fluid, the stratification ambient. The mechanism of these two fluid interactions can be, in principle, interpreted by two-fluid or two-phase flow theories.

Based on the physical model and the two-fluid theories, the governing equations of eddy-resolved variables ($^\wedge$), normalized according to the scales of velocity (U_c), space (L), and time ($T = L/U_c$) for each phase, can be expressed by applying Schumann's filtering technique, in the *Eulerian–Eulerian* scheme, as follows. For a continuous phase, in general:

$$\frac{\partial \bar{\rho}}{\partial t} + \frac{\partial \bar{\rho} \hat{u}_i}{\partial x_i} = \dot{w}_d, \tag{1}$$

$$\frac{\partial \bar{\rho} \hat{u}_i}{\partial t} + \frac{\partial \bar{\rho} \hat{u}_i \hat{u}_j}{\partial x_j} = \frac{\partial \hat{p}}{\partial x_i} + \frac{\partial \bar{\rho} \hat{\tau}_{ij}}{\partial x_j} + (\bar{\rho} - \rho_0)g_i + (\dot{F}_d)_i, \tag{2}$$

where $\hat{p}$ is a dynamic pressure and g_i is the gravitational acceleration.

$$\frac{\partial \bar{\rho} \hat{\phi}_k}{\partial t} + \frac{\partial \bar{\rho} \hat{\phi}_k \hat{u}_j}{\partial x_j} = \frac{\partial}{\partial x_j}\left(\bar{\rho} D_k \frac{\partial \hat{\phi}_k}{\partial x_j}\right) + \frac{\partial \bar{\rho} \hat{q}_k}{\partial x_j} + \dot{w}_d \delta_{kl}, \tag{3}$$

where $\delta_{kl} = 1$ when $l = k$ for mass fraction of dispersed phase solution, and $\delta_{kl} = 0$ when $l \neq k$ for scalars. $\dot{w}_d$ and $\dot{F}_d$ are the exchange rates of mass and momentum between the two phases, respectively.

The fundamental equations for the dispersed phase are

$$\frac{\partial \hat{n}_d}{\partial t} + \frac{\partial \hat{n}_d \hat{u}_{dj}}{\partial x_j} = 0, \qquad (4)$$

$$\frac{\partial \hat{\alpha}}{\partial t} + \frac{\partial \hat{\alpha} \hat{u}_{dj}}{\partial x_j} = -\frac{\dot{w}_d}{\rho_d}, \qquad (5)$$

$$\frac{\partial \bar{\rho}_d \hat{u}_{di}}{\partial t} + \frac{\partial \bar{\rho}_d \hat{u}_{di} \hat{u}_{dj}}{\partial x_j} = \hat{\alpha}(\bar{\rho} - \rho_d) g_i - (\dot{F}_d)_i, \qquad (6)$$

where $\hat{\alpha}$, $\hat{n}_d$, and $\hat{u}_{di}$ the volume fraction, number density, and velocity vector of droplets, respectively. $\bar{\rho}_d = \hat{\alpha} \rho_d$ and ρ_d the physical density of droplet/bubble.

In this study, $\dot{w}_d$ is set to zero (no mass exchange is assumed) and Equation (4) is neglected. To close this set of equations, the turbulent transport terms of Reynolds stress tensor $\hat{\tau}_{ij}$ and Reynolds scalar flux $\hat{q}_k$ in subscale smaller than the grid-size and the momentum-exchange rate $\dot{F}_d$ need to be modeled in terms of resolved variables. The turbulent transport terms are expressed by the so-called "structure-function" model[8] in order to take the effect of stratification into account. The momentum-exchange rate is estimated from

$$\dot{F}_d = 0.75 \left(\frac{\pi}{6}\right)^{1/3} \rho_d v_w \hat{\alpha}^{2/3} \hat{n}_{d_0}^{1/3} Cd |\hat{u}_j - \hat{u}_{dj}| (\hat{u}_j - \hat{u}_{dj}), \qquad (7)$$

and the drag coefficient Cd is calculated from

$$Cd = \frac{24}{Re}(1 + 0.173 Re^{0.657}). \qquad (8)$$

The Reynolds number Re is defined by the slip velocity and the diameter of the bubble.

The complete set of governing equations, (1)–(3), (5), and (6), is numerically solved by the SIMPLEC method. The algorithm is based on a finite control volume discretization on a staggered grid system to ensure the preservation of the conservation properties of the original differential equations. A second order central difference scheme was used to solve the momentum equations and a hybrid scheme of fourth order central difference "power-law" was used for scalar equations to avoid extremely false values and to reduce numerical diffusion. The solution is advanced in time by using implicit second-order Adams–Bashforth scheme.

SIMULATION VALIDATION

An experimental case[6] was simulated to test the physical model and the associated computer code developed in this work. The simulated domain was set as 1.0×1.0×0.6 m in horizontal and vertical directions, respectively, associated with a non-uniform grids number of 102×102×64. The minimum grid size was 0.5×0.5×1.0 cm within the center of the domain (0.44×0.44×0.6) where the bubble-release nozzle was located. The air bubbles were released into artificial seawater at

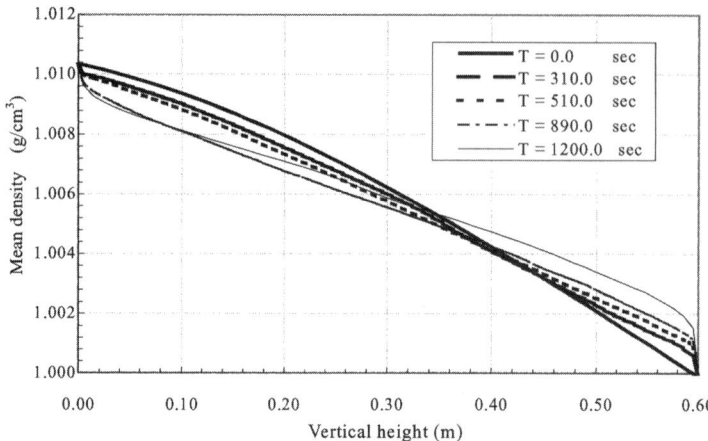

FIGURE 1. Mean density distribution in the vertical direction as it developed with time.

a height of 4.0 cm from the bottom with a release rate of 0.645×10^{-6} m^3/sec and an initial diameter of 1.05 mm. The initial ambient stratification was created by applying a presumed salinity distribution in the vertical direction to a standard state equation of seawater[9] to fit the experimental data. The initial buoyancy frequency $N^2 = 0.170$ (1/sec^2). Density was calculated by solving the transport equation of salinity to handle the dynamic stability of the stratification ambient. The boundary conditions at the top water free surface for both phases were treated as a rigid slip wall according to Lam and Banerjee,[10] which is a reasonable and simple way if no surface mass exchange occurs.

FIGURE 1 shows the time evolution of mean density distribution in the vertical direction, which is one of overall characteristics of an ambient plume and averaged for each horizontal section. Compared with experimental data[6] (not shown here), the numerical predictions are satisfactory, quantitatively and qualitatively. Another overall characteristic of the ambient plume, the multilayer outline, can be indicated by the difference of the ambient mass flux between peeling out (positive) and entraining into (negative) the continue phase plume. This mass flux difference is shown in FIGURE 2. It can be seen that, up to the time $T = 810$ sec the multiplayer plume formed with the first peeling layer at about 0.25 m from the bottom, the second layer at about 0.45 m from the bottom, and an upward entraining approaching the surface of the ambient. This result agrees qualitatively with experimental observations.[6] The mean density and mass flux differences in FIGURES 1 and 2 illustrate the evolution of the plume.

VISUALIZATION OF THE INNER STRUCTURE OF THE PLUME

After validation, the numerical model was used to investigate the inner structure and mechanism of plume development. The horizontal velocity vector and the

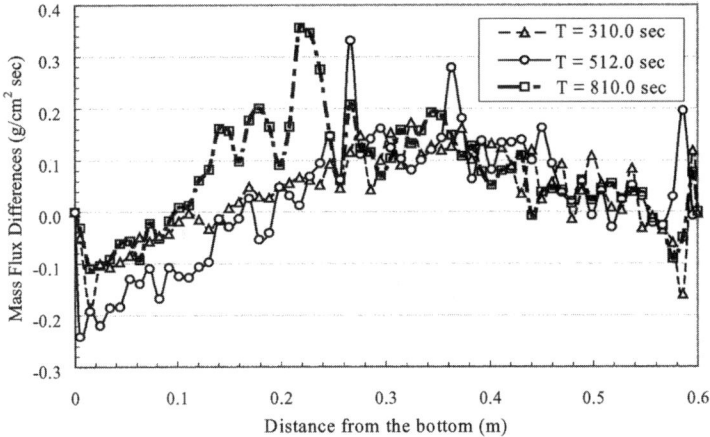

FIGURE 2. Mass flux differences between peeling-out and entraining-in.

density field were retreated to produce a mass flux field. Their centurial and centrifugal directions to the central line of computational domain define the entraining-in and peeling-out mass flux. The entraining-in mass flux (see FIGURE 3A) and peeling-out mass flux (FIG. 3B) at time of 810sec are visualized in three-dimensional by a contour surface with a value of $0.21 \text{g}/(\text{cm}^2\text{sec})$. The ambient mass driven by the buoyant bubble is entrained into the plume mostly within a vertical cylinder with a diameter not larger than 8.0cm and appears as a discrete property in space. In contrast to the entraining mass flux, the peeling mass flows mostly out of the plume asymmetrically at the middle and upper parts of the plume. The multilayer outline of the plume could also be presented by this peeling mass flux. To show the interaction characteristics of those two fluxes, the mass flux field on the central-vertical section, at the same time as in FIGURE 3A and B, is shown in FIGURE 3C in which the contours are the entraining flux and the image is the peeling. The discontinuity results from the coupled interaction between the bubble rising and the dynamic stability of the ambient stratification. The fluid element with larger density is lifted by the air bubble due to the action of drag up to a height where the driving force is balanced with the local negative buoyancy induced by the density difference. Consequently, it penetrates on the horizontal surface or is peeling down to a height where the density difference disappears.

FIGURE 3D gives the velocity field combined with streamlines in the central-vertical section. It further describes the mechanism of plume development. The stratification characteristics of the fluid are visualized in the three-dimensional density field by two density contour surfaces (FIG. 3E). The air bubble plume (contours) and density distribution (image) in the central-vertical section are given in FIGURE 3F. FIGURE 3E and F both demonstrate the reconstruction of the density field disturbed by air bubble plume and eddies created by turbulent flow.

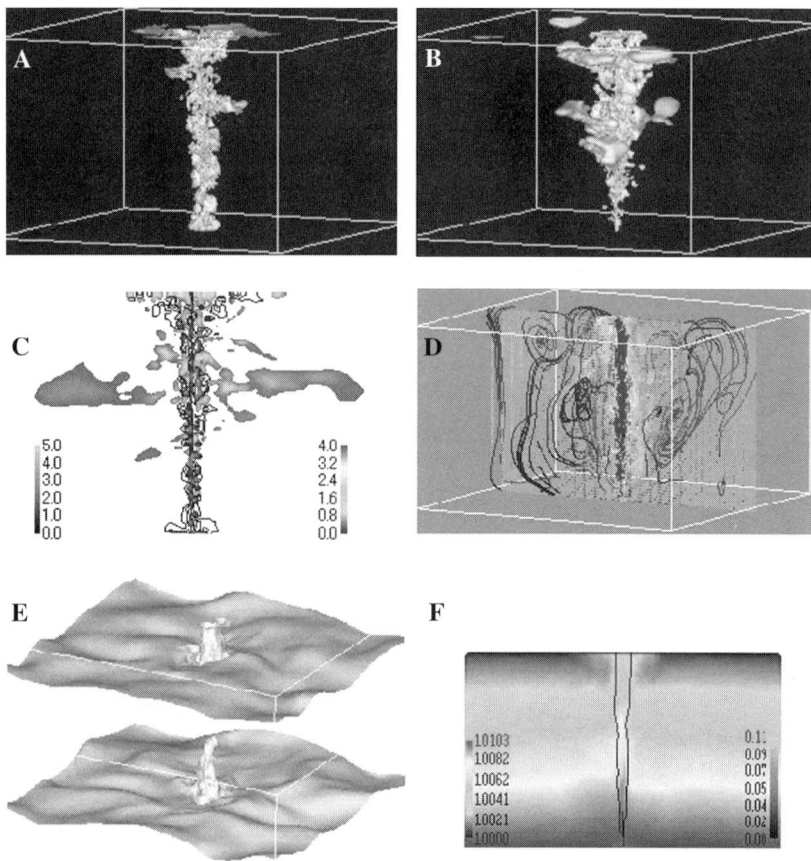

FIGURE 3. Visualization of inner structure of bubble and ambient plumes. **(A)** Three-dimensional view of entraining-in mass flux. **(B)** Three-dimensional view of peeling-out mass flux. **(C)** Coupling of entraining-in (contours) and peeling-out mass flux on central-vertical section. **(D)** Velocity vectors and associated streamlines on the central-vertical section. **(E)** Two contour-surfaces of the density field in three-dimensional view. **(F)** Air bubble plume (contours) and density field in the central-vertical section.

CONCLUSIONS

A three-dimensional unsteady large-eddy simulation model and associated CFD computer code developed in this study are capable of predicting the basic mechanism and the inner characteristics of two-phase plume formation in a stratification environment. The three-dimensional numerical visualizations indicate that the interactions among momentum exchange between the air bubble and ambient, the ambient dynamic stability, and stratification play the central role in two-phase plume development. In the inner part of plume, both mass entrainment and peeling are discrete in space. The former occurs within a cylinder-like regime whereas the latter is created by the density differences and appears in the middle and upper parts of continue phase plume. The simulations also show the potential for engineering applications, as for instance, the simulation of CO_2 sequestration in the ocean.

ACKNOWLEDGMENT

This study is a part of the investigation of the CO_2 Ocean Sequestration Project managed by the Research Institute of Innovative Technology for the Earth (RITE) and funded by New Energy and Industrial Technology Development Organization (NEDO), Japan.

REFERENCES

1. SCHLADOW, S.G. 1993. Lake destratification by bubble-plume systems: design methodology. J. Hydr. Eng. **199**(3): 350–368.
2. LIRO, C.R., E.E. ADAMS & H.J. HERZOG. 1992. Modeling the release of CO_2 in the deep ocean. Energy Conserv. Mgmt. **33**(5–8): 667–674.
3. HAUNGAN, P.M., F. THORKILDSEN & G. ALENDAL. 1995. Dissolution of CO_2 in the ocean. Energy Convers. Mgmt. **36**: 461–466.
4. JOHANSEN, Ø. 1999. Deep blow—a Lagrangian plume model for deep water blowouts. Proc. 3rd Int. Marine Environ. Modelling seminar, Lillebammer.
5. SOCOLOFSKY, S.A. 2000. Laboratory experiments of multi-phase plumes in stratification and cross flow. Ph.D, Thesis, Dept. of Civil and Environmental Eng, MIT, Cambridge, MA, U.S.A.
6. ASAEDA, T. & J. IMBERGER. 1993. Structure of bubble plumes in linearly stratified environments. J. Fluid Mech. **249**: 35–57.
7. MCDOUGALL, T.J. 1978. Bubble plumes in stratified environments. J. Fluid Mech. **85**: 655–672.
8. LESIEUR, M. & O. MÉTAIS. 1996. New Trends in large-eddy simulations of turbulence. Annu. Rev. Fluid Mech. **28**: 45–82.
9. UNESCO. 1981. Tenth report of the joint panel on oceanographic tables and standards. UNESCO Technical Papers in Marine Science **36**: 24–29.
10. LAM, K. & S. BANERJEE. 1992. On the condition of streak formation in a bounded turbulent flow. Phys. Fluids **A4**: 306–320.

Visualization Studies of an Acoustically Excited Liquid Sheet

VAYALAKKARA SIVADAS AND MANUEL V. HEITOR

Laboratory of Thermofluids, Combustion and Energy Systems, Center for Innovation, Technology and Policy Research, Instituto Superior Técnico, Technical University of Lisbon, Lisbon, Portugal

ABSTRACT: The interaction between atomizing air and a liquid sheet initiates interfacial instabilities that develop into large amplitude waves, which ultimately promote liquid disintegration. However, to reduce atomizing airflow, the dynamic influence of the surrounding gas is not sufficient to affect the inner core of the liquid. In other words, the characteristic length scales of perturbation promoting break-up are small in comparison with the initial liquid momentum. Therefore, to enhance the disintegration process, the intact liquid-core near the nozzle outlet must be influenced by artificial or external excitations. To accomplish this, the present investigation employed a plane laminar liquid-sheet that was perturbed by acoustic excitations at an optimum frequency through the impinging air-streams.

KEYWORDS: atomizing; liquid sheet; acoustic excitation; air blast

INTRODUCTION

High-speed flow visualization techniques are used to characterize the wave behavior at the air–liquid interface in the context of an air-assist atomizer under the influence of acoustic excitation. Research in this area has progressed considerably due to the recognized potential of air-blast atomizers for minimizing pollutant emissions from air-breathing engines.[1] Air-blast or twin-fluid atomizer has a number of advantages over a pressure atomizer, including lower requirement for fuel injection pressure and generating finer sprays. Among various injector configurations, planar liquid sheet studies have been given particular emphasis because this technique allows direct examination of the air–liquid interface without the added complexity of curvature effect.[2–4]

EXPERIMENTAL SYSTEM

Atomizer Configuration

The liquid film generator consists of a two-dimensional unit[5] built to generate thin and flat liquid sheets with large aspect ratios, allowing adequate control and

Address for correspondence: V. Sivadas, Ph.D., Center for Innovation, Technology and Policy Research, Technical University of Lisbon, Av. Rovisco Pais, 1049-001 Lisbon, Portugal. das@dem.ist.utl.pt

direct examination of the air–liquid interface. To eliminate fluctuations in the flow, water is supplied through a pressurized tank in closed circuit. The liquid emerges from a rectangular slit orifice with aspect ratio 200:1, an exit thickness of 0.4 mm, and is "sandwiched" between two impinging air jets. The atomizing edge is made sharp to prevent generation of flow disturbances and, hence, to ensure perfect laminar conditions for the liquid sheet. The liquid velocity is varied from 0.95 m/sec to 1.8 m/sec, corresponding to a Reynolds number (Re) in the range 375 to 710 based on film thickness. A calibrated rotameter continuously monitors the liquid flow rate.

Atomizing air enters through the rectangular channels, on either side of the liquid slot, and impinges on the liquid sheet with an angle of 30° at the nozzle-outlet. The air-channel thickness at the exit-port is 7 mm and the initial velocity of both air jets were kept identical at 10 m/sec. The atomizer is mounted vertically on a traversing-table having an accuracy of 0.5 mm in all directions. In order to impart acoustic perturbation through the atomizing air, both air-channels are equipped with a loud-speaker system. The speakers were kept at an equal distance upstream of the atomizer configuration. The system is excited by an amplified sinusoidal signal from a function generator, with tunable frequency, and the maximum output power of each loud speaker is 60 Watts. Frequency tuning is carried-out in such a way that the efficiency of electromechanical transudation for the combined loudspeaker system is maximized, which occurs at the resonant frequency of the system.[6] For the present experimental arrangement, resonance occurs at 750 Hz with a narrow bandwidth. Analysis has shown that at other frequencies the liquid sheet is unaffected by the modulation signal for input powers up to 60 Watts.

Measurement Techniques

Qualitative and quantitative analysis of the flow-field was carried-out by flow visualization techniques. To obtain high contrast images of the liquid sheet, a white background illumination was used that enhanced the recording quality because of the transparent nature of the liquid sheet. Images were acquired by a high-speed digital CCD-camera at a grabbing rate of 2,000 frames per second, and with an exposure time of 10 μsec. Correspondingly, the maximum resolution of the camera was 256 × 120 pixels. The recording parameters were chosen based on the expected maximum frequency of the flow-domain. Thus, instantaneous information about the flow structure could be inferred from each frame. For qualitative analysis a grabbing rate of 250 frames per second was used, which, in turn, boosts the image resolution to 512 × 480 pixels.

The camera views the object-plane normally, without distorting the image, so as to allow the direct measurement of the interfacial characteristics of the liquid sheet. In order to obtain the characteristic length scales of disintegration, both front-and-side view images of the flow-field were utilized. Image analysis was carried-out by identifying the major disturbances and their subsequent development. The respective scale factor for the analysis was obtained interactively using the scale function. To acquire reliable data, 8–10 pairs of frames were used for each flow condition. This ensured sampling of the highest amplitude waves, and hence, with reasonably low-levels of measurement errors, the associated break-up length and break-up frequency could be estimated. Maximum uncertainty in the break-up length was estimated to be below 10%, based on the repeatability of measurements.

RESULTS AND DISCUSSIONS

FIGURE 1 demonstrates the effect of acoustic perturbations on liquid sheets, as the flow rate (Q_{liquid}) varies from 8.1 to 6.7 g/sec. The images depict the front-view of the flow-field in stagnant ambient conditions. Acoustic excitations are imposed through an air column, on both sides of the liquid sheet, at the resonant frequency of 750 Hz. The resonant characteristics of the present atomizer configuration were established by correlating the pressure wave behavior at the loudspeaker and at the outlet of the injector.[5] In other words, measurements with a microphone at the sound source and outlet show that the minimum phase-difference and peak energy transfer occurs at the optimum frequency of 750 Hz.

FIGURE 1 revealed the pronounced effect of imposed perturbations that directly influence the thick rim of the liquid sheet. Consequently, the convergent liquid sheet, which is bounded by thick rims that are drawn together by surface tension forces, is not able to withstand the disruptive external excitation. The resulting imbalance of forces initiates the break-up of the liquid sheet. In the following sections, the scenario emerges by using mechanical energy applied in conjunction with atomizing air in the context of an air-blast atomizer configuration.

FIGURE 2 illustrates the influence of acoustic modulation on an air-blast liquid sheet. The images correspond to an air flow rate (Q_{air}) of 1.4 g/sec, while maintaining

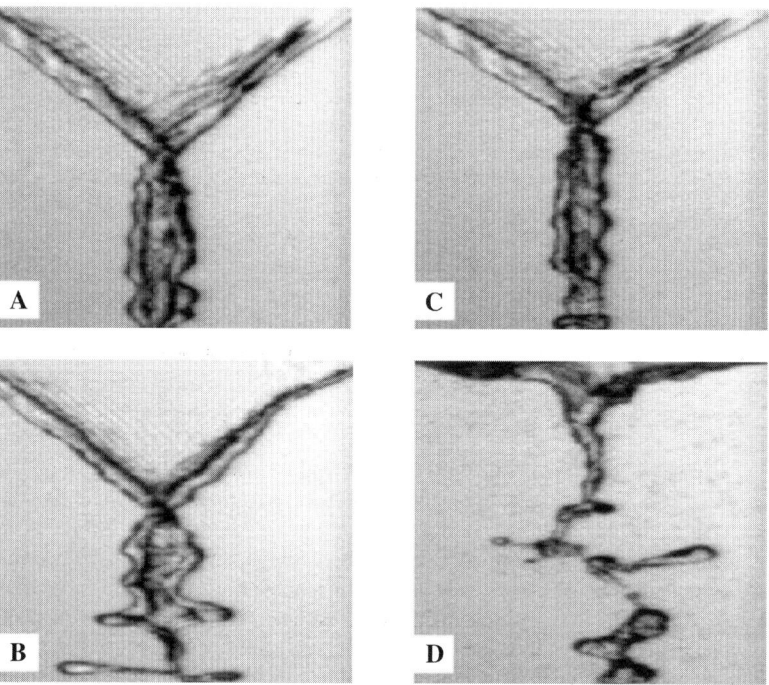

FIGURE 1. Front view images of liquid sheets: **A** and **C**, without perturbation; **B** and **D**, with perturbation; **A** and **B**, Q_{liquid} = 8.1 g/sec; **C** and **D**, Q_{liquid} = 6.7 g/sec.

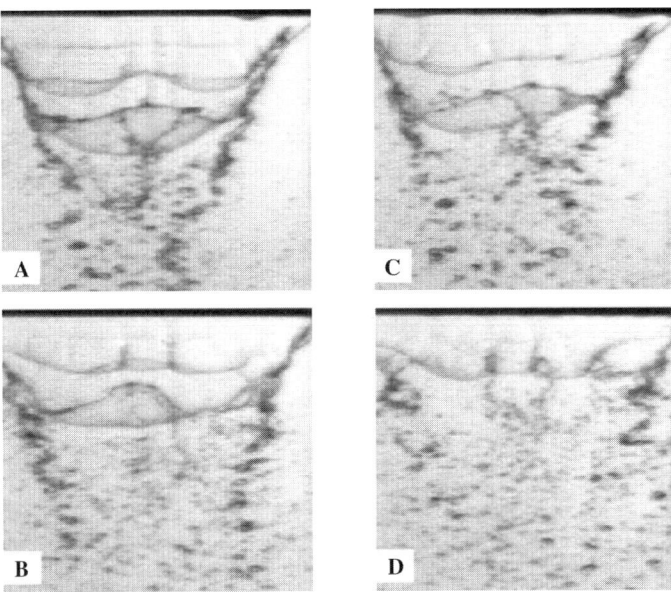

FIGURE 2. Front view images of air-blast liquid sheets: **A** and **C**, without perturbation; **B** and **D**, with perturbation; Q_{air} = 1.4 g/sec; **A** and **B**, Q_{liquid} = 8.1 g/sec; **C** and **D**, Q_{liquid} = 6.7 g/sec.

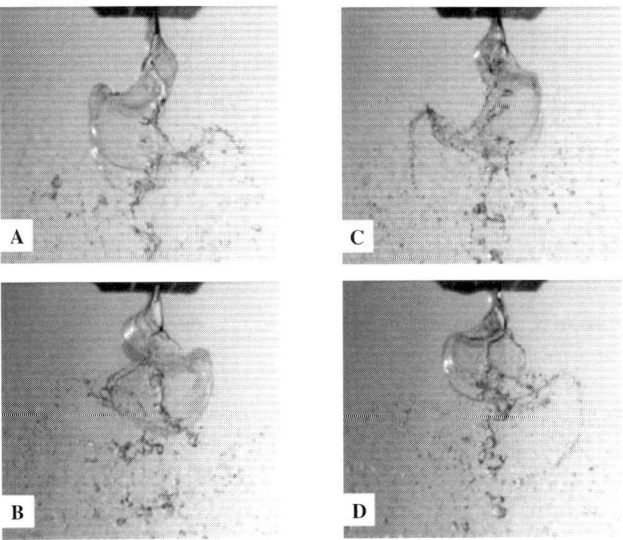

FIGURE 3. Side view images of air-blast liquid sheets: **A** and **C**, without perturbation; **B** and **D**, with perturbation; Q_{air} = 1.4 g/sec; **A** and **B**, Q_{liquid} = 8.1 g/sec; **C** and **D**, Q_{liquid} = 6.7 g/sec.

the liquid flow rate variation similar to the preceding case. The front view images show rapid rupturing of the liquid sheet due to increased transfer of energy from the atomizing air,[4] and the process is further augmented by the imposed acoustic excitations. A qualitative overview of the interfacial wave development and its breakdown can be extracted from the images. In order to characterize the critical waves at the liquid–air interface, side view images of the flow field for unexcited and acoustically excited conditions are presented in FIGURE 3.

The associated waves exhibit distinct characteristics; that is, sinusoidal instabilities connected with unexcited flows, (FIG. 3 A and C) that tend to become symmetric and attain the critical state quite rapidly under the influence of acoustic perturbations (FIG. 3 B and D). This can be attributed to the effect of electromechanical transudation that causes the thick rim of the liquid sheet to stretch symmetrically from either side, and thus gives rise to the bubble-like pattern. The images also show that early bursting of waves and, hence, short break up length is associated with the acoustically modulated liquid sheet.

Length Scale of Disintegration

FIGURE 4 quantifies the break-up length (L_b) of air-blast liquid sheets as a function of the absolute Weber number (We_{abs}), which represents the ratio of inertial to surface tension forces, for acoustically excited and unexcited cases. L_b is obtained from the images by measuring the axial distance from the nozzle outlet-plane to the point where the central part of the sheet is non-existent as a cohesive entity, and it is non-dimensionalised by the initial liquid sheet thickness, t_s. The liquid flow-rate varies

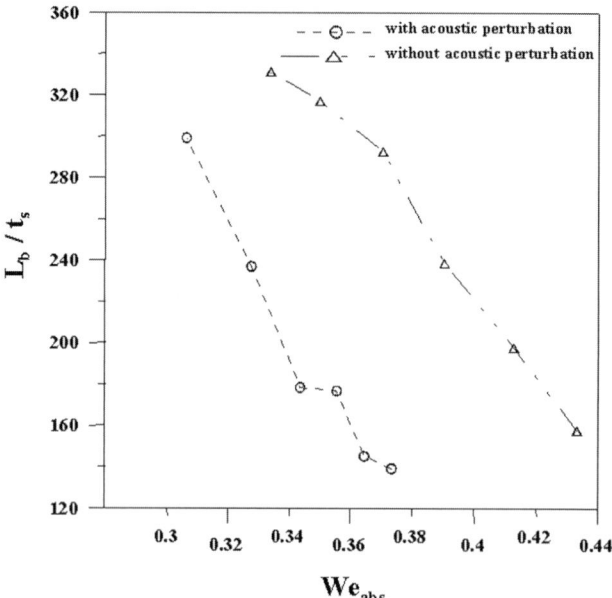

FIGURE 4. Break-up length as a function of Weber number.

over the range 6.7 to 12.5 g/sec, whereas the air flow rate is kept constant at 1.4 g/sec. For a given orifice dimension and fluid properties, the inertial and surface tension forces control the flow dynamics. Hence, We_{abs} can be used for proper characterization of the flow field. Due to the inclined atomizing-air configuration, the inertial force is the resultant of normal and tangential air momentum acting on the liquid sheet. Consequently, We_{abs} accounts for the resultant momentum and the enhanced energy transfer associated with impinging air blast atomizers.[7]

The results show that imposed perturbations cause steep decay in the break-up length due to the effective energy transfer from the acoustically excited air to the liquid sheet. This leads to an enhanced instability and, thus, to the disintegration process.

SUMMARY

The present visualization studies on air-assisted liquid sheets for Weber numbers below 0.4 reveal that properly tuned external perturbation may enhance the liquid break up and result in structurally different disintegration characteristics as compared with non-imposed flows. This can be attributed to the effective energy transfer from the atomizing air to the liquid sheet, which reduces the surface tension dominated instability.

The results of this investigation may find practical relevance because its effective utilization can reduce the required air pressure in air-blast atomizers. In addition, the symmetric structure exhibited by critical waves under acoustic excitations can enhance the radial dispersion of the liquid in the initial stages, which will be useful for direct injection systems.

ACKNOWLEDGMENTS

The authors acknowledge financial support from the Portuguese Science Foundation, project 34586/99, and the EC funded project DIME, "Direct Injection Spray Engine Processes—Mechanisms to Improve Performance", DIME – ENK6-2000-00101. V. Sivadas is a post-doctoral research fellow at IST under the sponsorship of Portuguese Science and Technology Foundation.

REFERENCES

1. LEFEBVRE, A.H. 1989. *In* Atomization and Sprays. Hemisphere Publishing Co. New York. 201–267.
2. MANSOUR, A. & N. CHIGIER. 1990. Disintegration of liquid sheets. Phys. Fluids **A2:** 706–719.
3. LOZANO, A., *et al.* 1996, Experimental study of the atomization of a planar liquid sheet. Atomization Sprays **6:** 77–94.
4. CARVALHO, I.S., M.V. HEITOR & D. SANTOS. 2001. Liquid film disintegration. Atomization Sprays. Submitted.
5. SIVADAS, V., E.C. FERNANDES & M.V. HEITOR. 2001. Acoustically excited air-assisted liquid sheets. Exp. Fluids. Submitted.

6. CHUNG, I.P., C. PRESSER & J.L. DRESSLER. 1998. Effect of piezoelectric transducer modulation on liquid sheet disintegration. Atomiz. Sprays **8:** 479–502.
7. SIVADAS, V. & M.V. HEITOR. 2001. Surface waves of air-assisted liquid sheets. ASME J. Fluids Eng. Submitted.

Flow Characteristics of Two Immiscible Liquid Layers Subjected to a Horizontal Temperature Gradient

SATOSHI SOMEYA,[a] TETSUO MUNAKATA,[a] MASAHIRO NISHIO,[a] KOJI OKAMOTO,[b] AND HARUKI MADARAME[b]

[a]*National Institute of Advanced Industrial Science and Technology, Tsukuba, Japan*

[b]*Nuclear Engineering Research Laboratory, University of Tokyo, Tokai-mura, Japan*

ABSTRACT: Marangoni convection, driven by an interfacial instability due to a surface tension gradient, presents a significant problem in crystal growth in normal microgravity environments. It is important to suppress and control the convection phenomenon for better material processing, especially in crystal growth by the liquid encapsulated Czochralski or liquid encapsulated floating zone techniques, in which the melt is encapsulated in an immiscible medium. Marangoni convection can occur on the liquid–liquid interface and on the gas–liquid free surface. Buoyancy driven convection can also affect and complicate the flow. In the study we report here, experiments were carried out with two liquid layers, silicone oil and fluorinert, in an open and enclosed rectangular cavity. The flow in the cavity was subjected to a horizontal temperature gradient. The interactive flow near the liquid–liquid interface was measured by the particle image velocimetry technique. The measured flow field is in agreement with numerical predictions. Free surface fluctuations with several dominant frequencies were also measured.

KEYWORDS: Marangoni convection; free interface; flow visualization

INTRODUCTION

Given the capability for crucibleless handling of melts, the floating zone (FZ) technique (see FIGURE 1 A) is preferred for growing crystals when chemical reactions between melts and crucible walls must be avoided. The FZ technique has inherent difficulties caused by hydrostatic forces on the diameter of the crystals, by vaporization loss at the free surface, and by Marangoni convection. Normally, the size of the melt zone is mainly determined by the equilibrium between the hydrostatic pressure and the capillary pressure jump on the contact line of the free surface and the solid wall of the growing single crystal. The size is practically restricted to a few millimeters for most of the melts. This restriction can be avoided in a microgravity environment, where a larger liquid column can be formed that yields bigger crystals are than normal on Earth. The absence of buoyancy driven convection in the microgravity

Address for correspondence: Satoshi Someya, Dr. Eng., Research Institute of Energy, National Institute of Advanced Industrial Science and Technology, 1-2-1 Namiki, Tsukuba, Ibaraki, 305-8564, Japan.

s.someya@aist.go.jp

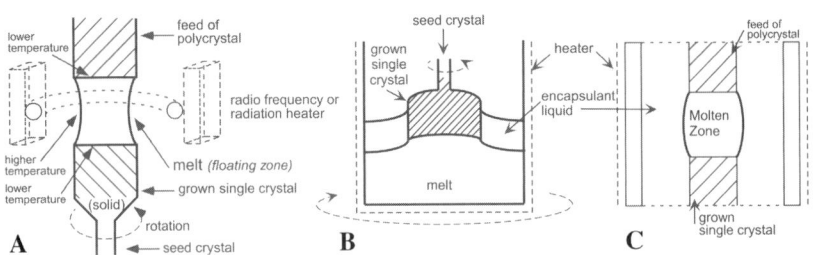

FIGURE 1. Schematics of floating zone (**A**), liquid encapsulated Czochralski (**B**), and liquid encapsulated floating (or melt) zone (**C**) crystal growth techniques.

environment strongly affects the quality of the grown crystal and the size of floating zone. However, the problems of surface tension driven Marangoni convection and volatilization remain. Marangoni convection often dominates convective motion under microgravity conditions. The driving force of the surface tension gradient is caused either by the temperature or the concentration distribution on the surface and the interface. To control the Marangoni convection, a liquid layer encapsulant, which has a free surface, can be used to coat the melt. Flow in melts can be suppressed or enhanced by choosing a liquid encapsulant with suitable physicochemical properties. It is, therefore, important to understand the flow phenomenon with interfacial Marangoni convection.

The liquid encapsulation technique under normal gravity conditions can suppress and control of the convective flow in melts without the vaporization losses. For example, in the liquid encapsulated Czochralski (LEC) crystal growth technique (FIG. 1B) the liquid encapsulant is put on the melts; for example, GaAs. In this case, buoyancy driven flow dominates the flow field. The convection characteristics in a double-layer system are generally much more complicated than in a single-layer system, due to a strong interaction between the motion in the two contiguous layers. Marangoni forces on Earth may have a little effect on the flow, but even a small effect can destabilize the flow. Understanding the interaction between buoyancy driven convection and Marangoni convection at the free surface and/or at the liquid–liquid interface in the encapsulation technique of crystal growth is highly desired.

Liu et al.[1] simulated flow in a horizontal two layered system subjected to a horizontal temperature gradient along the interface between two immiscible fluids. They assumed flat interfaces in a zero gravity field, and investigated effects of geometric factors and physicochemical properties. Azuma et al.[2] conducted experimental studies on combined effects of Marangoni and buoyancy forces in a double layer system, and have obtained typical flow patterns. Jing et al.[3] discussed in detail the effects of the liquid depth ratio, oscillatory instability, and interface contamination on the critical conditions for the incipience of thermal Rayleigh–Marangoni convection in horizontal benzene–water and water–CCl_4 two-liquid layers, assuming a flat nondeformable interface. Ar et al.[4] experimentally studied the velocity fields of oscillatory Marangoni convection in floating half zone by using PIV. The velocity distribution of the oscillatory flow fields in both a horizontal cross-section and a vertical cross-section were measured in a small liquid bridge 5 mm in diameter. Cheng et al.[5] also studied a floating liquid half zone in terms of temperature fluctuations. Lopez *et al.*[6] studied GaSb crystal processing by the liquid encapsulated melt zone (LEMZ) technique (FIG. 1 C) in a microgravity environment and under a 1 *g* environment (i.e., on Earth). They characterized the properties of GaSb crystals by the LEMZ method and determined the effects of gravity and other processing parameters that most affected the final properties of the crystals.

The measurement accuracy in previous studies was not too satisfactory, and only a few studies have investigated the velocity flow field experimentally and numerically with good agreement. In addition, some authors report that Marangoni convection was not observed.[7] Most numerical studies assumed relatively large *Ma* and small *Ra* values (i.e., microgravity conditions). In the present study, the experiments were carried out in a simple system with two liquid layers. The study of this system is very important for liquid encapsulated FZ (LEMZ) and LEC crystal growth techniques in normal gravity and microgravity environments. We observed numerical and experimental flow behavior and investigated the combined effects of Marangoni and buoyancy forces on the flow in a double-layered system in which the flow was dominated by buoyancy driven convection. Obviously, this kind of study is a first step toward obtaining a rudimentary understanding of the present phenomenon.

NUMERICAL AND EXPERIMENTAL METHODS

Numerical Procedure

Numerical simulation was carried out for an experimental system (see FIGURE 2A), consisting of two immiscible and incompressible viscous fluids in a two-dimensional cavity. The fluid motion was governed by the two-dimensional Navier–

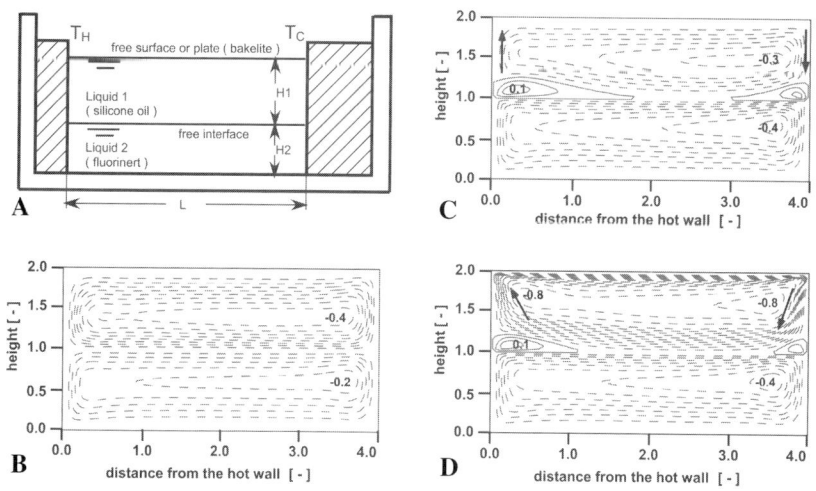

FIGURE 2. (**A**) Schematics of the horizontal two-layered system. (**B**) Numerical results for flow with the covered surface and without the interfacial tension gradient. (**C**) Result for the covered surface and interfacial tension gradient. (**D**) Result for the surface tension gradients at the free surface and at the interface.

Stokes equations, with the Boussinesq fluid approximations. The vertical heater and cooler side walls were maintained at constant temperatures, T_h and T_c. We assumed that the horizontal walls were thermally insulated. The free surface and the free interface were flat. The problem was rendered dimensionless by reference quantities corresponding to the lower layer. The scaling length was the height of the lower layer, H_2. Therefore, the Marangoni number and Rayleigh number were defined by

$$Ma = \frac{\sigma_t \Delta T H_2}{\mu \kappa}, \quad Ra = \frac{\beta \Delta T g H_2^3}{\nu \kappa}.$$

The dynamic and kinematic viscosities, the thermal diffusivity, and the thermal expansion coefficient are μ, ν, κ, β, respectively; g is the acceleration due to gravity, ΔT is the temperature difference, and σ_t is the absolute value of the surface tension gradient with temperature. The velocity boundary condition at the wall was the free-slip condition. The governing equations were solved by a finite difference method.

Experimental Setup

FIGURE 2A shows the schematics of a double layered system of two immiscible liquids. The experiments were made with fixed geometric parameters, except for the existence of the removable solid top plate of bakelite. The length of the cavity, L, was 20 mm and the heights, H_1 and H_2 were each 5 mm. The width of the system was 100 mm. We assumed that the flow was two-dimensional and unaffected by the walls and the thermal radiation from the side walls. The heater and a cooler were made of copper and the temperatures T_h and T_c were controlled by a wire heater and a temperature regulator bath. The temperature difference, $T_h - T_c$, was kept at 4.0°. The working fluids were silicone oil (KF96L-2cSt) and fluorinert (FC-40). Their thermophysical properties are well known and they are very stable. The surface tension gradient with temperature of the silicone oil was -0.067×10^{-3} N/(mK).[8] The interfacial tension between silicone oil and fluorinert was -0.044×10^{-3} N/(mK).[8]

Natural convective flow could be visualized inside the two-dimensional glass cavity with or without the bakelite top plate (i.e., without or with a free surface). The vertical cross-section of the system was illuminated by a light sheet of the second harmonic of YAG laser (double pulse system, 25 mJ/pulse). A high-speed and high-spatial resolution CMOS camera (1,024 × 1,024 pixels, fastcam-ultima1024, Photron Limited.) was used to capture images of tracer particles in the flow (i.e., 20.86 micrometers per pixel. The velocity field was calculated by using the PIV technique. The limitation of our measurement system was due to the frequency of a double-pulse, 15 Hz, whereas the camera could take a maximum of 1,600 pictures per second. In addition, it was a little difficult to measure the velocity near the interface/free surface due to reflection at the interfaces and the meniscus at the walls.

RESULTS AND DISCUSSION

Numerical Results

The values of the stream function are represented by contour lines in FIGURES 2B, C, and D. FIGURE 2B shows the results obtained without Marangoni convection at the interface and with a solid top plate. In this case, Ra was 7.8×10^4 and Ma was zero in the lower layer. As shown in FIGURE 2B, there was only a clockwise circulating flow in each layer. This pattern is a typical buoyancy driven flow calculated for the cavity of FIGURE 2A. The maximum velocity was faster in the upper layer due to the shear stress.

FIGURE 2C shows the numerical results of the flow field with the interfacial tension gradient and with the solid top plate. Ra and Ma in the lower layer were 7.8×10^4 and 6.25×10^3, respectively. The flow field was still dominated by the buoyancy driven convection under this condition. (If the horizontal length, L, is considered as a reference for these non-dimensional numbers, then Ra and Ma are 5.0×10^6 and 2.5×10^4, respectively.[9]) However, a couple of counterclockwise vortices were observed near the wall and the interface, suggesting an interfacial tension gradient. Thus, even in a flow field dominated by buoyancy driven convection, the interfacial Marangoni force can affect the flow in a two-liquid layer system. The interaction between Marangoni and buoyancy convections seems to cause the additional vortices in the flow field. In general, even a small vortex can induce instability of the flow. It is, therefore, very important to understand this instability in order to control the melt flow.

In practice, LEC and LEMZ systems have a free surface. A free surface is in general one of the most unstable boundary conditions of flow. In addition, Marangoni convection can occur especially under microgravity conditions. FIGURE 2D shows the results of a free surface with Marangoni convection. Ra and Ma values in the lower layer were same as in FIGURE 2C and the Ma value at the free surface was 10,823. The counterclockwise flow region is seen to narrow in FIGURE 2D. The flow velocity in the upper layer becomes much larger than that with the covered surface. The flow pattern shown in FIGURE 2D has also changed in the upper layer: the upward flow toward the free surface (arrow near the hot wall) and the downward flow from the free surface (arrow near the cold wall) sloped, while these were parallel to walls with the covered surface (arrows in FIG. 2C). These changes are caused by convection at the free surface, although the analysis includes many assumptions (e.g., a flat free surface/interface and a contact angle without meniscus).

Experimental Results

FIGURE 3A shows the measured velocities near the hot wall in the upper layer with the solid top plate. Due to reflection at the interface and at the solid plate, and due to the meniscus at the glass plate in front of the camera, it was impossible to visualize the flow very close to the interface. The velocities in the middle region in the cavity and near the wall were much larger than those near the interface. The large difference in velocities made it difficult to measure the flow near the interface due to a limited dynamic range of PIV the analyses. However, the rightward flow just above the interface and the counterclockwise circulating flow can be observed in FIGURE 3A, although the flow velocities near the contact point between the wall and the interface are quite different from the numerical results shown in FIGURE 2C.

The measured velocity map of the entire system with a free surface is shown in FIGURE 3B. The rightward flow just above the interface was observed near the hot wall in this case also. Another vortex flow near the cold wall shown in FIGURE 2D could not be observed in this case. The large meniscus at the hot/cold wall may have

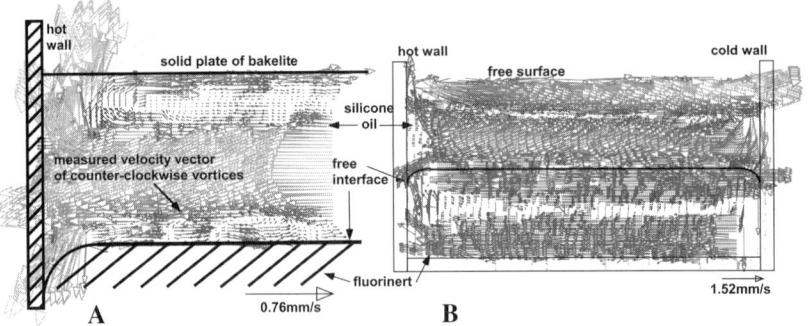

FIGURE 3. The visualized flow field in silicone oil layer near the hot wall with **(A)** a covered surface and **(B)** a free surface.

caused the different flow fields in the numerical and experimental studies. The upward flow toward the free surface near the hot wall was obviously different from that without a free surface.

FIGURE 4 summarizes the horizontal components of the measured and calculated velocities along the vertical axis at 6.0 mm from the hot wall, for a solid top plate and a free surface, respectively. As shown in FIGURE 4A, the measured velocities are in good agreement with the calculated values, except for those near the interface. The disagreement in this region may be caused by the complex and very slow flow, as well as by the intense reflection at the interface. The latter was particularly noted in the case of the solid non-transparent top plate, where the reflection of light at the interface was more intense than that with the free surface. The velocity near the interface gave relatively good agreement with the free surface, as shown in FIGURE 4B. The velocity distribution is a little different in the middle height region

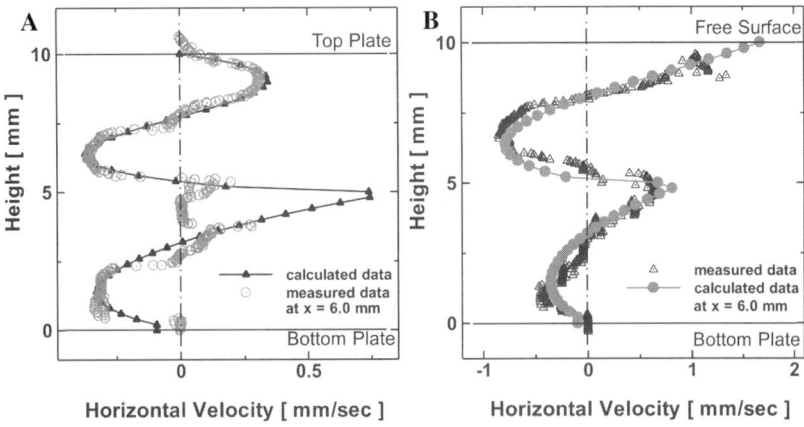

FIGURE 4. Measured and calculated horizontal velocities at $x = 6.0$ mm from the hot wall: **(A)** with the covered surface and **(B)** with a free surface.

of each layer. This may be due to the assumption of a flat free surface without a meniscus. The deformation of surface shape and capillary waves at the surface should be considered in future. The visualization and the PIV analyses may still have room for improvement. However, the absolute values of the measured velocities are quite consistent with calculated values and it is thought that the velocity field was measured successfully, for the first time, in the present experiments, with a two-immiscible layers system in a cavity.

Measurement of Free Surface Level Fluctuations

It is very difficult to simulate the flow in this system with a large temperature difference (i.e., large Ma and Ra values and large velocity gradients near the free surface/interface). To the best of our knowledge there is no report of a successful simulation of the combined convective flows. Here, we measured the velocity fields, especially in the upper layer, which is affected by Marangoni and buoyancy driven convection.

As a step toward understanding the characteristics of flow with a larger temperature difference between the hot and cold walls, we examined the free surface level fluctuation by using a laser focus level meter. The largest amplitude of level fluctuation at $x = 10.0$ mm in case with $\Delta T = 10.0$ degrees was about $15\,\mu m$, and its dominant frequency, calculated by FFT analysis, was 2.05 Hz. This suggests a three-dimensional flow system, since without three dimensions, the eigenfrequency of this system is about 6 Hz. The fluctuations were very complicated. The amplitudes were very slight and we could not rationalize the oscillation and its onset condition. However, we did find oscillations in this system, in the cavity with a large temperature difference. Future investigation of the oscillation are obviously important to understanding the flow in melts during a crystal growth process.

SUMMARY

Flow in two layers of immiscible liquids, with or without a free liquid–air surface, was investigated. Only in the presence of an interfacial tension gradient were a couple of counterclockwise vortices observed, both numerically and experimentally, in the upper layer, independently of the existence of the free surface. We believe that it is the first time that vortices were observed and measured under condition $Ra \gg Ma$, where the flow field was mainly dominated by the buoyancy driven convection. The velocity field with interfacial Marangoni convection was measured by using the PIV technique. The measured velocities gave good agreement with the numerical calculations. Preliminary data was also obtained on oscillations in the free surface.

Since accurate measurements of the flow field were very difficult to obtain in the large spatial velocity gradient of the present study, more reliable measurement techniques should be developed in order to investigate the flow field more carefully. However, the measured velocity field compared satisfactorily with the results to the numerical simulation, thus contributing to a better understanding of the phenomena involved.

REFERENCES

1. LIU, Q.S. & B. ROUX. 1994. Marangoni convection in immiscible double liquid layers. Micrograv. Sci. Tech. **VII/1:** 103–111.
2. AZUMA, H., S. YOSHIHARA, M. OHNISHI & T. DOI. 1991. Upper layer flow phenomena in two immiscible liquid layers subject to a horizontal temperature gradient. Microgravity Fluid Mechanics, IUTAM Symp. Bremen. 205–212.
3. JING, CH., T. SATO & N. IMAISHI. 1997. Rayleigh–Marangoni thermal instability in two-liquid layer systems. Micrograv. Sci. Tech. **X/1:** 21–28.
4. AR, Y., Z.M. TANG, J.H. HAN, *et al.* 1997. The measurement of azimuthal velocity field for oscillatory thermocapillary convection of floating half zone. Micrograv. Sci. Tech. **X/3:** 129–135.
5. CHENG, M. & S. KOU. 2000. Detecting temperature oscillation in a silicon liquid bridge. J. Crystal Growth **218:** 132–135.
6. LOPEZ, C.R., J.R. MILEHAM & R. ABBASCHIAN. 1999. Microgravity growth of GaSb single crystals by the liquid encapsulated melt zone (LEMZ) technique. J. Crystal Growth **200:** 1–12.
7. KIMURA, T., N. HEYA, M. TAKEUCHI & H. ISOMI. 1986. Natural convection heat transfer phenomena in an enclosure filled with two stratify fluids. JSME (B) **52:** 617–625. (Japanese)
8. OTSUBO, F., K. KUWAHARA & T. DOI. 2001. Effect of temperature and storage time on the surface tension of the silicone oil and fluorinert. J. Jap. Soc. Micrograv. Appl. **18**(1): 29–34.
9. BETHANCOURT, A.M., L.M. HASHIGUCHI, K. KUWAHARA & J.M. HYUN. 1999. Natural convection of a two layer fluid in a side-heated cavity. Int. J. Heat Mass Transfer **42:** 2427–2437.

Quantification of Myocardial Microcirculatory Function with X-ray CT

STEFAN MÖHLENKAMP,[a] LILACH O. LERMAN,[b] ŽELJKO BAJZER,[c] PATRICIA E. LUND,[d] AND ERIK L. RITMAN[d]

[a]*Department of Cardiology, University Clinic Essen, Essen, Germany*

[b]*Department of Internal Medicine, Division of Hypertension, Mayo Clinic, Rochester, Minnesota, USA*

[c]*Department of Biochemistry & Molecular Biology, Mayo Clinic, Rochester, Minnesota, USA*

[d]*Department of Physiology & Biophysics, Mayo Clinic, Rochester, Minnesota, USA*

ABSTRACT: A mathematical model of the intramyocardial coronary microcirculation is used to explore the validity of a fast CT imaging method for characterizing the myocardial microcirculatory functional status. The fast CT method depends on the demonstrated CT-based estimation of myocardial perfusion (F) and the intramyocardial blood volume (B_v). The observed curvilinear myocardial blood volume-to-flow relationship, empirically fitted to $B_v = a \cdot F + b \cdot F^{0.5}$, is a signature of the underlying early pathophysiologic processes thought to be involved in systemic disease processes, such as atherosclerosis, hypertension, and diabetes mellitus. The sensitivity and specificity of the CT-based estimate of this characteristic relationship is explored by altering the characteristics of the vascular diameter-to-flow relationship and the variation in the fraction of capillaries perfused at different coronary flows. The simulation results also indicate that if the vascular diameters change so that the vascular resistance corresponds to the change in flow, then the empirically observed myocardial B_v-to-F relationship holds well.

KEYWORDS: microcirculation; CT; myocardium; blood volume; flow

INTRODUCTION

Coronary arterial disease is currently diagnosed and treated primarily on the basis of its impact on the large diameter epicardial arteries. A structural change, usually a localized narrowing (stenosis) of a coronary artery lumen, is generally detected and quantified by selective coronary arteriography. However, by the time that the epicardial arteries result in reduced myocardial perfusion reserve, it is generally too late to arrest (much less reverse) the disease process in that artery in a clinically meaningful manner. Since coronary artery atherosclerosis affects the intramyocardial arterial microcirculation early in the disease process, well before the epicardial vessels are hemodynamically compromised, it makes sense to characterize the function of

Address for correspondence: E.L. Ritman, M.D., Ph.D., Department of Physiology and Biophysics, Mayo Clinic, 200 First Street SW, Rochester, MN 55905, USA.
elran@mayo.edu

the intramyocardial microcirculation as a bellwether of the presymptomatic disease process.[1,2] Moreover, once detected, noninvasive treatment for the arrest and reversal of the microvascular disease process would appear to be more likely to be effective in a clinically meaningful manner. For this reason we are exploring a method to quantify intramyocardial microcirculatory function in a minimally invasive manner.

The intramyocardial microcirculation consists of arteries that are up to 1.0 mm in lumen diameter and progressively branch to the 5 μm diameter capillaries. Unfortunately these small microvessels cannot be individually visualized by clinically applicable imaging methods. Consequently, indirect estimates of vessel diameters must be made. We propose to do this from the intramyocardial intravascular blood volume—a value that we obtain from the analysis of clinical whole-body CT images using intravascular contrast medium dilution curves. In this study we use a mathematical simulation of the myocardial microcirculation, with dimensions based on published experimental data. The goal is to evaluate (1) if the dilution curves of the intramyocardial microcirculation provide the needed blood volume and flow data with adequate sensitivity and specificity to permit a meaningful analysis, and (2) the extent to which changes in the integrated intravascular volume of all the intramyocardial microcirculation reflects altered diameters of the *large* or *small* microvessels.

METHODS

The CT Scanner

An Imatron (C-150, South San Francisco CA) electron beam computed tomography (EBCT) scanner[3] was used. The scan sequence was repeated every, or every other, heart cycle (i.e., approximately at 1 second intervals) for up to 40 pairs of CT images that were recorded over a period of one breathhold (i.e., approximately 40 seconds, depending on heart rate). The CT image computed from these scans was generally of an 18 cm diameter field centered on the heart with pixels $0.5 \times 0.5\,\text{mm}^2$ in area and a 7 mm slice thickness (i.e., voxel volume $1.75\,\text{mm}^3$).

Animal Experiments

As described in detail elsewhere,[4,5] three-month old pigs (approximately 30 kg) were anesthetized and placed in the scanner so that the imaged planes were at right angles to the long axis of the atrioventricular valve plane and located between the apex and base of the left ventricle. The scan was started at the same time as the start of the injection of a 0.33 mL/kg bolus of nonionic contrast agent (iohexol 350 mg iodine/mL) over a two second period into the right atrium. The scans were repeated, at 15-minute intervals, during infusion of adenosine solution (at selected delivery rates) through a 2F catheter into one of the coronary arteries. These different rates of infusion resulted in proportionately increased myocardial blood flow and blood volume.

CT Image Analysis

A region of interest was "outlined" on the imaged myocardium and over the left ventricular chamber. The average grey scale value within each region of interest was then calculated. This was applied for each sequential CT image, so that a dilution

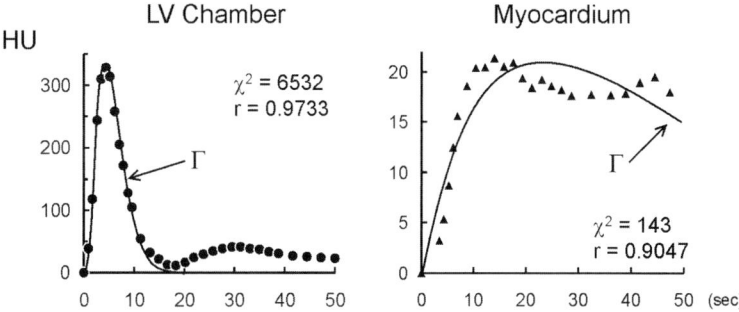

FIGURE 1. Typical contrast dilution curves measured over the LV chamber (input) and myocardium.

curve could be generated from this sequence of average grey scale values. The curve over the left ventricular (LV) chamber was used as the "input" curve. As illustrated in FIGURE 1, each time series of CT image-derived contrast dilution curves was fitted with a gamma variate curve. From these fitted curves, intramyocardial blood volume (B_v, mL blood/cm^3 myocardium) and perfusion (F, mL/cm^3 myocardium/minute) were calculated as described in detail elsewhere.[6–8]

The gamma variate fit (and the χ^2 and correlation coefficients) show excellent correspondence for the LV curve (excluding the recirculation of greater than 20 seconds) but much poorer fit in the myocardium. This is because the measured myocardial curve is a superposition of the intravascular, extravascular (due to permeability of endothelium), and recirculation of the contrast agent. To obtain the intravascular component, we assume that it and the extravascular curves are also gamma variates.[7]

MODELING MICROVASCULAR BLOOD VOLUME DISTRIBUTION: THE DYE DILUTION CURVE

We assume that a region of interest within the myocardium has representation of all the generations of the microvasculature in proportion to their numbers in a single tree.

Branching Geometry Assumptions

The arterial vascular tree is modeled by dichotomous branching: the mother branch has two equal daughter branches. Their diameters are governed by Murray's law:[9–11] $D_{mo}^3 = 2D_d^3$, where D_{mo} is the diameter of the mother branch and D_d is the diameter of daughter branches. For the vascular tree with n levels of branching, one obtains the recursion $D_{n+1} = 2^{-1/3}D_n$, which implies

$$D_n = 2^{-n/3}D_0, \quad n = 0, 1, 2, ..., N, \tag{1}$$

where D_0 is the diameter of the *root* vascular segment, assumed to be 1.0 mm. For $N = 17$ it follows that $D \approx 20\,\mu\text{m}$, which is the diameter of a terminal arteriole that branches into many capillaries.

Flow in each daughter branch is half of the flow in the mother branch, $F_d = F_{mo}/2$. The length L_n of a segment at the nth level of branching is, as shown elsewhere,[12] $L_n = kD_n^\mu$, $\mu \approx 1$ and $k \approx 10$. The length is fixed for all flows at the value computed at maximum flow (see below). The proportionality of length to diameter is a consequence of Murray's law and Poiseuille's formula for viscous flow. Indeed, according to Poiseuille, the flow is given by

$$F = \frac{\pi D^4 \Delta p}{2^7 \eta L}, \qquad (2)$$

where D is diameter of a tube, L its length, Δp the pressure drop, and η the viscosity. Applying $2F_d = F_{mo}$ Equation (2) yields $D_{mo}^4/L_{mo} = D_d/L_d$, if the pressure drop is considered constant. Murray's law then implies $D_{mo}/L_{mo} = D_d/L_d$, which means that the length of a vessel along the tree is proportional to its diameter.

Each terminal artery branches into m identical capillaries of which m_o capillaries are open to flow: $m = 30$, at rest only 20% of capillaries receive flow;[13,14] that is, $m/5 \leq m_o \leq m$. Diameter and the length of capillaries[15] are, respectively, $D_c = 0.005$ mm and $L = 1.0$ mm. The venous tree is a mirror image of the arterial tree; it progressively coalesces until a single branch segment is 1.0 mm in diameter.

For the purposes of this model, the blood flow is considered time-independent, but dependent on the number of open capillaries $F = (m_o/m)F_m$, where the maximal flow F_m (when all m capillaries are open) is estimated to be 6.55 mL/min. This is derived from the deduction[16] that each terminal arteriole perfuses 0.01 mm³ of tissue and that maximum myocardial perfusion is approximately 5 mL/cm³/min of myocardium/minute. Diameters of arteries and veins change with the flow according to the equation

$$\frac{F}{F_m} = \frac{D_n^4(m_o)}{D_n^4} = \frac{m_o}{m}, \quad D_n \equiv D_n(m). \qquad (3)$$

This relation can be derived from the Poiseuille's formula for flow under the assumption that only diameter, but not the length of vessels, changes with the flow as m_o changes, and that the pressure drop is not affected by the flow change. The size of the capillaries does not change with the flow.

Based on these assumptions, the volume V_n of a branch at level n of branching is

$$V_n = k\pi D_0^3 2^{-n-2} \left(\frac{m_o}{m}\right)^{1/2}.$$

The volume of m_o open capillaries is

$$V_c = \pi L D_c^2 2^{N-2} m_o,$$

and the volume of the whole vascular tree (note that B_v is V within a known volume of myocardium) is then

$$V = 2\sum_{n=0}^{N} 2^n V_n + V_c = \pi\left[kD_0^3 \left(\frac{m_o}{m}\right)^{1/2} \frac{N+1}{2} + LD_c^2 2^{N-2} m_o\right] \qquad (4)$$

or, as a function of blood flow in the whole myocardium,

$$V = aF + bF^{1/2}, \qquad (5)$$

where the numerical constants are $a = 0.0118$ min, $b = 0.1133$ [mL/min/cm³]$^{1/2}$, when the flow is given in mL/min/cm³ of myocardium.

Under the condition of constant flow, the transit time through a branch for the nth level is

$$t_n = \frac{2^n V_n}{F} = \frac{k\pi D_0^3}{4F_m}\left(\frac{m}{m_o}\right)^{1/2}, \quad (6)$$

and the total transit time for the whole vascular tree is given by

$$\bar{t} = 2\sum_{n=0}^{N} t_n + t_c = 2Km_o^{-1/2}(N+1) + t_c = \frac{V}{F}, \quad (7)$$

with $t_c = L\pi D_c^2 2^{N-2} m/F_m = 0.354 \sec$ and $K = \pi k D_0^3 m^{1/2}/4F_m$.

Indicator Dilution Response of the Model

We assume that the probability density $f_n(t)$ that a contrast particle, which entered the nth level branch at zero time, will leave this branch at time t is known and is equal for all level n branches. The output $O_n(t)$ from a branch (the amount of contrast leaving per unit of time) at the nth level can be modeled by the linear response model, that is,

$$O_n(t) = [f_n \circ I_{n-1}](t) = \int_0^t f_n(t-x)I_{n-1}(x)dx = \int_0^t f_n(x)I_{n-1}(t-x)dx, \quad (8)$$

where I_{n-1} is the input to the branch (the amount of contrast entering the branch per unit of time). Because of dichotomous branching I_{n-1} is half of the output from the mother branch, $I_{n-1}(t) = O_{n-1}(t)/2$, and, therefore, (7) yields the recursion formula

$$O_n = f_n \circ O_{n-1}, \quad n = 1, 2, \ldots, N, \quad O_0 = f_0 \circ I, \quad (9)$$

where $I(t)$ is the input to the root of the tree. By applying this recursion to the whole vascular tree (arteries, capillaries, and veins) one obtains the output $O(t)$ from the largest vein considered

$$O = f \circ I, \quad f = f_0 \circ f_0 \circ f_1 \circ \ldots \circ f_N \circ f_c \circ f_N \circ \ldots \circ f_1 \circ f_0, \quad (10)$$

where $f_c(t)$ is the transit time probability density for capillaries. The amount of contrast $C(t)$ within the region of the whole vascular tree at time t is given by the equation

$$\frac{dC(t)}{dt} = I(t) - Q(t) = I(t) - [f \circ I](t). \quad (11)$$

By taking the Laplace transform[17] of this expression, one obtains, with $C(0) = 0$,

$$\mathcal{L}[C](s) = \frac{1 - \mathcal{L}[f](s)}{s}\mathcal{L}[I](s), \quad (12)$$

where, because of Equation (9) and by means of simplifying notation $\mathcal{L}[g](s) \equiv \tilde{g}(s)$, one obtains

$$\tilde{f} = \tilde{f}_c \tilde{f}_0^2 \tilde{f}_1^2 \ldots \tilde{f}_N^2 \quad (13)$$

To make this model analytically solvable, the simplest assumption is that the contrast in vascular branches is carried with blood in such a way that all contrast particles have the same transit time t_n given by Equation (4). Then the probability density is $f_n(t) = \delta(t - t_n)$ and similarly, for capillaries $f_c(t) = \delta(t - t_c)$. In this case $\tilde{f}_n(s) = e^{-t_n s}$, $\tilde{f}_c(s) = e^{-t_c s}$, and one obtains $\tilde{f}(s) = e^{\bar{t}s}$. The use of well known properties of the Laplace transform then yields

$$C(t) = \int_0^t I(x)dx - u(t-\bar{t})\int_0^{(t-\bar{t})} I(x)dx, \qquad (14)$$

where $u(t)$ is the Heaviside unit step function. The first moment of $C(t)$ can be expressed through Laplace transforms and related to the mean transit time $\bar{t}$ and to the first moment of the input function.

$$M_1[C] = \frac{\int_0^\infty tC(t)dt}{\int_0^\infty C(t)dt} = -\left[\frac{d\tilde{C}(s)}{ds}\frac{}{\tilde{C}(s)}\right]_{s=0} = \frac{\bar{t}}{2} + M_1[I]. \qquad (15)$$

When the input function is given by the gamma variate,[18] $I(t) = at^b e^{-ct}$, one obtains

$$C(t) = ac^{-b-1}[\gamma(b+1, ct) - u(t-\bar{t})\gamma(b+1, c(t-\bar{t}))], \qquad (16)$$

where $\gamma(y,t)$ is the incomplete gamma function.

RESULTS

CT Image Capacity to Measure Iodine Concentration

A fundamental aspect of CT-based dilution curves is the linearity of the relationship of the curve to the amount and rate of contrast medium injected. In a previous study,[19] performed under virtually identical conditions as our current studies, the iodine (the radiopaque component of the contrast medium) concentration less than 100 mg iodine/cubic centimeter, the CT number changed linearly with iodine concentration.

The Characteristics of the CT Dilution Curves

Using a gamma variate curve (time to peak 1.5 sec and first moment of 3 sec) as the input to the simulated myocardial circulation, the simulation shows that the myocardial curve is close to a gamma variate curve, which at maximum flow has a time to peak of 1.9 seconds and a first moment of 3.34 seconds. The first moment of the gamma variate fitted to the simulation curve (Y) relates to the first moment of the actual computed curve (X) as $Y = 1.01X - 0.26$ with $R^2 = 0.9997$ over the range of resting to maximum flow and $L = 1$ mm, but $Y = 0.83 + 0.37$, $R^2 = 0.995$ for $L = 0.5$ mm.

To prepare the CT dilution curves for analysis we assume that the left ventricular CT dilution curves is well represented by a gamma variate curve.[20] The validity of this consumption is illustrated in FIGURE 1. However, as also illustrated in FIGURE 1, the myocardial dilution curve is not well represented by a single gamma variate curve because of the concurrent extravascular accumulation, and washout, of some of the contrast agent as well as the much less significant, delayed, recirculation effects. However, the intramyocardial intravascular curve could be extracted.[8]

As illustrated in FIGURE 2, the intramyocardial, intravascular, dilution curve is well represented by a gamma variate curve is indicated by our simulation, and it is well represented by a *single* gamma variate curve.

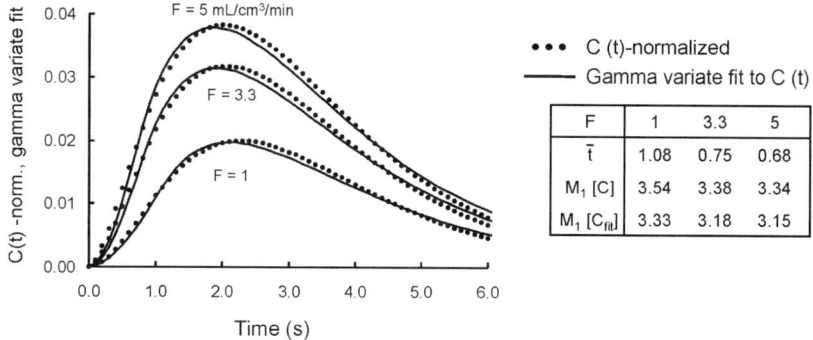

FIGURE 2. A dilution curve over the simulated intramyocardial vascular tree is close to a gamma variate curve when the input curve is a gamma variate. The three curves show the intravascular curve at three levels of myocardial perfusion: maximum, medium, and resting (i.e., 20% of maximum) flow.

FIGURE 3 shows the similarity of the model-based compared to the experimental blood volume-to-flow relationships characteristics.

The model-based blood volume-to-flow relationship shows characteristics similar to those obtained from experimental data. The reason(s) for the model's under estimation of B_v for a given F is probably primarily due to the choice of microcirculatory dimensions, which are based on average values reported in the literature.

DISCUSSION

A previously developed, catheter sampling, indicator dilution theory[21] was applied here to imaging-based dilution curves to quantify local myocardial perfusion

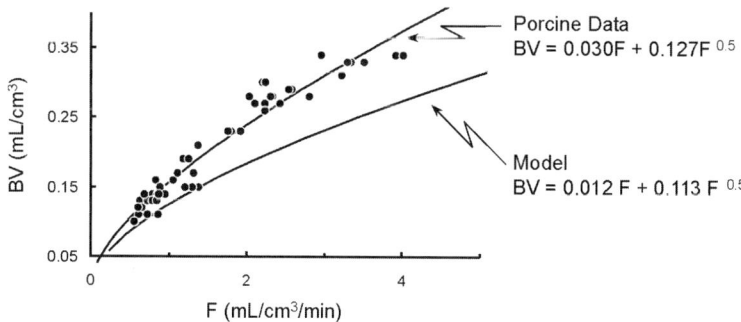

FIGURE 3. A typical myocardial blood volume-to-flow relationship generated from repeated scans performed at different levels of adenosine infusion into a coronary artery and the same relationship computed for our model.

and intravascular blood volume in multiple locations within the heart wall. In principle, this involves estimating intramyocardial intravascular blood volume (B_v) and mean transit time (MTT) of blood flowing through the B_v in that region of interest. Blood flow is then calculated as $F = B_v/\text{MTT}$. There are a number of problems in the application of catheter-based theory[22,23] to the imaging approach. In addition, there are a number of practical issues in using contrast medium as an intravascular indicator. Thus, we know in the catheter-based method what the amount of indicator entering the organ is, whereas for the imaging application this is only true if all the indicator has entered the imaged region of interest before any indicator leaves that region of interest.[24,25] Moreover, the concentration of indicator is measured in catheter-based methods by flow-sampling (i.e., effectively all of the blood flowing out of the organ is acquired), whereas in imaging we have cross-sectional sampling, which has the problem that in the absence of plug flow in an artery the slowest flow results in disproportionate measurement of the indicator. Because the indicator theory generally assumes steady flow, the impact of cardiogenic periodic flow can best be overcome by having the duration of the indicator dilution curve be a much longer duration than the cardiac cycle duration.[26] This is counterproductive to the need for the entire bolus of indicator entering the region of interest before any leaves it.

SUMMARY

Several attempts at modeling the microvascular branching geometry as a means to providing insight into blood flow distribution,[27] blood pressure distribution,[28] and X-ray based indicator dilution curves[29] have been reported. Our simulation has much in common with these models but differs in that it aims to provide insight into the impact of changes in the vessel dimensions and capillary recruitment on the observed X-ray CT contrast dilution curves and if this impact is specific and unique for that functional change of vessel diameters. Our model is relatively simplistic in its assumptions relative to some[30,31] in that it assumes steady flow, whereas it is periodic with reflections and phase lags occurring within the vessels, and the vessels are viscoelastic rather than rigid. Our model suggests that dilution curves provide the needed B_v and F data to permit a meaningful analysis of the intramyocardial microcirculation, and that changes in the integrated volume of the intramyocardial microcirculation reflect altered diameters of small microvessels.

REFERENCES

1. MARCUS, M.L., W.M. CHILIAN, H. KANATSUKA, et al. 1990. Understanding the coronary circulation through studies at the microvascular level. Circulation **82:** 1–7.
2. NASERI, A., K.F. CREA & D. CRAINFLONE. 1992. Myocardial ischemia caused by distal coronary vasoconstriction. Am. J. Cardiol. **7:** 1602–1605.
3. BOYD, D.B. & M.J. LIPTON. 1983. Cardiac computed tomography. Proc. IEEE **71:** 298–307.
4. MOHLENKAMP, S., L.O. LERMAN, A. LERMAN, et al. 2000. Minimally invasive evaluation of coronary microvascular function by electron beam computed tomography. Circulation **102:** 2411–2416.

5. MOHLENKAMP, S., T.R. BEHRENBECK, L.O. LERMAN, et al. 2001. Quantitation of long-term changes in coronary microvascular functional reserve using electron beam tomography—preliminary results in a porcine model. Radiology **221**: 229–236.
6. WANG, T., S. WU, N. CHUNG & E.L. RITMAN. 1989. Myocardial blood flow estimated by synchronous, multislice, high-speed computed tomography. IEEE. Trans. Med. Imaging **8**: 70–77.
7. LIU, Y.-H., R.C. BAHN & E.L. RITMAN. 1992. Dynamic intramyocardial blood volume: evaluation with a radiological opaque marker method. Am. J. Physiol. **263**: H963–H967.
8. LERMAN, L.O., S. SIRIPORNPITAK, N. LUNA-MAFFEI, et al. 1999. Measurement of in vivo myocardial microcirculatory function with electron beam CT. J. Comp. Assist. Tomogr. **23**: 390–398.
9. MURRAY, C.D. 1926. The physiological principle of minimum work: I. The vascular system and the cost of blood volume. Proc. Nat. Acad. Sci. **12**: 207–214.
10. GRIFFITH, T.M. & D.H. EDWARDS. 1990. Basal EDRF activity helps to keep the geometrical configuration of arterial bifurcations close to Murray optimum. J. Theor. Biol. **146**: 545–573.
11. MAYROWITZ, H.N. & J. ROY. 1983. Microvascular flow: evidence indicating a entire dependence on arteriolar diameter. Am. J. Physiol. (Heart Circ. Physiol.) **245**: H1031–H1038.
12. KASSAB, G.S., C.A. RIDER, N.J. TANG & Y.B. FUNG. 1993. Morphometry of pig coronary arterial trees. Am. J. Physiol. (Heart Circ. Physiol) **265**: H350–H365.
13. GROVER, G.J., M. ROSOLOWSKI, J.M. KEDEM & H.R. WEISS. 1986. Effect of hypoxia and adenosine on the mivcrovascular reserve in the rabbit heart. Microcirc. Endothelium Lymphatics **3**: 359–382.
14. HENQUELL, L. & C.R. HONIG. 1976. Intercapillary distances and capillary reserve in right and left ventricles: significance for control of tissue pO_2. Microvas. Cardiol. Res. **12**: 35–41.
15. RAKUSAN, K. 1971. Quantitative morphology of capillaries in the heart. Number of capillaries in animal and human hearts under normal and pathological conditions. Method. Archiev. Exp. Pathol. **5**: 272–281.
16. BASSINGTHWAIGHTE, J.B., R.B. KING & S.A. ROGER. 1989. Fractal nature of regional myocardial blood flow heterogeneity. Circ. Res. **65**: 578–590.
17. RAINVILLE, E.D. 1963. The Laplace Transform. An Introduction. MacMillan Mathematics Paperback. London.
18. KORN, G.A. & T.M. KORN. 1961. Mathematical Handbook for Scientists and Engineers. 106. McGraw-Hill New York.
19. RITMAN, E.L. 1994. Scan timing considerations in indicator dilution analysis of dynamic CT image sequences. Proc. SPIE. **2168**: 2–12.
20. THOMPSON, H.K., JR., C.F. STARMER, R.E. WHALEN & H.D. MCINTOSCH. 1964. Indicator transit time considered as a gamma variate. Circ. Res. **14**: 502–515.
21. MEIER, P. & K.L. ZIERLER. 1954. A theory of indicator dilution method for measurement of blood flow and volume. J. Appl. Physiol. **6**: 731.
22. VISCHER, M.B. & J.A. JOHNSON. 1953. The Fick principle: analysis of potential errors in its conventional application. J. Appl. Physiol. **5**: 635.
23. STOW, R.W. 1954. Systematic errors in flow determination by the fick method. Minnesota Med. **37**: 30–35.
24. MULLANI, N.A. & K.L. GOULD. 1983. First-pass measurements of regional blood flow with external detectors. J. Nucl. Med. **24**: 577–581.
25. CLOUGH, A.V., A.W. MANUEL, S.T. HAWORTH & C.A. DAWSON. 2000. Application of indicator dilution theory to time-density curves obtained from dynamic contrast images. Proc. SPIE **3978**: 457–465.
26. BASSINGTHWAIGHTE, J.B., R.E. STURM & E.H. WOOD. 1970. Advances in indicator dilution techniques applicable to studies of the acutely ill patient. Mayo. Clin. Proc. **45**: 563–572.
27. VAN BEEK, J.H.G.M., S.A. ROGER & J.B. BASSINGTHWAIGHTE. 1989. Regional myocardial flow heterogeneity explained by fractal networks. Am. J. Physiol. **257**(Heart Circ. Physiol 26): H1670–H1680.

28. CHILIAN, W.M., S.M. LAYNE, E.C. KLAUSNER, et al. 1989. Redistribution of coronary microvascular resistance produced by dipyridamole. Am. J. Physiol. **256**(Heart Circ. Physiol. 25): H383–H390.
29. CLOUGH, A.V., J.H. LINEHAN & C.A. DAWSON. 1997. Regional perfusion parameters from pulmonary microfocal angiograms. Am. J. Physiol. **272**: H1537–H1548.
30. STETTLER, J.C., P. NIEDERER & M. ANLIKER. 1987. Nonlinear mathematical models of the arterial system: effects of bifurcations, wall viscoelasticity, stenoses and counter pulsation on pressure and flow pulses. R. Skalak, S. Chien, Eds.: Handbook of BioEngineering, Chpt. 17, pp. 17.1-17.26. McGraw Hill Book Company, New York.
31. KASSAB, G.S., J. BERKLEY & Y.-C.B. FUNG. 1997. Analysis of pig's coronary arterial blood flow the detailed anatomical data. Ann. Biomed. Engr. **25**: 204–217.

Renal Handling of X-ray Contrast Media

Imaging and Exploration with Electron Beam CT

ANDREW D. RULE,[a] ŽELJKO BAJZER,[b] ERIK L. RITMAN,[c] AND LILACH O. LERMAN[d]

[a]*Department of Internal Medicine, Mayo Clinic, Rochester, Minnesota, USA*

[b]*Department of Biochemistry and Molecular Biology, Biomathematics Resource, Mayo Clinic, Rochester, Minnesota, USA*

[c]*Department of Physiology and Biophysics, Mayo Clinic, Rochester, Minnesota, USA*

[d]*Division of Hypertension, Mayo Clinic, Rochester, Minnesota, USA*

ABSTRACT: Physiologic changes in renal hemodynamics and function reflect its role as a regulatory organ in maintaining homeostasis, whereas other alterations may mirror development of renal injury and often precede overt signs of morphologic changes. Furthermore, intrarenal alterations may be discreet and manifest only in the renal cortical, medullary, or papillary zones. The high spatial and temporal resolution of electron-beam computed tomography enables external detection and quantification of cortical, medullary, and papillary tissue density changes following an intravenous bolus injection of X-ray contrast media. These changes reflect flow of contrast media in these renal zones through the successive renal vascular, glomerular, and tubular compartments that can be individually plotted as time–density curves (TDC). Mathematical modeling then allows calculation of unique parameters of renal function from these TDC. This ability to quantify renal regional attributes may not only shed light on the physiologic mechanisms that the kidney controls, but also assist in detecting subtle impairment in its function.

KEYWORDS: kidney; X-ray contrast media; renal function; electron beam; CT; blood flow

PHYSICS AND PHYSIOLOGY

The kidney plays a pivotal role in physiologic processes, such as regulation of blood volume and arterial blood pressure, and it is a vulnerable target organ to common systemic diseases like hypertension and hyperlipidemia. Because of the involvement of the kidney in homeostasis and the potential ramifications of renal impairment, diverse techniques have been developed and used for measurement of renal hemodynamic and functional attributes, including renal perfusion and renal blood flow (RBF), glomerular filtration rate (GFR), and the tubular reabsorption processes. However, no single technique has reliably quantified these characteristics simultaneously and non-invasively across a wide range of physiologic conditions,

Address for correspondence: Lilach O. Lerman, M.D., Ph.D., Division of Hypertension, Mayo Clinic, 200 First Street SW, Rochester, MN 55905, USA.
  lermanlilach@mayo.edu

or in a manner readily applicable in humans. Furthermore, the region specificity of intrarenal structure and function has made evident the need to study them separately in various regions of the intact kidney, such as the renal cortex and medulla (see FIGURE 1). The intermingling and often parallel vascular and tubular compartments contribute to the spatial and functional complexity of the kidney, and make a distinction between them challenging. This has been facilitated by X-ray techniques incorporated into clinical and experimental medicine, which provide potential insight into regional renal function. Such techniques were initially introduced in the 1920s with excretory urography, in which sequential projection images of the kidney were recorded over several minutes to time contrast media excretion.[1] From this evolved high resolution, cross-sectional CT techniques that dynamically record sub-second changes in regional renal density and eliminate superimposition of intrarenal structures.[2]

In particular, electron beam computed tomography (EBCT; Imatron Inc., South San Francisco, CA), an ultra-fast scanner, provides reliable,[3,4] reproducible,[3,5] and non-invasive quantifications of single kidney volume,[6] perfusion,[4] GFR,[3,7] and segmental tubular function[3,8,9] that are difficult to obtain with comparable spatial and temporal resolution using other techniques. The high temporal resolution of EBCT (up to 17 frames/sec at 50 msec/image) enables registration of dynamic tissue density changes during the transit of a contrast media bolus and, thus, provides an opportunity to investigate renal functional traits.

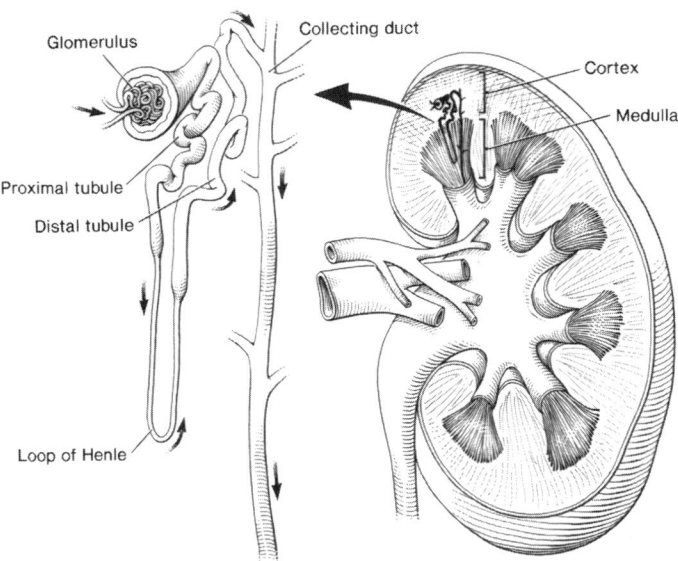

FIGURE 1. Schematic section of the kidney, illustrating the cortex and the medullary pyramids, as well as the basic functional unit of the kidney, the nephron. The nephron is shown enlarged on the left. The glomeruli, proximal tubules, and distal tubules are located primarily in the cortex. The *arrows* indicate the direction of tubular fluid flow along the nephron as water, electrolytes, and other substances are reabsorbed or secreted during the process of urine formation.

Exploration of intrarenal hemodynamics and function using CT or other X-ray techniques is facilitated by the manner with which the kidney handles the X-ray contrast media. The urographic contrast media commonly used for cardiovascular x-ray imaging are iodinated soluble compounds with small molecular size. Under physiologic conditions, 15–30% of an injected contrast media bolus may leak into the extravascular compartment during its first pass through the vasculature.[10,11] Furthermore, their clearance from the body is achieved primarily via renal glomerular filtration in a manner similar to inulin—that is, they are neither secreted nor reabsorbed in the renal tubules.[2] Hence, the determination of their elimination rate can be used to estimate the rate of glomerular filtration.

Contrast media can be given as a single bolus from which the first pass through the kidney can be temporally observed. The contrast-filled blood initially transverses the renal vasculature, inducing a marked increased in renal tissue x-ray density. The majority of the bolus then washes out through the renal vein with about 20% of the contrast bolus undergoing filtration in the glomerular capillaries and accumulating in the proximal tubule,[8] a nephron segment contained in the renal cortex. This results in residual tissue density that remains in the cortex following the vascular phase and venous washout of the contrast media.[12] Using EBCT we have shown that subsequent flow of the contrast-filled filtrate along the cortical and medullary nephron segments (FIG. 1) leads to distinct changes in the cortical and medullary X-ray densities (see FIGURE 2). In addition, concomitant reabsorption or secretion of intratubular fluid as it flows along the nephron contributes to changes in fluid concentration or dilution, which thereby determines intratubular contrast concentration (ITC).

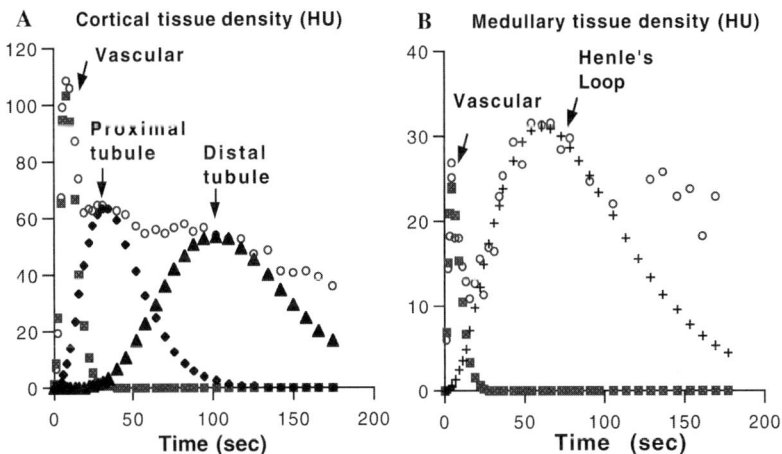

FIGURE 2. Time density curves obtained with EBCT from (**A**) the swine renal cortex and (**B**) medulla after a central venous bolus injection of contrast media. The labels and arrows indicate the tubular location of the bolus during its transit along the nephron. Note that the scales of the TDC obtained from the two regions are different, because the cortex receives the majority of renal blood flow. *HU*, Hounsfield units (CT numbers); ○, measured tissue density (raw data); ■, intravascular transit; ♦, proximal tubule; ▲, distal tubule; +, Loop of Henle.

Because of the linear relationship between contrast media concentration and CT-derived tissue attenuation, measurements of tissue density reflect the tissue concentration of contrast media contained in the renal vasculature, interstitium, and/or tubular fluid. These measurements can be normalized by the concurrent measurement of the contrast density in the abdominal aorta (that serves as an input function), and subsequently provide absolute quantifications of contrast flow or concentration per unit tissue volume. External detection of the spatial and temporal course of these density changes allows generation of renal TDC, from which vascular, glomerular, and tubular flow can be quantified using appropriate models.

MODELING RENAL TRANSIT OF CONTRAST MEDIA

The renal TDCs that result from contrast media flow are partially overlapping and require mathematical stripping. The reasons for this include dispersion of the bolus following intravenous injection and during vascular and tubular flow, the spatially intermixed vascular and tubular compartments, and the heterogeneity in the lengths of the tubular segments, particularly the medullary Loops of Henle. To enable this stripping we have recently developed mathematical models that allow separate delineation of renal vascular and tubular fluid flow in individual nephron segments.[3,7] This approach permits calculation of curve parameters for each segment to be used for assessment of regional perfusion, ITC, tubular transit times (represented by the first moment of the curve), and GFR. These models represent extensions of the standard gamma-variate curve-fitting algorithm[13] that are sequentially applied to each raw-data TDC in accordance with the region of interest from which the data was originally obtained (e.g., cortex, medulla, or papilla).

The renal cortical TDC typically exhibits three sequential peaks (FIG. 2A) corresponding to the distribution pattern of the contrast bolus in the cortical vascular compartment, proximal tubule, and distal tubule, respectively.[7] These data can, therefore, be fitted with a tricompartmental curve-fitting algorithm (vascular, proximal tubular, and distal tubular)[3,7]

$$C(t) = ct^a \{e^{-t/b} + de^{-t/h}\} + it^j e^{-t/k}, \qquad (1)$$

where $C(t)$ is the contrast concentration as a function of time t (sec), and the remaining constants are curve-fitting parameters describing the magnitude and spread of the curve. Equation (1) assumes dependence of the proximal tubular on the vascular term, since the contrast medium accumulating in the proximal tubule is filtered in the glomerular capillaries and is, therefore, directly derived from vascular transit. Distal tubular flow of tubular fluid, on the other hand, is significantly delayed relative to the first two compartments, and is modeled by an independent gamma function. In addition, these data can also be similarly fitted with a model that represents the sum of three independent gamma functions, where $p \approx a$:

$$C(t) = ct^a e^{-t/b} + dt^p e^{-t/h} + it^j e^{-t/k}. \qquad (2)$$

The advantage of Equation (2) is that it does not make any mathematical assumption about the relationship between the vascular and proximal compartments. On the other hand, the addition of a parameter (p) makes curve fitting computationally more demanding.

The medullary TDC exhibit two peaks (FIG. 2B), corresponding to the transit of the contrast medium in the vascular compartment, followed by independent tubular fluid arrival via the Loop of Henle. The model applied to the medullary TDC is, therefore, bicompartmental[3,7]

$$C(t) = ct^a e^{-t/b} + it^j e^{-t/k}. \quad (3)$$

The papillary TDC (which is derived from the tip of the medullary pyramid draining into a calyx) resembles the medullary TDC, and is treated similarly. However, the second papillary peak results from the transit of the contrast in the collecting duct, and is therefore delayed relatively to the second medullary peak (Loop of Henle).

The aortic input TDC takes the form of a simple gamma-variate curve describing a contrast agent bolus through a vessel and has been well described in the literature.[13]

The data and the curve-fit parameters are used to define various properties of the fitted gamma-variate curves for each peak detected in the TDC observed in these regions of interest. Thus, for a gamma-variate component in the form $f(t) = ct^a e^{-t/b}$, the first moment that estimates the transit time is given by $(a + 1)b$, the area under the curve is $c\Gamma(a + 1)b^{a+1}$, the peak height is given by $c(ab/e)^a$, and the maximal ascending slope is $c(sb)^{a-1}e^s a^{-1/2}$ with $s = a(1 - a^{-1/2})$.[13] These properties are then used to estimate various measures of renal function. The vascular gamma-variate curve in each region (FIG. 2) is used to calculate regional (e.g., cortical, medullary, and papillary) perfusion (mL/min/cc tissue), whereas data obtained from all subsequent peaks are used to calculate tubular dynamics (ITC and transit times).[7] Furthermore, the maximal slope of the ascending arm of the proximal tubular curve, which represents the rate of contrast accumulation secondary to glomerular filtration, is used to calculate unit GFR (mL/min/cc tissue)[3] using the peak height or area under the aortic curve for normalization.[3,7]

To explore the function of the whole kidney, the regional renal perfusion and unit GFR need to be multiplied by the renal volume to obtain single-kidney RBF and GFR (mL/min), respectively, which are clinically useful parameters of renal function. EBCT measurements of RBF obtained with this technique correlate well with those obtained with an intravascular Doppler wire (which measures blood velocity in the main renal artery), and EBCT-derived GFR correlates well with the clearance of the reference standard inulin.[3] Importantly, this methodology has been shown to be highly reproducible in both humans[5] and pigs.[3]

APPLICATION OF MODELING IN RENAL PHYSIOLOGY AND PATHOPHYSIOLOGY

The analysis of renal TDCs permits derivation of unique measures of intrarenal regional hemodynamics and function, and has been successfully applied to investigate alterations in the kidney under both physiologic and pathophysiologic conditions. For example, systemic administration of renal diuretics and vasodilators, such as furosemide,[8] acetylcholine, and sodium nitroprusside[14] in dogs or pigs demonstrated physiologic changes in cortical and medullary perfusion and tubular dynamics that were detectable with EBCT. Examination of intrarenal distribution of blood flow and tubular dynamics during an acute increase of renal perfusion pressure within the range of RBF autoregulation revealed decrements of fluid reabsorption in the proximal tubule, Loop of Henle, and distal tubules.[9,15] Furthermore, in pigs with unilateral renal artery stenosis, chronic alterations of renal perfusion pressure demonstrated reciprocal changes in the kidneys exposed to low, as compared to high, renal perfusion pressure.[16] The stenotic kidney exposed to low renal perfusion pressure showed an increase in ITC (tubular fluid reabsorption),[16] whereas in the contralateral kidney, exposed to renovascular hypertension, intratubular fluid was diluted.[17]

In the early phase of renal disease, these measurements may not be sensitive enough to detect differences in renal functional attributes under basal conditions. However, they can be accentuated by examining subtle alterations in renal response to a vasodilator or diuretic challenge. For example, in the intact kidney of pigs with diet-induced hypercholesterolemia, EBCT-derived regional renal perfusion and tubular function were shown to be normal under basal conditions, but the increase in renal perfusion in response to both endothelium-dependent (acetylcholine) and endothelium–independent (sodium nitroprusside) vasodilators was significantly attenuated. Conversely, renal tubular response to acetylcholine was enhanced, whereas tubular response to sodium nitroprusside was relatively preserved.[14] This abnormality was augmented in the presence of concurrent hypertension,[17] in association with cellular and metabolic alterations indicative of increased oxidative stress,[17,18] but was normalized with chronic dietary anti-oxidant intervention.[19]

Similar principles have also been applied to investigate intrarenal perfusion in humans. Humans with a family disposition[20] or with overt essential hypertension[21] were found to have decreased cortical perfusion compared to normal. Furthermore, in renovascular hypertensive patients, alterations in intrarenal perfusion were related to the underlying etiology of renal artery stenosis. In patients with fibromuscular dysplasia, a local vascular disease, which is mainly confined to the renal artery, cortical and medullary perfusion correlated inversely with degree of stenosis. On the other hand, in patients with renal artery stenosis due to atherosclerosis, a disseminated vascular disease that also involves the intrarenal microvasculature, the decrease in cortical perfusion exceeded in severity the degree of the stenosis.[21] This demonstrated the feasibility of this methodology for detection of subtle pathophysiologic intrarenal alterations. Future studies of renal tubular dynamics in human hypertension may potentially shed light on the nephron sites, extent, and potential reversibility of the impairment in renal function.

Finally, this technique may not be limited to measurement of renal perfusion and function. A similar approach has also been useful to quantify myocardial perfusion[22–25] and microvascular permeability,[25] and with organ-specific adjustments may potentially be adapted to explore the hemodynamics and function of other organs as well.

SUMMARY

Depiction of contrast-media transit in the renal vascular and tubular compartments allows simultaneous measurements of regional renal perfusion, tubular dynamics, and GFR in intact bilateral kidneys. The X-ray contrast media can be injected in the central venous circulation and externally detected within the renal tissue without requiring arterial catheterization. This technique can be useful for quantification of regional hemodynamics and function in the intact kidneys, and is potentially applicable to humans.

ACKNOWLEDGMENTS

The authors are grateful to Professor Avinoam Libai, Ph.D., from the Faculty of Aerospace Engineering in the Technion, Israel Institute of Technology, Haifa, Israel, for his valuable advice regarding the methodology described in this paper.

REFERENCES

1. CARELLI, H.H. & E. SORDELLI. 1921. A new procedure for examining the kidney. Rev. Assoc. Med. Argent. **34:** 18–19.
2. LERMAN, L.O., M. RODRIGUEZ PORCEL & J.C. ROMERO. 1999. The development of X-ray imaging to study renal function. Kidney Int. **55:** 400–416.
3. KRIER, J.D., E.L. RITMAN, Z. BAJZER, et al. 2001. Noninvasive measurement of concurrent, single-kidney perfusion, glomerular filtration, and tubular function. Am. J. Physiol. Renal Physiol. **281:** F630–638.
4. LERMAN, L.O., M.R. BELL, V. LAHERA, et al. 1994. Quantification of global and regional renal blood flow with electron beam computed tomography. Am. J. Hypertens. **7:** 829–837.
5. LERMAN, L.O., A.L. FLICKINGER, P.F. SHEEDY & S.T. TURNER. 1996. Reproducibility of human kidney perfusion and volume determinations with electron beam computed tomography. Invest. Radiol. **31:** 204–210.
6. LERMAN, L.O., M.D. BENTLEY, M.R. BELL,et al. 1990. Quantitation of the *in vivo* kidney volume with cine computed tomography. Invest. Radiol. **25:** 1206–1211.
7. LERMAN, L.O., J.D. KRIER, E.L. RITMAN, et al. 2000. Quantification of single-kidney glomerular filtration rate with electron-beam computed tomography. SPIE Proc. **3978:** 539–546.
8. LERMAN, L.O., M. RODRIGUEZ-PORCEL, P.F.I. SHEEDY & J.C. ROMERO. 1996. Renal tubular dynamics in the intact canine kidney. Kidney Int. **50:** 1358–1362.
9. RODRIGUEZ PORCEL, M., L.O. LERMAN, P.F. SHEEDY II. & J.C. ROMERO. 1997. Pressure dependency of renal tubular flow. Am. J. Physiol. **273:** F667–F673.
10. CANTY, J.M., JR., R.M. JUDD, A.S. BRODY & F.J. KLOCKE. 1991. First-pass entry of nonionic contrast agent into the myocardial extravascular space. Effects on radiographic estimates of transit time and blood volume. Circulation **84:** 2071–2078.
11. WU, X., D.L. EWERT, Y.H. LIU & E.L. RITMAN. 1992. *In vivo* relation of intramyocardial blood volume to myocardial perfusion. Circulation **85:** 730–737.
12. BENTLEY, M.D., L.O. LERMAN, E.A. HOFFMAN, M.J. FIKSEN-OLSEN, et al. 1994. Measurement of renal perfusion and blood flow with fast computed tomography. Circ. Res. **74:** 945–951.
13. THOMPSON, H.K., F. STARMER, R.E. WHALEN & H.D. MCINTOSH. 1964. Indicator transit time considered as a gamma variate. Circ Res. **14:** 502–515.
14. FELDSTEIN, A., J.D. KRIER, M. HERSHMAN SARAFOV, Lerman, et al. 1999. *In vivo* renal vascular and tubular function in experimental hypercholesterolemia. Hypertension **34:** 859–864.
15. LERMAN, L.O., M.D. BENTLEY, M.J. FIKSEN-OLSEN, et al. 1995. Pressure dependency of canine intrarenal blood flow within the range of autoregulation. Am. J. Physiol. **268:** F404–F409.
16. LERMAN, L.O., R.S. SCHWARTZ, J.P. GRANDE, et al. 1999. Noninvasive evaluation of a novel swine model of renal artery stenosis. J. Am. Soc. Nephrol. **10:** 1455–1465.
17. RODRIGUEZ-PORCEL, M., J.D. KRIER, A. LERMAN, et al. 2001. Combination of hypercholesterolemia and hypertension augments renal function abnormalities. Hypertension **37:** 774–780.

18. LERMAN, L.O., K.A. NATH, M. RODRIGUEZ-PORCEL, et al. 2001. Increased oxidative stress in experimental renovascular hypertension. Hypertension **37:** 541–546.
19. STULAK, J.M., A. LERMAN, M. RODRIGUEZ PORCEL, et al. 2001. Renal vascular function in experimental hypercholesterolemia is preserved by chronic antioxidant vitamin supplementation. J. Am. Soc. Nephrol. **12:** 1882–1891.
20. FLICKINGER, A.L., L.O. LERMAN, P.F. SHEEDY & S.T. TURNER. 1996. The relationship between renal cortical volume and predisposition to hypertension. Am. J. Hypertens. **9:** 779–786.
21. LERMAN, L.O., S.J. TALER, S. TEXTOR, et al. 1996. CT-derived intra-renal blood flow in renovascular and essential hypertension. Kidney Int. **49:** 846–854.
22. LERMAN, L.O., S. SIRIPORNPITAK, N. LUNA MUFFEI, et al. 1999. Measurement of in vivo myocardial microcirculatory function with electron beam CT. J. Comput. Assist. Tomogr. **23:** 390–398.
23. MÖHLENKAMP, S., T.R. BEHRENBECK, A. LERMAN, et al. 2001. Coronary microvascular functional reserve: quantification of long-term changes with electron-beam CT. Preliminary results in a porcine model. Radiology **221:** 229–236.
24. MÖHLENKAMP, S., L.O. LERMAN, A. LERMAN, et al. 2000. Minimally invasive evaluation of coronary microvascular function by electron beam computed tomography. Circulation **102:** 2411–2416.
25. RODRIGUEZ-PORCEL, M., A. LERMAN, P.J. BEST, et al. 2001. Hypercholesterolemia impairs myocardial perfusion and permeability: role of oxidative stress and endogenous scavenging activity. J. Am. Coll. Cardiol. **37:** 608–615.

Visualization of Blood Microcirculation Parameters in Human Tissues by Time-Integrated Dynamic Speckles Analysis

MARIA M. GONIK, ALEXANDER B. MISHIN, AND DMITRY A. ZIMNYAKOV

Saratov State University, Astrakhanskaya, Saratov, Russia

> ABSTRACT: Statistical analysis of images of time-integrated non-stationary speckle patterns is a tool for diagnostics and imaging of *in vivo* dynamics of blood microcirculation in superficial layers of tissues and organs. The approach to monitoring blood microcirculation using the contrast analysis of time-averaged speckle images is known as the laser speckle contrast analysis (LASCA) technique. This paper presents a modified version of LASCA, based on the application of a localized light source in combination with the speckle contrast analysis of time-integrated dynamic speckle patterns. This method adds possibilities for the analysis of depth distributions of the blood microcirculation parameters. A theoretical background for the depth-resolved analysis of blood microcirculation parameters is considered here on the basis of the concept of effective optical path distributions for a multiply scattered probe light.
>
> KEYWORDS: blood; microcirculation; speckle image; dynamic patterns; contrast analysis; superficial tissues

INTRODUCTION

Various applications of laser Doppler methods for blood flow monitoring have become part of the universally adopted and rapidly developing collection of techniques in laser medicine during the past two decades. In a typical configuration for diagnostic applications, the optical unit of a laser Doppler flowmeter can be considered as an optode with open optical channel; this channel is traced through the probed tissue volume with expressed motion of erythrocytes. In particular, such an optode can be designed as a diode laser and a photodetector coupled with a pair of spatially separated light-delivering and light-collecting fibers. Coherent light travels through the delivering optical channel, reaches the probed tissue volume, is multiply scattered and, as a result, undergoes a Doppler frequency modulation due to a sequence of scattering by the moving elements of the tissue structure, such as erythrocytes. Part of the backward or forward scattered frequency-modulated light is collected by the second optical fiber and, when observed in the detection plane, induces the random interference pattern or dynamic speckle pattern, with a correlation or spectral properties that depend on the dynamic parameters of an ensemble of

---

Address for correspondence: Maria M. Gonik, Ph.D., Saratov State University, Astrakhanskaya str. 8, flat 43, Saratov, 410028, Russia.

mgmaria@san.ru *or* gonikmm@info.sgu.ru

erythrocytes. Thus, by studying the spectral moments of *temporal* speckle intensity fluctuations, one can evaluate the average velocity of erythrocyte motion through the microcapillary net in the probed volume.[1–3]

Another approach is based on the statistical analysis of *spatial* fluctuations of speckle pattern images induced by scattered light and captured with a given exposure time. It can be shown that, for statistically homogeneous and ergodic spatial-temporal speckle intensity fluctuations, the estimates of statistical moments give identical results for spatial averages of instantaneous values of speckle intensity across the pattern area and for temporal averages of the time-dependent intensity fluctuations at the fixed detection point. Thus, measurements of the contrast of speckle pattern images captured with varying exposures allow one to analyze the decorrelation of temporal fluctuations of speckle intensity and to evaluate the dynamic properties of the scattering system.[4–6] This approach is known as the LASCA technique. Contrast estimates of the fragments of the dependence of time-integrated images on exposure time permit reconstructing of two-dimensional distributions of the average level of blood perfusion (*microcirculation maps*).

Here, we consider the modification of the full-field speckle technique, which is based on the application of the localized light source in combination with the speckle contrast analysis of time-integrated dynamic speckle patterns. This approach gives certain opportunities for depth-resolved analysis of blood microcirculation parameters.

PHYSICAL BACKGROUND

If a multiple scattering non-stationary *homogeneous* medium is probed using a full-field speckle technique, such as the LASCA method, then temporal fluctuations of backscattered light amplitude in an arbitrary point of the object plane (which usually coincides with the interface) can be characterized by the field correlation function, $G_I(\tau)$:

$$G_I(\tau) \sim \int_0^\infty \exp\left(-\frac{k_0^2 \langle \Delta^2 r(\tau) \rangle s}{3l^*}\right) \rho(s) ds, \qquad (1)$$

where $\rho(s)$ is the probability density function of effective optical paths of diffusion photons of the scattered field, k_0 is the wave number of the probe light in the scattering medium, l^* is the transport mean free path, and $\langle \Delta^2 r(\tau) \rangle$ is the variance in scattering particle displacement for the observation time τ. The probability density $\rho(s)$ is determined by observation and detection conditions (e.g., by source-detector separation in the case of localized light source) and the optical properties of the scattering medium. An increase in the modal value of effective optical path due to an increase of the source-detector separation or a decrease of the transport mean free path, leads to faster decay of temporal fluctuations of scattered light and, consequently, diminishes the correlation time.

The most probable trajectory of the probe light propagation between localized light source and point-like detector can be obtained using the following diffusion approximation:[7]

$$z_{\text{mod}}(x) \approx \sqrt{\frac{1}{8}[\sqrt{(x^2 + (R-x)^2)^2 + 32x^2(R-x)^2} - x^2 - (R-x)^2]}, \quad (2)$$

where x is the coordinate along the source-detector axis. For such scattering and detection geometry, the normalized temporal correlation function of diffusing light amplitude fluctuations can be written

$$g_l(\tau) = \frac{G_l(\tau)}{G_l(0)} = \frac{\exp\left(-\sqrt{k_0^2 \langle \Delta^2 \bar{r}(\tau) \rangle + 3\mu_a l^*} \frac{\rho_1}{l^*}\right) - \exp\left(-\sqrt{k_0^2 \langle \Delta^2 \bar{r}(\tau) \rangle + 3\mu_a l^*} \frac{\rho_2}{l^*}\right)}{\exp\left(-\sqrt{3\mu_a l^*} \frac{\rho_1}{l^*}\right) - \exp\left(-\sqrt{3\mu_a l^*} \frac{\rho_2}{l^*}\right)}, \quad (3)$$

where μ_a is the absorption coefficient of the probed medium,

$$\rho_1 = \sqrt{R^2 + l^{*2}}, \quad \rho_2 = \sqrt{R^2 + (l^* + 2l_{\text{ext}})^2},$$

and l_{ext} is the so-called extrapolation length. In particular, in the case of probe light propagation for large distances within weakly absorbing Brownian medium, the correlation time of speckle intensity fluctuations can be roughly estimated from $\tau_c \approx l^*/24R^2 D_B k_0^2$, where D_B is the self-diffusion coefficient of the scattering particles.

In the case of probe light propagation in an *heterogeneous* scattering medium, the temporal correlation function of speckle intensity fluctuations at the detection point can be obtained by summation over the parts of the effective optical paths that are associated with the propagation of each partial component (i.e., each diffusing photon) of the scattered optical field in regions with differing optical or dynamic properties.

$$G_l(\tau) \sim \int_s \prod_i^N \exp\left(-\frac{k_0^2 \langle \Delta^2 \bar{r}(\tau) \rangle_i \Delta s_i^s}{3l_i^*}\right) \rho(s) ds, \quad (4)$$

where N is the number of heterogeneous regions along the propagation trace from source to detector for each partial component characterized by the pathlength s, Δs_i^s is the part of the propagation trace with length s that is inside the ith heterogeneity. Using standard notation, the following normalization condition should be satisfied:

$$\int_0^\infty \rho\left(\sum_i^N \Delta s_i^s\right) d\left(\sum_i^N \Delta s_i^s\right) = \langle I \rangle, \quad (5)$$

where $\langle I \rangle$ is the average intensity of the scattered probe light of the detection point.

FIGURE 1 illustrates the effect of underlying *dynamic* layer when probing a two-layered scattering medium for the correlation decay with an increase of the source-detector separation between the localized light source (focused laser beam) and light-collecting system (optical fiber).

Monte Carlo simulations and analytic calculations, using the diffusion approximation, show that for such illumination geometry the traces of the propagating photons pass through a "banana-shaped" region of scattering tissue.[7] This region can be considered a region of localization of photon trajectories inside the probed tissue. When the banana-shaped region penetrates into the modulating layer with the expressed motion of erythrocytes, it causes a fast decorrelation of the detected light fluctuations and results in a blurred speckle pattern. This technique makes it possible to estimate the thickness of the *static* layer and can be successfully used, for example, for burn depth diagnostics.[8]

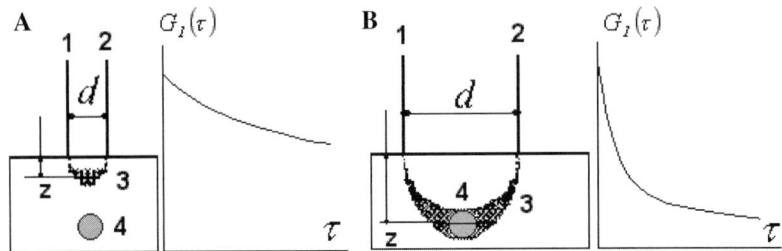

FIGURE 1. Location of the dynamic heterogeneity using measurements of the correlation decay of speckle amplitude fluctuations: 1, light-collecting fiber; 2, light-delivering fiber; 3, "banana-shaped" region; 4, dynamic heterogeneity. $G_I(\tau)$ is the normalized temporal correlation function of the amplitude fluctuations, d is the source-detector separation. **(A)** The correlation decay in the case of probing the static layer of two-layered scattering medium. **(B)** The correlation decay in the case of probing underlying *dynamic* layer.

Additional information about the dynamic properties of an underlying *modulating* layer can be obtained from an analysis of the radial distribution of the speckle contrast in the dependence on exposure time. In this case, the speckle contrast for time-integrated image of the probed object surface at a given observation point can be estimated from

$$V(R, T) = \frac{\sqrt{\frac{1}{T}\int_0^T |\tilde{G}_I(\tau)|^2}}{\langle I \rangle}, \tag{6}$$

where T is the integration time used to capture speckle-modulated image of the object surface and $|\tilde{G}_I(\tau)|$ should be estimated using Equation **(1)**, with a modified pathlength probability density that depends on R and the superficial layer thickness.

Information about the dynamic properties of the underlying *modulating* layer can be obtained from an analysis of radial distribution of the speckle contrast $V(R,T)$ for the dependence of the time-integrated image of the probed object surface on the exposure time T.

EXPERIMENTAL TECHNIQUE

A focused beam of He–Ne laser was used as a localized light source. The diameter of laser beam on the object surface was 100μ. Outgoing light from localized region of the object surface was collected by a multimode fiber. EDC-1000L monochrome CCD camera (Electrim Corp., USA) was used to capture dynamic speckle patterns from the fiber output with a given value of exposure time for further computer processing. Teflon plate with cylindrical channel filled by strongly scattering liquids (e.g., milk solution) was used as the phantom object. Experiments were carried out with phantoms as well as *in vivo* human tissue, such as human skin.

RESULTS

Statistical properties of the time-integrated speckle patterns, obtained with *in vivo* tissue and phantom scatterers, were studied using the setup arranged as shown in FIGURE 1. The value of a residual contrast for a given exposure time was determined from the relative fractions of Doppler-shifted and non-shifted components of the backward scattered light, and correspondingly, by the localization and scattering of stationary and moving scatterers (milk particles) in the probed volume. Thus, the dependence of the residual contrast on exposure time could be used to reconstruct the depth distribution and localization of the dynamic heterogeneity.

In the case of weak absorption limit, the mean value of penetration depth of a probe light can be estimated from $z_{pen} \approx (2)^{0.5} R/4$, where R is the source–detector separation. Thus, changes in R that are related to the mean value of the penetration depth of a probe light change the speckle contrast (see FIGURE 2).

For real tissue, the presence of motionless scatterers in the probed volume leads to the formation of a stationary component in the observed speckle pattern. As a result, spatial modulation of the time-integrated speckle-modulated image of the object surface or the speckle pattern, which is recorded in the diffraction plane, is not suppressed even in the case of very large exposure times. Recording the dynamic speckle patterns, with an exposure time of 25 msec or larger, produced images with a significant value of the residual contrast. For comparison, the correlation time of speckle intensity fluctuations, which are induced by blood microcirculation in human derma, typically does not exceed 3–5 msec. The effect of the residual contrast is closely related to the limitation of the modulation depth of the speckle intensity fluctuations detected at a fixed point. Superposition of Doppler-shifted and non-shifted components of the scattered light induces a detected signal with modulation depth $M = \delta I/\langle I \rangle$, which depends on the fraction of Doppler-shifted photons f as follows:[9]

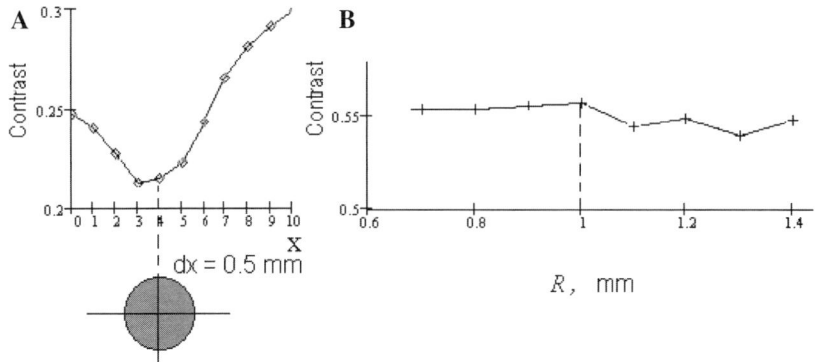

FIGURE 2. Localization of the dynamic heterogeneity: (**A**) speckle contrast as a function of the dynamic heterogeneity displacement in the transverse direction; (**B**) speckle contrast as a function of the source–detector separation R, mm.

$$M = \frac{1}{N} f(2-f), \qquad (7)$$

where N is the number of speckles within the detector aperture.

CONCLUSIONS

A modified LASCA technique based on the application of localized light source, in combination with speckle contrast analysis of time-integrated dynamic speckle patterns, can be used to analyze axial distributions of blood microcirculation parameters. One of the simplest approaches to depth distribution analysis of blood microcirculation parameters is based on estimating the contrast of speckle patterns recorded with given exposure times and the dependence on localization of the dynamic heterogeneity. Further development of this technique should be directed at improving its spatial resolution.

ACKNOWLEDGMENTS

This work was supported by the Civilian Research and Development Foundation through Grant # REC-006.

REFERENCES

1. BRIERS, J.D. 1996. Laser Doppler and time-varying speckle: a reconciliation. J. Opt. Soc. Am. A **13:** 345–350.
2. NILSSON, G., A. JAKOBSSON & K. WARDELL. 1991. Tissue perfusion monitoring and imaging by coherent light scattering. Proc. SPIE **1524:** 90–109.
3. ESSEX, T.J.H. & P.O. BYRNE. 1991. A laser Doppler scanner for imaging blood flow in skin. J. Biomed. Eng. **13:** 189–194.
4. FERCHER, A.F. & J.D. BRIERS. 1981. Flow visualization by means of single-exposure speckle photography. Opt. Commun. **37:** 326–329.
5. FERCHER, A.F., M. PEUKERT & E. ROTH. 1986. Visualization and measurement of retinal blood flow by means of laser speckle photography. Opt. Eng. **25:** 731–735.
6. BRIERS, J.D. & S. WEBSTER. 1996. Laser speckle contrast analysis (LASCA): a non-scanning, full-field technique for monitoring capillary blood flow. J. Biomed. Opt. **1:** 174–179.
7. FENG, S., F. ZENG & B. CHANCE. 1993. Monte Carlo simulations of photon migration path distributions in multiple scattering media. Proc. SPIE **1888:** 78–89.
8. SADHWANI, A., K.T. SCHOMACKER, G.J. TEARNEY & N.S. NISHIOKA. 1996. Determination of Teflon thickness with laser speckle. I. Potential for burn depth diagnosis. Appl. Optics **35:** 5727–5735.
9. SEROV, A., W. STEENBERGEN & F. DE MUL. 2000. A method for estimation of the fraction of Doppler-shifted photons in light scattered by mixture of moving and stationary scatterers. Proc. SPIE **4001:** 178–189.

Measurement of a Velocity Field in Microvessels Using a High Resolution PIV Technique

YASUHIKO SUGII,[a] SHIGERU NISHIO,[b] AND KOJI OKAMOTO[a]

[a]*Nuclear Engineering Research Laboratory, University of Tokyo, Tokai-mura, Japan*
[b]*Kobe University of Mercantile Marine, Higashinada, Kobe, Japan*

ABSTRACT: Because endothelial cells are subject to flow shear stress, it is important to determine the velocity distribution in microvessels during studies of the mechanical interactions between the blood and the endothelium. Particle image velocimetry (PIV) is a quantitative method for measuring velocity fields instantaneously in experimental fluid mechanics. The authors have developed a high-resolution PIV technique that improves the dynamic flow range, spatial resolution, and measurement accuracy. The proposed method was applied to images of the arteriole in the rat mesentery, using an intravital microscope and high-speed digital video system. Taking the mesentery motion into account, the PIV technique was improved to measure red blood cell (RBC) velocity. Velocity distributions with spatial resolutions of $0.8 \times 0.8 \mu m$ were obtained even near the wall in the center plane of the arteriole. The arteriole velocity profile was blunt in the center region of the vessel cross-section and sharp in the near-wall region. Typical flow features for non-Newtonian fluid are shown.

KEYWORDS: blood flow; RBC velocity; microcirculation; micro-PIV; arteriole velocity

INTRODUCTION

Microcirculation, such as occurs in arterioles, capillaries, and venules with diameters from 5 to 50μm, is the mechanism by which oxygen and other substances are delivered to the tissues and organs of the body. Because microcirculation is so important in the process of maintaining healthy tissues and organs, it is essential that velocity fields be measured and the characteristics of microcirculation described, including abnormal microcirculation in a diseased system. To this end, a number of measurement techniques have been developed, such as the electromagnetic blood flow meter, the ultrasonic Doppler flow meter, laser Doppler velocimetry, and the dual slit method. The most practical and commonly used method is the dual slit method,[1-2] which measures the passage of a blood flow signal between two predetermined points by the dual window method. However, this technique is based on the assumption that blood flow passes both points under the same conditions. In laser Doppler velocimetry (LDV),[3-4] a measurement at any probe area and depth is limited

---

Address for correspondence: Yasuhiko Sugii, Ph.D., Nuclear Engineering Research Laboratory, University of Tokyo, Tokai-mura, Ibaraki 319-1188, Japan.
sugii@tokyo.t.u-tokyo.ac.jp

to a few millimeters, which can give spurious signals and reduces the accuracy of the measurement. Particle image velocimetry (PIV) is a quantitative method for measuring velocity fields instantaneously in experimental fluid mechanics system.[5] A number of PIV methods, such as cross-correlation, particle tracking, and iterative correlation, have been proposed. Tsukada et al.[6] applied the conventional cross-correlation technique to *in vivo* blood flow images recorded by using a high-speed camera. However, they are not suitable for microcirculation because of the measurement accuracy and spatial resolution.

We have developed a high-resolution PIV technique[7] that can improve the dynamic range, spatial resolution and measurement accuracy. *In vivo* blood images of an arteriole in a rat mesentery were recorded using an intravital microscope and high-speed digital video system. The PIV method was applied to determine mesentery motion, and it improved the accuracy of RBC velocity by taking the motion into account. Since blood vessels usually move during blood velocity measurement, the measurement accuracy is low.

EXPERIMENTAL SETUP AND METHODOLOGY

A male Wister rat (8 weeks, 310 g body weight) was anesthetized with thiobutabarbital sodium intraperitoneously and allowed to respire spontaneously. An intestinal loop was mounted on the stage of an intravital microscope with a water-immersion objective lens with a magnification $M = 60$ and a numerical aperture $NA = 1.0$. The mesentery was placed on an observation window and perfused with Krebs–Ringer solution maintained at 37 degrees. Blood flow images were recorded into a computer for two seconds using a high-speed CCD camera at a rate of 1,000 frames/sec. The images consisted of 512×512 pixels with 8 bit gray levels. The measurement region was illuminated from the bottom using back light illumination by metal halide. The basis flow features were initially investigated by examining a relatively straight length of arteriole. FIGURE 1 shows a picture of blood flow in an arteriole with internal diameter of about 24–26 μm at the center of the image. The observed region is 136×136 μm in size. One pixel is equivalent to 0.27 μm. The vessel curved slightly to the right at about $x = 50$ μm. Generally, the diameter of endothelial cells is approximately 5 μm. However, the apparent diameter of these cells in the present images is 2 to 4 μm. This is considered to be due to the out-of-plane alignment of cells in the optical plane. A plasma layer, where erythrocytes hardly pass, is clearly observed near the wall.

The PIV technique[7] was applied to the first blood flow image, at $t = 0$, as a reference, and then sequentially to subsequent images at $t = k$ ($k = 1, 2, \ldots, 2{,}047$). In this method, a pixel unit displacement was obtained by using the iterative cross-correlation method whereas a sub-pixel displacement was calculated by the use of the gradient method. The error was analytically assessed by means of Monte Carlo simulations. PIV images with known a displacement were generated synthetically and the root-mean-square (RMS) error was of the order of 0.01 pixels, even with the small interrogation window size of 8×8 pixels, or less. Thus, the method achieved high sub-pixel accuracy and high spatial resolution compatibly. The aim in this analysis was to measure the movement of the mesentery rather than the RBC velocity.

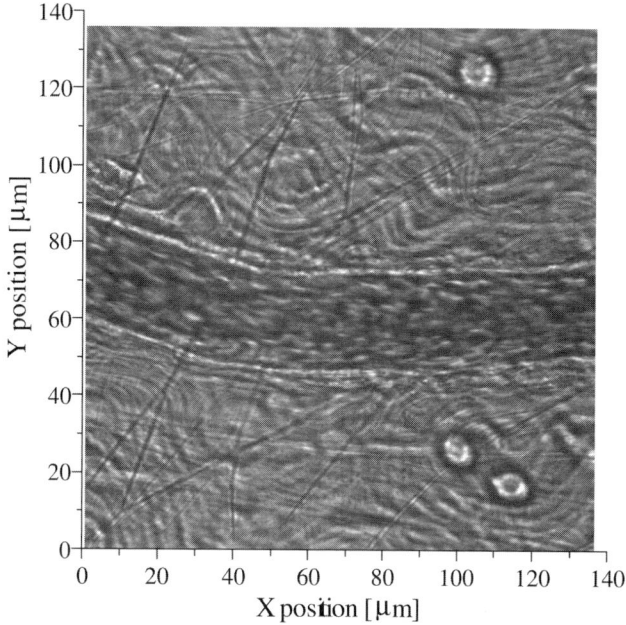

FIGURE 1. Image of blood flow in an arteriole.

The motion of patterns within the mesentery image, such as musculi and vessel walls, was then obtained. The displacement was determined as the relative distance from the reference image. The displacements were approximately -0.52 pixels in the x direction and -1.73 pixels in the y direction, corresponding to 0.14 and 0.45 μm, respectively. The gradients and variation in displacement in the upper and lower areas of the arteriole were small. It is considered that the motion involved only parallel translation, without higher-order displacements, such as deformation or rotation. Compared to the displacements in the upper and lower side of arteriole wall in several cross sections, the differences in displacements were smaller than 0.1 pixels, or 0.027 μm. These results indicate that the contraction and relaxation of blood vessels in the arteriole were too small to measure.

A strong correlation between x displacement and y displacement was confirmed from the time series of spatially averaged displacement, indicating that the mesentery periodically vibrated upper-right to lower-left. The amplitude of vibration varied significantly, and the maximum displacement was about 8.9 pixels, or 2.4 μm. A temporal derivation of the displacement gave the instantaneous velocity of the mesentery. The peak frequency of motion was about 16 Hz, obtained by spectrum analyses. Since the cardiac cycle of a rat is usually about 6–7 Hz, it is considered that this mesentery vibration was caused by intestinum motion.

RBC velocity vector was obtained in all of the previous studies by applying the PIV technique to two successive images, without taking mesentery motion into account. However, the relative position of the RBC velocity vector to vessel wall moves with time. Bias error due to mesentery motion interfered with the analysis of instantaneous RBC velocity. An improved technique is proposed here to reduce the effect of mesentery motion. Both the blood image and RBC velocity were modified using the measured mesentery motion. The PIV technique was applied to two successive shifted images in order to obtain the RBC velocity distributions. The measurement accuracy of RBC velocity improved after the relative positions of the arteriole in all images were arranged consistently and the effect of the mesentery motions eliminated.

RESULTS

FIGURE 2 shows time-averaged velocity distributions of 2,008 images for about 2 sec, calculated using the highly accurate iterative PIV technique. An interrogation window of 7×7 pixels was taken, with a 50% overlap rate. This corresponds to a spatial resolution of $0.8 \times 0.8\,\mu m$. The velocities in the horizontal direction were thinned out in order to be displayed clearly. A velocity distribution with high spatial resolution and highly measurement accuracy was obtained. The velocity vectors very close to the wall were measured and it was found that the wall-normal component of the velocity vectors was close to zero.

The maximum velocity is about 11.0 pixel/frame or 3.0 mm/sec at the center of the arteriole. The flow around inner side of the bent corner of arteriole was lower. The lower velocity areas near the wall were in the plasma layer. Averaged velocity profiles show the blood flow volume to be constant at the vessel cross-sections. The measured velocities represent integrated values of velocities through the depth of the vessel. In principle, the depth of focus δz of a microscope objective lens is obtained as follows:[8]

$$\delta z = \frac{\lambda}{NA^2} + \frac{e}{M \cdot NA}, \qquad (1)$$

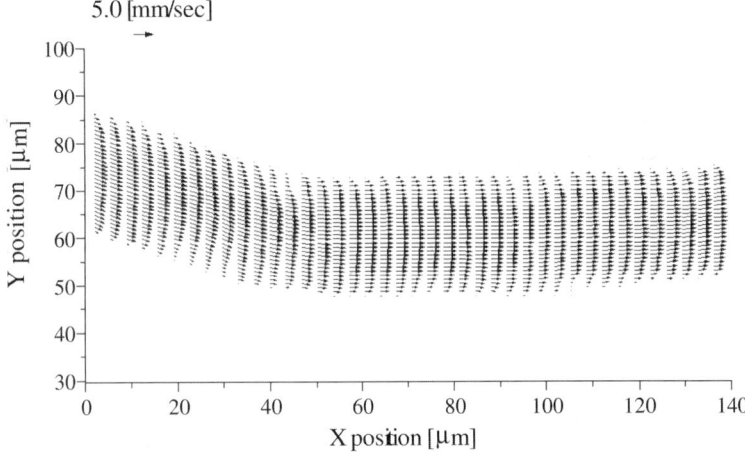

FIGURE 2. Time-averaged velocity distribution on blood flow in an arteriole.

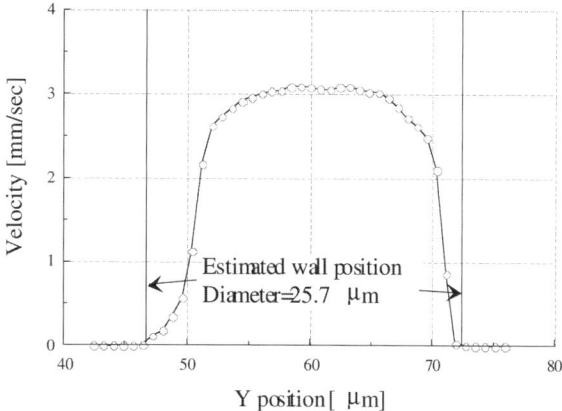

FIGURE 3. Cross-sectional time-averaged axial velocity profile at $x = 101\,\mu m$ on blood flow in an arteriole.

where N is the refractive index of the fluid between the subject flow and the objective lens, λ is the wavelength of light, NA is the numerical aperture. Since the depth of focus, δz, is 0.91 mm in the experiment, which is about 3% of the vessel diameter, the effect can be ignored.

FIGURE 3 shows cross-sectional time-averaged axial velocity profile at $x = 101$ mm. Thirty velocity values were obtained along the capillary diameter at a spacing of $0.8\,\mu m$. The wall positions and capillary diameter in each section are displayed in the figure. The wall position was identified via the luminance of the cross section in a

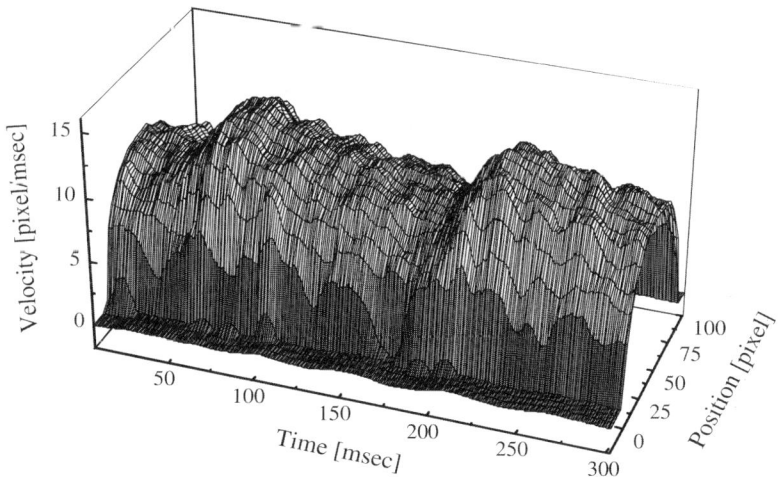

FIGURE 4. Time series of cross-sectional blood velocity profile at the center of the image.

time-averaged image after mesentery motion compensation. The velocity of all profiles was maximal near the center of the vessel and decreased to zero near the wall. The arteriole velocity profiles were broad at the center of the vessel and sharp near the wall, compared to a parabolic flow profile. This suggests that the shear stress on the vessel wall was higher than expected.

FIGURE 4 shows time series of the axial blood velocity profiles of the vertical section in the center of the images. Thirty velocity values were recorded along the capillary diameter at spacing of 0.9 µm. Blood flow velocities at the center of vessel fluctuated within the range of about 8 to 16 pixels/frame. The velocity increased sharply toward peak systole, and then decreased moderately toward late diastole. Finally, it was found that the velocity profile synchronized with the cardiac cycle of 6.8 Hz even in the microvessels.

SUMMARY

A highly accurate PIV technique was used to obtain *in vivo* blood images of the arteriole in the rat mesentery. The images were recorded using the intravital microscope with a water immersion objective lens and illumination by metal halide, and a high-speed digital video system. The velocity distributions with spatial resolutions of 0.8×0.8 µm were measured even near the wall in the center plane of arteriole. The arteriole velocity profiles were blunt at the center region of the vessel cross-section and sharp near the wall. The results show that the proposed method can be used to measure the blood flow velocity profiles with highly accurate temporal and spatial resolution even in the microvessels.

REFERENCES

1. WAYLAND, H. & P.C. JOHNSON. 1967. Erythrocyte velocity measurement in microvessels by two-slit photometric method. J. Appl. Physiol. **22:** 333–337.
2. NAKANO, A. & M. MINAMIYAMA. 1996. The regulation mechanism between the red blood cell velocity and the skin blood flow during the vasomotion in the rabbit ear chamber. Proc. Sixth World Congress for Microcirculation. 833–837.
3. COCHRANE, T., J.C. EARNSHAW & A.H.G. LOVE. 1981. Laser Doppler measurement of blood velocity in microvessels. Med. Biol. Eng. Comput. **19**(15): 589–596.
4. SEKI, J. 1999. Pulse wave propagation in microvessels studied by a dual laser-Doppler anemometer microscope. Microcirc. Ann. **15:** 37–38.
5. RAFFEL, M., C.E. WILLERT & J. KOMPENHANS. 1998. Particle Image Velocimetry. Springer.
6. TSUKADA, K., H. MINAMITANI, E. SEKIZUKA & C. OSHIO. 2000. Image correlation method for measuring blood flow velocity in microcirculation: correlation 'window' simulation and *in vivo* image analysis. Physiol. Meas. **21:** 459–471.
7. SUGII, Y., S. NISHIO, T. OKUNO & K. OKAMOTO. 2000. A highly accurate iterative PIV technique using gradient method. Meas. Sci. Technol. **11:** 1666–1673.
8. BORN, M., *et al.* 1997. Principles of Optics. Pergamon Press, Oxford.

Biosimulation and Visualization

Effect of Cerebrovascular Geometry on Hemodynamics

MARIE OSHIMA,[a] TOSHIO KOBAYASHI,[a] AND KIYOSHI TAKAGI[b]

[a]*Institute of Industrial Science, University of Tokyo, Tokyo, Japan*
[b]*Teikyo University School of Medicine, Tokyo, Japan*

ABSTRACT: Hemodynamics plays an important role in cardiovascular disorders, and the authors are applying numerical and experimental studies of cerebrovascular blood flow to the creation and rupture of cerebral aneurysms. In particular, this study aims to investigate the effects of cerebrovascular geometry on hemodynamics, such as flow pattern, wall shear stress distribution, and pressure. This report consists mainly of two parts: numerical study of blood flow in the artery extracted from computer tomography data, and numerical and experimental studies of a curved pipe model. The simulation was conducted by using a finite element method; the experiment was conducted by particle imaging velocimetry. Numerical and experimental results are compared and both show similar secondary flow behavior.

KEYWORDS: computational hemodynamics; cerebral aneurysm; PIV; CT images

INTRODUCTION

It is known that more that 90% of subarachnoideal hemorrhage is caused by rapture of a cerebral aneurysm.[1] The subarachnoidal hemorrhage is quite fatal, so that 35% of patients die immediately after occurrence, 18% suffer from paralysis or disability, and only 22% show full recovery. However, according to a 1998 paper published in the *New England Journal of Medicine,* the rate of rupture of aneurysms is relatively low.[1,2] In general, the surgical operation for an aneurysm is conducted either by clipping or coiling. Both operations present a difficult and intensive procedure for patients. Thus, it is important to predict risk factors associated with aneurysms rupture in order to avoid unnecessary surgery.

Clinical statistics show that a cerebral aneurysm tends to be created at three preferential locations, among the preferential groups aged 40–60 years.[3] This evidences leads to the hypothesis that hemodynamics plays an important role in the creation and rupture of aneurysms. Numerical and experimental studies were conducted to investigate effects of the geometry of the carotid siphon on hemodynamic factors.

For the numerical study, an integrated numerical simulation system was developed by the authors.[4] The simulation system consisted of preprocessing, simulation, and postprocessing. In the preprocessing step, an analysis model was constructed

Address for correspondence: Marie Oshima, Ph.D., Institute of Industrial Science, University of Tokyo, De 503, 4-6-1 Komaba Meguro-Ku, Tokyo, Japan.
marie@iis.u-tokyo.ac.jp

from computer tomography (CT) data, and finite element generation was employed. The simulation was performed using the finite element method (FEM) under pulsatile flow conditions and measured by the Doppler ultrasound method. Postprocessing was performed to visualize the results to better understand the flow characteristics.

Particle imaging velocimetry (PIV), a flow measurement technique wherein the light scattered by tracer particles in the flow is recorded on a sequence of frames, was used in this study. The information on velocity fields was obtained by processing the recorded particle images.[6] The PIV technique can measure the velocity of two-dimensional (2D) or three-dimensional (3D) flow fields instantaneously, without disturbing the flow. Using PIV, the global structure of a complicated unsteady flow field can be obtained quantitatively.

The paper presents the numerical results of flow in the internal carotid artery (ICA)–posterior communicating artery (PcomA), extracted from CT data, and a comparison of numerical and experimental results from the curved pipe model. To evaluate the geometry effects, the location and size of branching PcomA were varied. The curved model was used to evaluate curvature effect of the ICA on the flow field. Results of the simulation and experiments using curved pipe models were compared and are discussed to validate the methodology. Consequently, the knowledge obtained in this study can be applied to improve simulation models for patient-specific diagnosis.

THE EXPERIMENTAL PIV SYSTEM

The experimental system consisted of a flow circuit and pulse generator. To imitate cardiac cycles, pulsatile motion was superimposed on steady mean flow by using a piston motor. The compression chamber was set up to suppress water hammer caused by the closure of the valve. The cross-correlation method was applied to extract information of the velocity vectors from the particle visualization images.[7]

NUMERICAL SIMULATION

Vascular flow in the artery is considered to be Newtonian, incompressible, and represented by the continuity and Navier–Stokes equations. The governing equations for a computational domain Ω are as follows:

$$\frac{\partial u_i}{\partial x_i} = 0, \tag{1}$$

$$\rho\left(\frac{\partial u_i}{\partial t} + u_j\frac{\partial u_i}{\partial x_j}\right) = -\frac{\partial P}{\partial x_j}\delta_{ij} + \mu\frac{\partial^2 u_i}{\partial x_j \partial x_j} + f_i, \tag{2}$$

where u_i is the velocity in the ith direction, ($i = 1, 2,$ or 3, denoting the x, y, or z direction, respectively), P is the pressure, f_i is the body force, ρ is the density, μ is the viscosity, and δ_{ij} is the Kronecker delta function. The Newtonian assumption is reasonable, since the diameter of the internal carotid artery is about 0.5 cm on average and is large enough compared to the size of a red cell, which is about 8μm. The initial and boundary conditions governing Equations (1) and (2) are

$$u_i = g_i(x, t) \text{ on } \Gamma_g, \tag{3}$$

$$t_i = h_i(x, t) \text{ on } \Gamma_h, \tag{4}$$

$$h_i = \left(-P\delta_{ij} + \mu \frac{\partial u_i}{\partial x_j}\right) n_j, \tag{5}$$

where n_j is the jth component of a vector normal to the Neumann boundary, Γ_h, and Γ_g is the Dirichlet boundary.

The finite element method is used for spatial discretization. The weighted residual method is employed to derive the weak form of Equations (1) and (2). To satisfy the constraint condition of incompressibility, a mixed finite element method was used, so that the order of interpolation for the pressure was one order smaller than that for velocity. Since analyses were conducted in 3D, a brick element with a trilinear function for velocity and a piecewise constant for pressure was used.

Integrating the product of the weighting function and the governing equations by parts yielded the weak form. Next, it was approximated using the Galerkin finite element method, in which the same basis function was used for both the variable and the corresponding weighting functions. In order to avoid numerical instability, one-point quadrature was applied to the convective terms.[5] Implementation of the finite element equations was performed in a conventional semidiscrete formulation. The numerical algorithm was based on the marker and cell (MAC) method, and the second-order Adams–Bashforth method was used for time discretization.[4]

NUMERICAL SIMULATION OF BLOOD FLOW IN THE CEREBRAL ARTERY

The Physiological Model

The junction between ICA and PcomA was used to investigate the effects of cerebrovascular geometry on hemodynamics. The geometry of ICA was extracted from CT data. The artificial PcomA was modeled and bifurcates from ICA as shown in FIGURE 1.

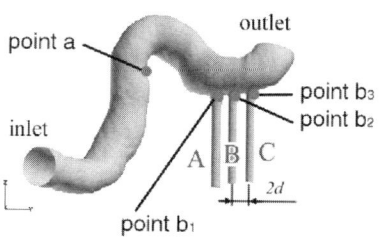

FIGURE 1. ICA-PcomA model.

TABLE 1. Analysis conditions: location and diameter of PcomA

Case	d (mm)	Location
1	—	—
2	1.0	A
3	1.0	B
4	1.0	C
5	0.5	B
6	1.5	B

To investigate the geometry effect, the location and diameter of the modeled PcomA were varied. A total of six cases, as summarized in TABLE 1, were studied. Case 1 had no PcomA and only ICA was considered. The location of the modeled PcomA was varied (points A, B, and C in FIG. 1) in Cases 2–4, while keeping a constant diameter of $d = 1.0$ mm (1/4 of that of ICA). In Cases 5 and 6, the diameters of the modeled PcomA were $d = 0.5$ mm and 1.5 mm, respectively, at a fixed location, point B.

For the inflow, pulsatile boundary conditions were prescribed, so that the velocity measured by the ultrasound Doppler technique was modeled using the Womersley condition. FIGURE 2 shows the profiles of inflow boundary conditions. The resulting number of finite elements was 49,684 and the number of finite nodes was 53,991.

Traction free conditions were prescribed for the outflow boundary conditions of the ICA and PcomA. To minimize effects of the outflow boundary conditions, a straight duct, with a length five times the diameter of ICA, was attached to ICA. In these analyses, the wall was considered to be rigid. The physical properties of the fluid were the actual values of blood, with a viscosity $v = 2.0$ cP.

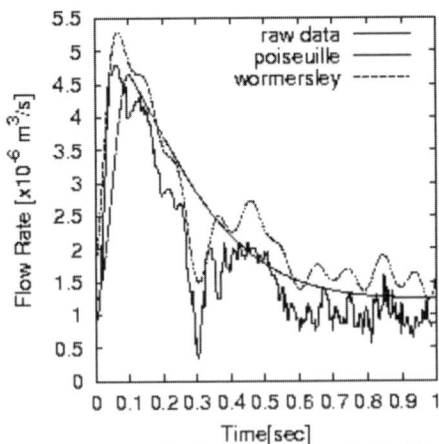

FIGURE 2. Profiles of inflow boundary conditions.

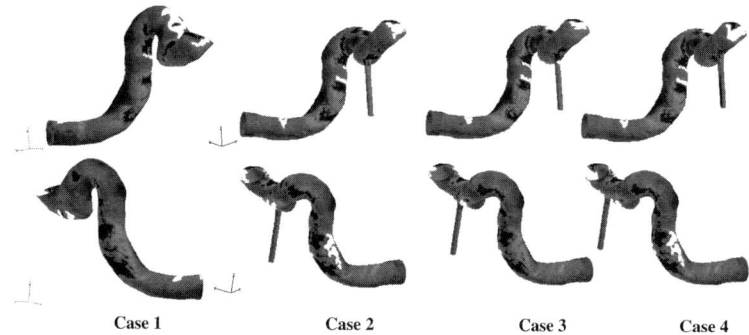

FIGURE 3. Distribution of averaged wall shear stress over one cardiac cycle. (COLOR PLATE 12.)

Numerical Results

FIGURE 3 compares four distributions of the averaged wall shear stress over one cardiac cycle, for various locations of the branching modeled PcomA, but with the same diameter. As shown in FIGURE 3, Case 1, with no branching of PcomA, the wall shear stress concentrates on curvature of ICA. On the other hand, if there is a bifurcation, the high shear stress also appears around the bifurcation area. The magnitude of the wall shear stress for all cases was 60 dyne/cm^2, which is within the measured values of 10–70 dyne/cm^2.[8]

When the location of the branching PcomA moves, the distribution of the wall shear stress changes as described in FIGURE 4. This figure shows that as the location of the branching PcomA moves downstream, the high magnitude of wall shear stress moves from the side to the center of PcomA. However, the diameter of the branching PcomA does not affect the distribution of the wall shear stress.

This tendency is initiated by the curvature of ICA upstream. FIGURE 5 compares the transient behaviors of wall shear stress for various locations of PcomA. FIGURE 6 compares these behaviors for various sizes of PcomA. The monitored points are designated in FIGURE 1. The first area, denoted by Point a, is near the curvature of ICA. The other area, denoted by Point b_i, is the junction between ICA and PcomA and i denotes the case number. The highest magnitude of wall shear stress is obtained at systole since the inflow velocity reaches a maximum. As the location of the branching

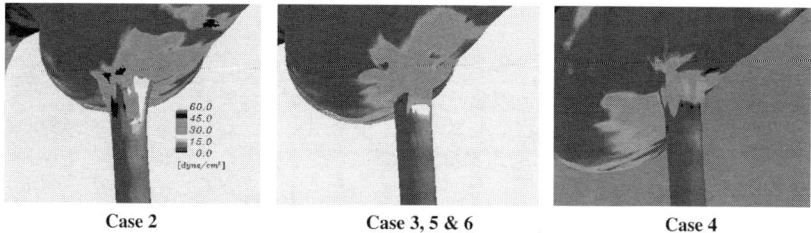

FIGURE 4. The wall shear distribution near bifurcation. (COLOR PLATE 13.)

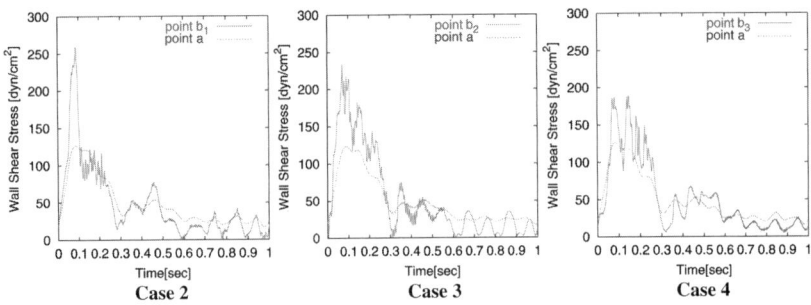

FIGURE 5. Transient behavior of wall shear stress for various branching locations.

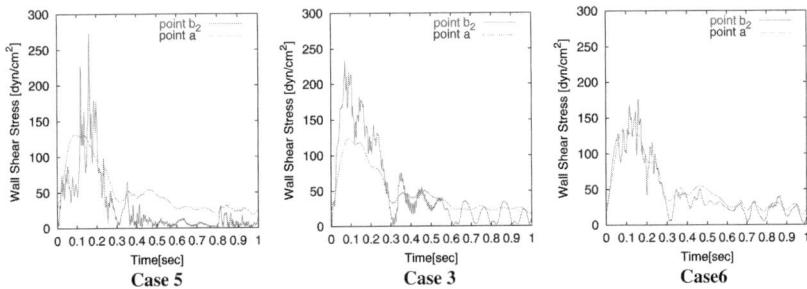

FIGURE 6. Transient behavior of wall shear stress for various PcomA diameters.

PcomA moves closer to ICA, the magnitude of shear stress becomes larger and peaky, as shown in FIGURE 6. On the other hand, as the diameter of PcomA becomes smaller, the velocity gradient of incoming flow of PcomA becomes steep, which leads to a high wall shear stress at bifurcation. As the diameter becomes larger, the wall shear stress becomes smaller because, despite an increase in flow rate, the velocity gradient of the incoming flow becomes lower due to the increased diameter.

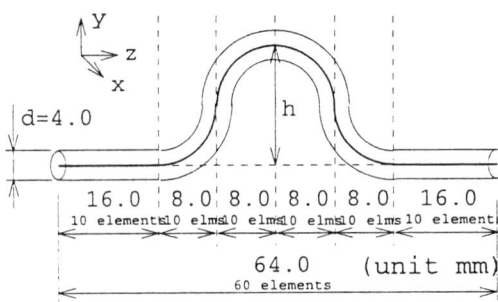

FIGURE 7. Curved pipe model.

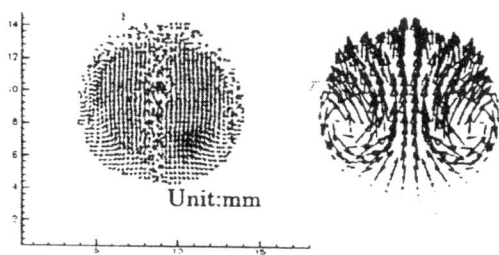

FIGURE 8. Comparison between experiment (*left*) and simulation (*right*) at the apex of the curved pipe model.

SIMULATION OF THE CURVED PIPE MODEL

To investigate the curvature effect of the carotid siphon, the simplified model shown in FIGURE 7 was used for both numerical and experimental studies. In the simulation, the curved model assumes that the diameter of the artery, D, is 0.5 cm and the length is about 3.2 cm. The test section is twice as large as the simulation model and is made of an acrylic resin. The curvature of the artery is defined here as the height h with respect to diameter D, and $h = 4D$ in both studies.

Identical physical properties of blood were used in the simulation with the ICA-PcomA model. In the experiment, a saturated aqueous solution of sodium iodide with a refraction index of 1.495 was used for the curved tube in order to avoid refraction between the model and the working fluid.

COMPARISON OF NUMERICAL AND EXPERIMENTAL RESULTS

FIGURE 8 shows a comparison between the experiment and the simulation at the apex of the curved tube. The simulation and experiments were conducted at $Re = 500$. Secondary flows are evident in both cases, and the two procedures show qualitatively similar flow behavior.

SUMMARY

Numerical and experimental studies of blood flow were conducted in the cerebral artery. This paper presents the results of two studies: (1) a numerical study of blood flow in the artery, with a geometry as extracted from CT data, and (2) numerical and experimental studies of flow in the curved pipe model.

In the first study, the location and a diameter size of the section PcomA were varied to examine effects of cerebrovascular geometry on flow dynamics in the artery. The results suggest that the curvature of ICA determines overall flow characteristics. It was also found that the location and the size of branching artery affect the distribution of wall shear stress. In the second, curved tube model study, a similar secondary flow behavior was observed. However, it seems necessary to improve the experimental procedure for quantitative comparison, particularly for the wall shear stress.

REFERENCES

1. WIEBERS, D.O. 1998. Unruptured intracranial aneurysms—risk of rupture and risks of surgical intervention. N. Engl. J. Med. **339:** 1725–1733.
2. KANO, T., S. TAKEUCHI, *et al.* 1993. Fluid dynamics in cerebrovascular diseases. Neurosurgeon **12:** 15–24. (Japanese.)
3. KASSELL, N.F. & J.C. TORNER. 1990. The international cooperative study on the timing of aneurysm surgery. J. Neurosurg. **73:** 18–36.
4. OSHIMA, M., R. TORII, *et al.* 2001. Finite element simulation in the cerebral artery. *In* Computational Methods in Applied Mechanics and Engineering. To be published.
5. OSHIMA, M., T.J.R. HUGHES & K. JANSEN. 1998. Consistent finite element calculation of boundary and internal fluxes. Int. J. Comput. Fluid Dynam. **9:** 227–235.
6. KOBAYASHI, T. 2000. High performance computation and visualization of fluid flows. 9th Int. Symp. on Flow Visualization 172.1–172.14.
7. RAFFEL, M., C. WILLERT & J. KOMENHANS. 1998. Particle Imaging Velocimetry. Springer, Berlin.
8. MALEK, A.M., S.L. ALPER & S. IZUMO. 1999. Hemodynamic shear stress and its role in atherosclerosis. JAMM **21:** 2035–2042.

Index of Contributors

Affeld, K., 200–205, 247–253
Aguilera, M.E., 242–246
Akai, M., 206–212, 285–291
Aleksic, J., 158–163
Anbar, M., 111–118
Aruga, Y., 81–86
Astarita, T., 177–186

Bajzer, Ž., 307–316, 317–324
Barequet, G., 10–18
Bauters, T.W.J., 103–110
Baveye, P., 103–110
Baysal, U., 127–132
Bruckstein, A., 10–18

Cardone, G., 177–186
Carlomagno, G.M., 177–186
Charbonnier, J.M., 271–277
Chen, B., 206–212, 285–291
Cho, S.-H., 260–264
Choi, H.M., 235–241
Choi, J.-H., 260–264
Chong, T.-P., 95–102

Dabiri, D., 1–9
Darnault, C.J.G., 103–110
Debaene, P., 200–205
Delfos, R., 247–253
Desbat, L., 87–94
DiCarlo, D.A., 103–110
Dizene, R., 271–277
Dorignac, E., 271–277
Dzyatkovskaya, N.N., 144–150

Fuchiwaki, M., 61–66
Fujii, K., 265–270
Fujimatsu, N., 265–270

Gharib, M., 1–9
Gonik, M.M., 325–331
Goubergrits, L., 200–205, 247–253
Guzović, Z., 67–72

Hamabe, K., 187–192
Hanjalić, K., 19–28
Hassan, Y.A., 223–228
Haueisen, J., 127–132, 133–138
Heitor, M.V., 292–298
Hervieu, E., 87–94
Hirochi, T., 171–176
Hodson, H.P., 95–102
Hu, H., 254–259
Hua, T.-S., 151–157

Inada, S., 164–170

Jouet, E., 87–94

Kenjereš, S., 19–28
Kertzscher, U., 200–205, 247–253
Kimoto, H., 187–192
Kobayashi, T., 254–259, 337–344
Korzyńska, A., 139–143
Kowalewski, T.A., 213–222

Lablanc, R., 271–277
Landesberg, A., 119–126
Landesberg, Y., 119–126
Leder, U., 133–138
Lee, I.-S., 260–264
Lerman, L.O., 307–316, 317–324
Loc, T.P., 73–80
López de Ramos, A., 242–246
Lund, P.E., 307–316

Machacek, M., 36–42
Madarame, H., 299–306

Makhervaks, V., 10–18
Martinis, V., 67–72
Matijašević, B., 67–72
Matsui, G., 235–241
Meier, R., 277–284
Mel'nik, Y.I., 144–150
Mishin, A.B., 325–331
Modarress, D., 1–9
Mogilenskikh, D.V., 43–52
Möhlenkamp, S., 307–316
Monji, H., 235–241
Montemagno, C.D., 103–110
Munakata, T., 299–306

Nam, Y.-S., 260–264
Neau, L., 73–80
Nishio, M., 206–212, 285–291, 299–306
Nishio, S., 331–336
Nuntadusit, C., 187–192

Ojeda, A., 242–246
Okamoto, K., 299–306, 331–336
Orel, V.E., 144–150
Oshima, M., 337–344
Ota, M., 53–60

Padet, C., 29–35
Padet, J., 29–35, 193–199
Parlange, J.-Y., 103–110
Pavlov, I.V., 43–52
Pereira, F., 1–9
Polidori, G., 193–199
Pruvost, J., 73–80

Qi, Y., 53–60

Rachek, A., 29–35
Richard, J., 229–235
Ritman, E.L., 307–316, 317–324
Rodriguez, O., 73–80
Romanov, A.V., 144–150
Rondón, C., 242–246

Rösgen, T., 36–42
Rule, A.D., 317–324

Saga, T., 254–259
Schäfer, O., 277–284
Schober, P., 277–284
Seeger, A., 247–253
Shintate, T., 171–176
Shirakashi, M., 81–86, 171–176
Sideman, S., xi–xiv, 119–126
Sivadas, V., 292–298
Someya, S., 206–212, 285–291, 299–306
Song, Y., 206–212, 285–291
Steenhuis, T.S., 103–110
Stoian, M., 29–35
Sugii, Y., 331–336
Susset, A., 229–235
Szymczyk, J.A., 158–163

Takagi, K., 337–344
Takahashi, T., 81–86
Tanaka, K., 61–66
Taniguchi, N., 254–259
Tao, L.-R., 151–157
ter Keurs, H.E.D.J., 119–126
Terauchi, T., 235–241
Tsuboi, N., 265–270

Uchida, T., 206–212

Vantelon, J.-P., 229–235

Watanabe, A., 81–86
Wellnhofer, E., 247–253
Wittig, S., 277–284

Yamada, S., 171–176

Zhong, S., 95–102
Zielke, P., 158–163
Zimnyakov, D.A., 325–331
Ziolkowski, M., 133–138

COLOR PLATES

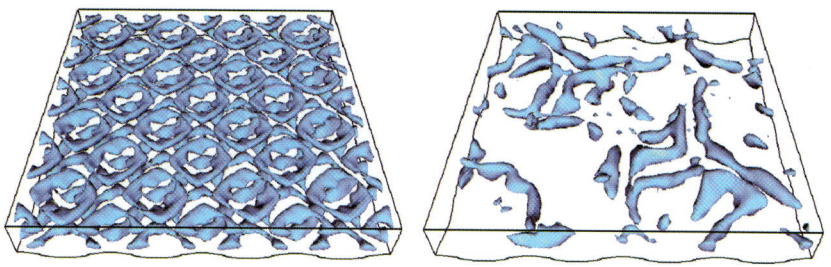

PLATE 1. FIGURE 2, Hanjalić & Kenjereš, page 25.

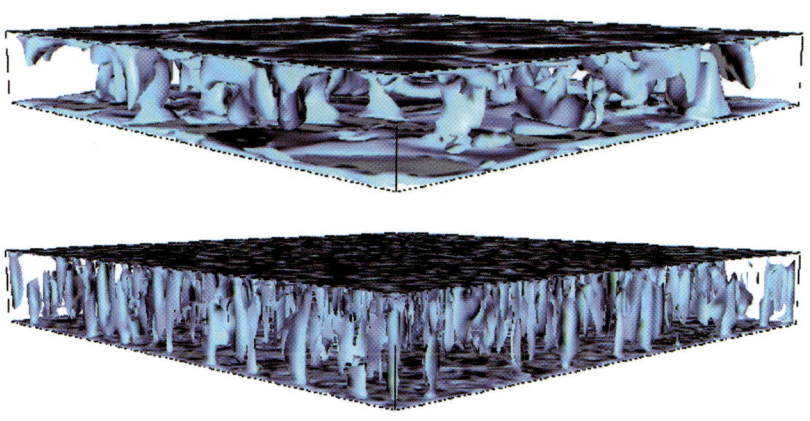

PLATE 2. FIGURE 6, Hanjalić & Kenjereš, page 27.

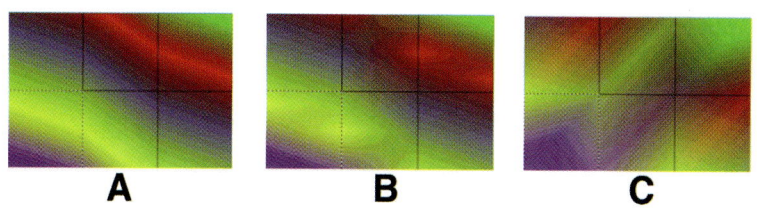

PLATE 3. FIGURE 3, Mogilenskikh & Pavlov, page 46.

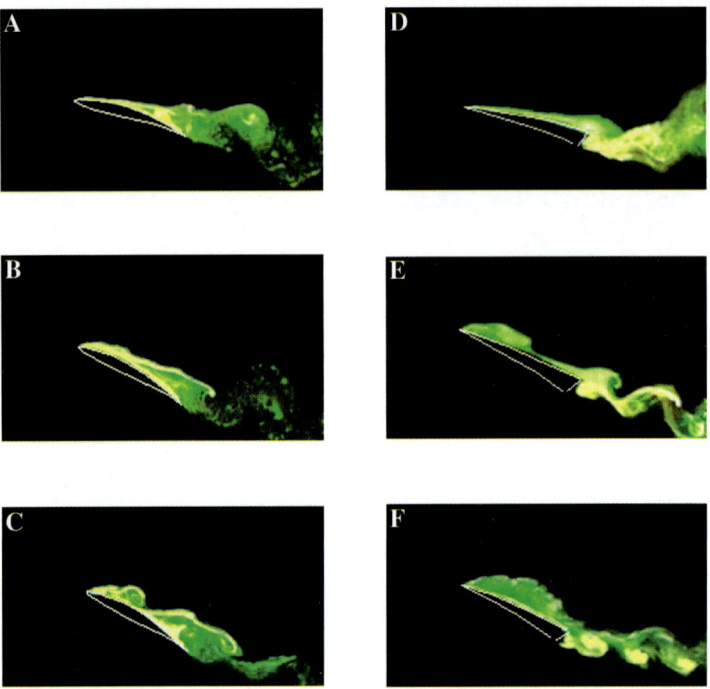

PLATE 4. FIGURE 1, Fuchiwaki & Tanaka, page 63.

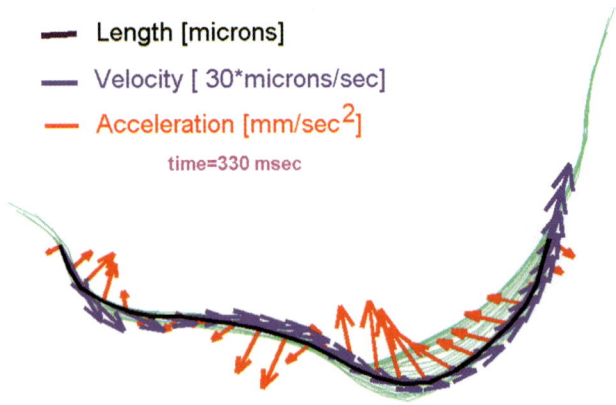

PLATE 5. FIGURE 2, Landesberg *et al.*, page 122.

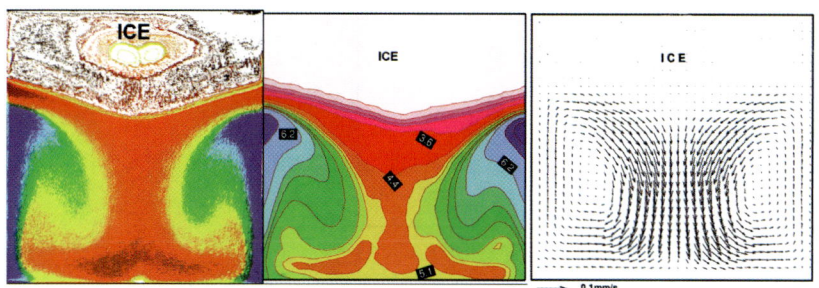

PLATE 6. FIGURE 2, Kowalewski, page 218.

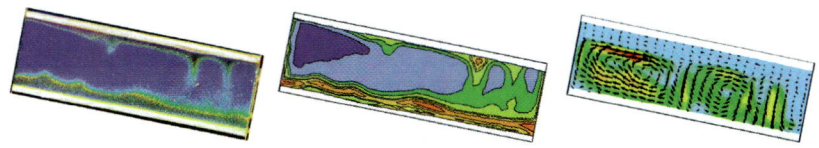

PLATE 7. FIGURE 4, Kowalewski, page 219.

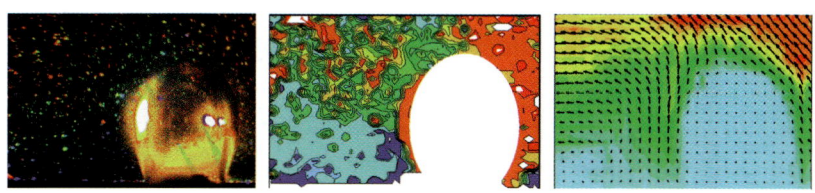

PLATE 8. FIGURE 5, Kowalewski, page 220.

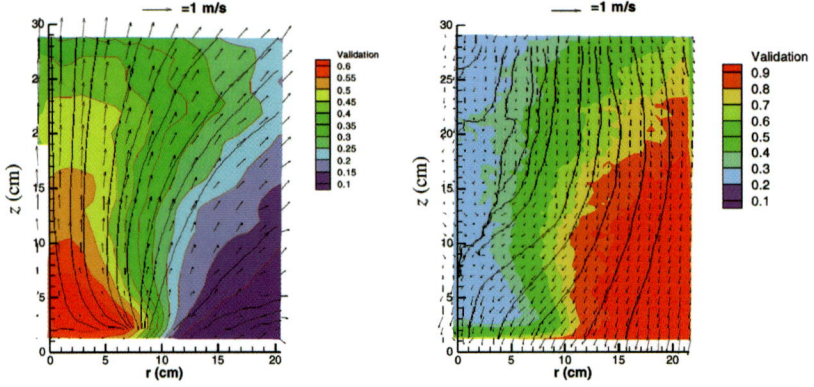

PLATE 9. FIGURE 3, Richard *et al.*, page 232.

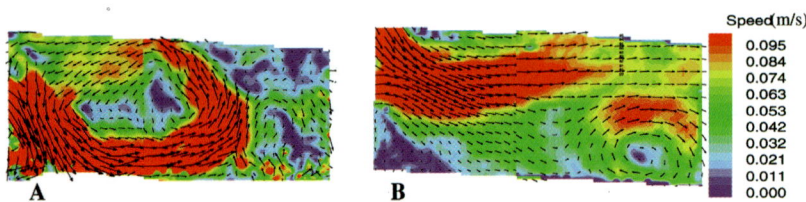

PLATE 10. FIGURE 2, Cho *et al.*, page 262.

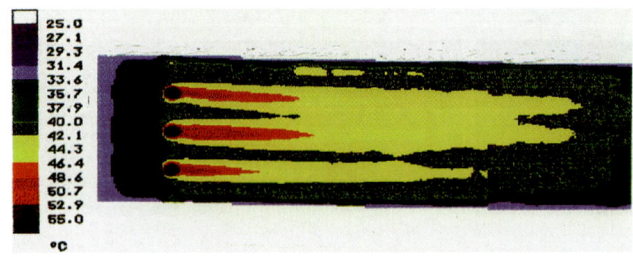

PLATE 11. FIGURE 2, Dizene *et al.*, page 275.

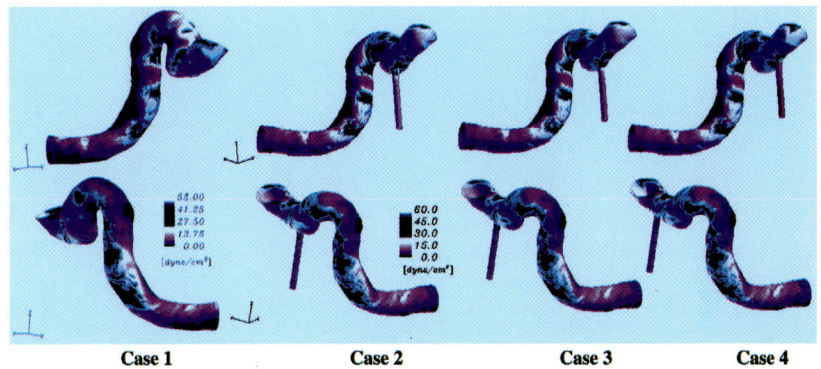

PLATE 12. FIGURE 3, Oshima *et al.*, page 341.

PLATE 13. FIGURE 4, Oshima *et al.*, page 341.